Induced Resistance for Plant Defence

Induced Resistance for Plant Defence

A Sustainable Approach to Crop Protection

Edited by

Dale Walters
Crop and Soil Systems Research Group,
Scottish Agricultural College, Edinburgh, UK

Adrian Newton
Scottish Crop Research Institute, Invergowrie, Dundee, UK

Gary Lyon
Scottish Crop Research Institute, Invergowrie, Dundee, UK

Blackwell
Publishing

Blackwell Publishing editorial offices:

Blackwell Publishing Ltd, 9600 Garsington Road, Oxford OX4 2DQ, UK
Tel: +44 (0)1865 776868
Blackwell Publishing Professional, 2121 State Avenue, Ames, Iowa 50014-8300, USA
Tel: +1 515 292 0140
Blackwell Publishing Asia Pty Ltd, 550 Swanston Street, Carlton, Victoria 3053, Australia
Tel: +61 (0)3 8359 1011

First published 2007 by Blackwell Publishing Ltd

ISBN: 978-1-4051-3447-7

Library of Congress Cataloging-in-Publication Data
Induced resistance for plant defence: a sustainable approach to crop protection / edited by Dale Walters, Adrian Newton, Gary D. Lyon. – 1st. ed.
p. cm.

Includes bibliographical references and index.
ISBN: 978-1-4051-3447-7 (hardback : alk. paper)
1. Plants–Disease and pest resistance–Genetic aspects. 2. Plants–Disease and pest resistance–Molecular aspects. I. Walters, Dale. II. Newton, Adrian C. III. Lyon, Gary D.

SB750.I4745 2007
632′.9–dc22
2006026449

A catalogue record for this title is available from the British Library

Set in 10/12.5 pt Times by Charon Tec Ltd (A Macmillan Company), Chennai, India
www.charontec.com

The publisher's policy is to use permanent paper from mills that operate a sustainable forestry policy, and which has been manufactured from pulp processed using acid-free and elementary chlorine-free practices. Furthermore, the publisher ensures that the text paper and cover board used have met acceptable environmental accreditation standards.

For further information on Blackwell Publishing, visit our website:
www.blackwellpublishing.com

Contents

List of contributors

Dr Tim Daniell	Scottish Crop Research Institute, Invergowrie, Dundee DD2 5DA, UK E-mail: tim.daniell@scri.ac.uk
Dr Elizabeth Dann	Department of Primary Industries and Fisheries, 80 Meiers Road, Queensland 4068, Australia E-mail: elizabethdann@dpi.qld.gov.au
Professor Brian Deverall	Faculty of Agriculture, Food and Natural Resources, University of Sydney, NSW 2006, Australia E-mail: bdeveral@mail.usyd.edu.au
Dr Christophe Garcion	Departemente de Biologie, Université de Fribourg, 1700 Fribourg, Switzerland E-mail: christophe.garcion@unifr.ch
Professor Ray Hammerschmidt	Department of Plant Pathology, 107 CIPS Building, Michigan State University, East Lansing, MI 48824-1311, USA E-mail: hammersr@anr.msu.edu
Professor Martin Heil	Department of General Botany – Plant Ecology, University of Duisburg-Essen, Universitatsstr. 5, D-45117 Essen, Germany E-mail: martin.heil@uni-duisburg-essen.de
Dr Kemal Kazan	Commonwealth Scientific and Industrial Research Organisation, Plant Industry, Queensland Bioscience Precinct, St Lucia, Queensland 4069, Australia E-mail: kemal.kazan@csiro.au
Dr Olivier Lamotte	Departemente de Biologie, Université de Fribourg, 1700 Fribourg, Switzerland olivier.lamotte@unifr.ch
Dr Andy Leadbeater	Syngenta Crop Protection AG E-mail: andy.leadbeater@syngenta.com
Dr Gary Lyon	Scottish Crop Research Institute, Invergowrie, Dundee DD2 5DA, UK E-mail: gary.lyon@scri.ac.uk
Professor Jean-Pierre Métraux	Departemente de Biologie, Université de Fribourg, 1700 Fribourg, Switzerland E-mail: jean-pierre.metraux@unifr.ch

Dr Adrian Newton Scottish Crop Research Institute, Invergowrie, Dundee DD2 5DA, UK
E-mail: adrian.newton@scri.ac.uk

Professor Corné M.J. Pieterse Phytopathology, Institute of Environmental Biology, Utrecht University, PO Box 80084, 3508 TB Utrecht, The Netherlands
E-mail: c.m.j.pieterse@bio.uu.nl

Dr Jörn Pons-Kühnemann Biometry and Population Genetics, Giessen University, Heinrich-Buff-Ring 26-32, 35392 Giessen, Germany
E-mail: joern.pons@agrar.uni-giessen.de

Dr Tony Reglinski Bioprotection Group, HortResearch, Ruakura Research Centre, East Street, Private Bag 3123, Hamilton, New Zealand
E-mail: treglinski@hortresearch.co.nz

Dr Philippe Reignault Mycologie/Phytopathologie/Environnement, Université du Littoral Côte d'Opale, 17, avenue Louis Blériot BP 699, F-62228 Calais Cedex, France
E-mail: philippe.reignault@univ-littoral.fr

Dr Peer Schenk School of Integrative Biology, University of Queensland, St Lucia, Queensland 4072, Australia
E-mail: p.schenk@uq.edu.au

Dr Theo Staub Syngenta Crop Protection AG

Professor Mike Stout Department of Entomology, Louisiana State University Agricultural Center, 402 Life Sciences Building, Louisiana State University, Baton Rouge, LA 70803, USA
E-mail: mstout@agctr.lsu.edu

Professor L.C. Van Loon Phytopathology, Institute of Environmental Biology, Utrecht University, PO Box 80084, 3508 TB Utrecht, The Netherlands
E-mail: l.c.vanloon@bio.uu.nl

Professor Dale Walters Crop & Soil Systems Research Group, Scottish Agricultural College, King's Buildings, West Mains Road, Edinburgh EH9 3JG, UK
E-mail: dale.walters@sac.ac.uk

Preface

Plant diseases have been a problem for mankind since the very beginnings of agriculture. As we write this preface, some 12,000 years later, plant diseases are still a problem. We have learned a great deal about plant diseases and how to control them in the intervening millennia, but disease still takes its toll on our crops every year. The problem is the result, in large part, of the genetic adaptability of the pathogens responsible for causing plant diseases: they develop resistance to our crop protection chemicals and rapidly overcome the resistance bred into our new crop varieties. In the fight against plant disease, it is essential therefore that we keep one (or preferably several) steps ahead of the pathogens.

In their review of global food security, Strange & Scott (2005; Annual Review of *Phytopathology* **43,** 83–116) point out that more than 800 million people worldwide do not have sufficient food, and some 1.3 billion people survive on less than $1 a day. Further, a survey by The Economist in 2000 (*The Economist*, **March 25**) estimated that there will be an additional 1.5 billion people to feed by 2020, requiring farmers to produce 39% more grain. Since it is estimated that some 12% of global crop production is lost to plant disease annually, it is clear that the need for efficient, reliable and affordable disease control measures has never been greater. Equally important from the modern perspective is the need to ensure that any new disease control measures maintain crop yield and quality, without harming our fragile and long suffering environment.

Although the first recorded observations of induced resistance date back to the 19th century, the phenomenon was largely ignored until the late 1950s and early 1960s. Even then, the concept of induced resistance was largely ignored, despite the very solid foundation being laid by Joe Kuć and his colleagues. There was a gradual awakening of interest, and induced resistance has attracted increasing attention in the last 15 years or so. This interest is not surprising, since induced resistance offers the prospect of broad spectrum, long lasting and, hopefully, environmentally benign disease control. However, this prospect will not be realized unless we are able to translate our ever increasing understanding of the cellular basis of induced resistance to the practical, field situation. This requires integration of molecular biology and biochemistry, with crop science and ecology. In this book, our aim is to provide plant pathologists, crop protectionists, agronomists and others with an update of the broad and complex topic that is induced resistance and to highlight the efforts being made to provide the understanding necessary to allow induced resistance to be used in practice. The various chapters in the book cover the cellular aspects of induced resistance, including signalling and defence mechanisms, the trade-offs associated with the expression of induced resistance, work on integrating induced resistance into crop protection practice and induced resistance from a commercial perspective. Our hope is that this book will excite the interest of plant and crop scientists and encourage the collaboration between molecular biologists, plant pathologists and ecologists that will be necessary to realize the great potential offered by induced resistance.

Dale Walters, Adrian Newton and Gary Lyon

Chapter 1
Introduction: definitions and some history

Ray Hammerschmidt

Department of Plant Pathology, Michigan State University, East Lansing, USA

1.1 Induced resistance: an established phenomenon

It is very well established that certain types of infection or other treatments can induce disease resistance (e.g. Kuć, 1982; Hammerschmidt & Kuć, 1995; Sticher *et al.*, 1997; Vallad & Goodman, 2004; da Rocha & Hammerschmidt, 2005). The induced plant is able to resist attack by virulent pathogens and other pests because of an enhanced ability to rapidly express defences upon infection and, in some cases, an increase in defences that were expressed in response to the inducing treatment. Although well established and studied, it is important to consider why induced resistance occurs. How can a plant that is known to be susceptible to a disease or several diseases be physiologically or biochemically changed so that it can now resist those infections?

Two basic assumptions must be considered to explain the overall phenomenon of induced resistance. First of all, plants must have all genes necessary to mount an effective defence. Second, the inducing treatment is capable of activating some of the defences directly and, more importantly, that the inducing treatment primes or sensitizes that plant in such a way that allows rapid expression of a broad set of defences upon infection by a pathogen.

The first assumption is easy to support. It is a well known plant pathology concept that plants resist the vast majority of pathogens that exist in nature, and that this phenomenon (non-host resistance) is associated with the expression of defences (Heath, 2000). Most plants, however, are susceptible to some pathogens or isolates or races of those pathogens. This does not mean that the plant lacks the defence to fend off the pathogen, but rather that the plant does not have the means to rapidly detect the presence of the pathogen (e.g. a major gene for resistance). The second assumption also has significant support: plants that are induced have enhanced capacity to express defences after an infection challenge (Conrath *et al.*, 2002).

1.2 Terminology and types of induced resistance

Plant resistance to pathogens and pests can be active and/or passive (Hammerschmidt & Nicholson, 1999). Passive resistance depends on defences that are constitutively expressed in the plant, while active resistance relies on defences that are induced after infection or attack. Induced resistance is an active process that can describe resistance at two levels. First, active defence to an incompatible race or isolate of a pathogen is a form of induced resistance that is characterized by highly localized expression of defences such as phytoalexins and the hypersensitive response (Hammerschmidt & Nicholson, 1999). Second, induced resistance can also describe plants that express resistance to a broad range of compatible pathogens after some initial inducing treatment (Kuć, 1982). It is this latter form of induced resistance that is the focal point of this book. The term 'induced resistance' in itself only describes the general phenomenon and does not imply any specific type of defence expression or regulation.

1.2.1 Local and systemic induction of resistance

Induced resistance can be local or systemic. Local induced resistance refers to those cases where the inducing treatment is applied to the same tissue as the subsequent challenge by a pathogen. Systemic induced resistance describes resistance that is induced in a part of the plant that is spatially separated from the point of induction. Although spatially different, both local and systemic resistances are characterized by needing time to develop after the inducing treatment and the non-specific nature of the resistance. The mechanisms of stopping pathogen development in locally induced resistance may be due to the production of defences such as pathogenesis-related (PR) proteins and cell wall alterations that are thought to be involved in stopping the development of the inducing inoculum as well as propagules of the challenge pathogen that have the unfortunate luck of landing directly on the site occupied by the inducing inoculum (Hammerschmidt, 1999). In the case of systemic resistance, the inducing or resistance activating treatments result in a change in cells at a distance from the induction site that allows those cells to rapidly deploy defences upon challenge. This is the part of systemic resistance that is now known as 'priming' (Conrath *et al.*, 2002). In addition to being primed, the systemically induced tissues may also have some degree of defence established by the induction process that is there prior to any challenge. An obvious example is the systemic expression of PR proteins in certain forms of systemic induced resistance (Van Loon, 1997).

1.2.2 SAR and ISR

Over the last 10 years, it has become clear that induced resistance to disease is not one phenomenon. At least two forms of induced resistance, known as systemic acquired resistance (SAR) and induced systemic resistance (ISR), have been characterized as distinct phenomena based on the types of inducing agents and host signalling pathways that result in resistance expression (Sticher *et al.*, 1997; Van Loon *et al.*, 1998).

A major characteristic of SAR is the need for the expression of localized necrosis caused by the inducing pathogen. This necrosis can be either a hypersensitive response or a local necrotic lesion caused by a virulent pathogen. SAR is also dependent on salicylic acid signalling and the systemic expression of pathogenesis related protein genes (Sticher *et al.*, 1997; Hammerschmidt, 1999). ISR is induced by certain strains of plant growth promoting rhizobacteria (PGPR) (Van Loon *et al.*, 1998). Unlike SAR, ISR is not associated with local necrotic lesion formation. ISR also differs in that it depends on perception of ethylene and jasmonic acid, and is not associated with expression of PR genes. Both SAR and ISR do result in broad spectrum resistance. The differences in mechanisms and signalling leading to SAR and ISR as well as potential trade-offs between these different forms of induced resistance are described in Chapters 4 and 9.

It should also be noted that many of the features that have been used to distinguish ISR from SAR are based in studies with Arabidopsis in which specific genetic analyses have been coupled with biochemical and pathological analyses (see Chapter 4). Because the phenotypes of SAR and ISR are similar if not identical in terms of reducing the effects of pathogen challenge, distinguishing between ISR and SAR should be approached with caution when dealing with plant–pathogen interactions other than genetically well defined systems, such as those utilizing Arabidopsis. With the many types of inducing agents that have been identified and the great number of microbes that can also induce resistance (see Chapters 2, 8 and 10), it is likely that other forms of induced resistance may occur. Use of the tools of genomics to understand the molecular basis and regulation of induced resistance, such as those outlined by Kazan and Schenk in this book (Chapter 3), will be invaluable in sorting out types of induced resistance in model systems as well as those crops in which induced resistance may be applied in the future.

1.2.3 Protection

Certain reports from 1970s used the term 'protection' to describe induced resistance (e.g. Skipp & Deverall, 1973; Kuć *et al.*, 1975; Hammerschmidt *et al.*, 1976). These reports on induced resistance in both cucumber and green bean plants described the ability of incompatible fungal pathogens to induce resistance. Although the term 'protection' adequately describes what is happening in terms of the end result, the term is really too generic to be of use in describing induced resistance.

1.2.4 Cross protection

It has been known for many years that prior infection of plants with milder strains of a virus can result in reduced disease development by a subsequent infection by a more severe strain of the same virus (Price, 1940; Pennazio *et al.*, 2001). This phenomenon is known as cross protection and is really very different from the induced resistance phenomena that are discussed throughout this book. Unlike induced resistance where defences or the potential to express defences are activated by the inducing treatment, cross protection is mechanistically very different and relies more on interference of the mild viral stain with the more severe strain than by defensive action (Fulton, 1986). Cross protection also differs from induced resistance in that the protection is only effective against strains of the same virus, whereas induced resistance is much broader spectrum (Fulton, 1986). However, there is true induced resistance against viruses, as will be discussed later in this chapter and throughout the book.

1.3 A little history

The general concept that plants can actively defend themselves and have resistance induced against virulent pathogens has been known for over 100 years. Much of the early work was summarized in the classic review by Chester (1933) who sorted through numerous reports from the early 20th century. Other reviews have also detailed many of the early observations, and the readers are directed to these sources for other details (Matta, 1971; Kuć, 1982; Sequeira, 1983; Hammerschmidt & Kuć, 1995). Rather than be comprehensive, a few representative examples of induced resistance systems and their origins will be discussed.

1.3.1 Early reports

The earliest reports of what appears to be induced resistance to disease come from the first part of the 20th century, when Bernard demonstrated that prior infection of orchid embryos with a mycorrhizal *Rhizoctonia* of low pathogenicity resulted in an increased ability of the embryo to resist infection by a more pathogenic isolate of *Rhizoctonia* (described in Gäumann, 1950; Allen, 1959). In 1940, Müller and Börger reported that prior inoculation of the cut surface of a potato tuber with an avirulent race of *Phytophthora infestans* resulted in the local induction of resistance to virulent races of the same pathogen (described in Gäumann, 1950; Allen, 1959; Müller, 1959). If the necrotic, hypersensitively responding tissues were carefully removed, the healthy tissue that was immediately beneath the necrotic tissue was also resistant to infection by a virulent isolate of *P. infestans*. Although these experiments are best known for the development of the phytoalexin hypothesis, Müller and Börger also provided evidence that would be readily recognized as features characteristic of induced resistance as is known today: the need for pathogen induced necrosis as part of the defence triggering process, a time delay between the application of the inducing pathogen and the expression of resistance against a virulent pathogen.

1.3.2 Developments leading towards today's state of knowledge

In the 1950s, initial biochemical evidence for inducible defences was being reported (e.g. Kuć, 1957; Allen, 1959; Müller, 1959) and this included induced resistance. Kuć *et al.* (1959) found that application of D- or DL-phenylalanine induced resistance in apple leaves to *Venturia inaequalis*. Within a few years, Hijwegen (1963) demonstrated that phenylserine would induce resistance in cucumber, and by the end of the 1970s, salicylic acid was shown to be an inducer of resistance (White, 1979). Many synthetic and natural compounds subsequently have been shown to induce resistance (Kessmann *et al.*, 1994; Cohen, 2002). The first synthetic resistance activator (acibenzolar-*S*-methyl) was commercialized in the 1990s, and many other materials that induce resistance have been identified (see Chapters 2 and 12). With the discovery of resistance activators or elicitors that can be easily applied via conventional production tools, the potential for practical applications has increased greatly, as discussed by Reglinski *et al.* and by Leadbeater & Staub later in this book.

Most of the basis of our understanding of induced resistance has come from the use of pathogens or other microbes to induce resistance. Cruickshank & Mandryk (1960) found that injecting stems of tobacco plants with sporangia of *Peronospora tabacina* induced resistance in the foliage to further infection by the same pathogen. Although the resistance was clearly induced, there was an obvious cost to the plant as the induced plants were visibly stunted. Because induced resistance is an active process that is associated with new transcription and translation, the stunting effect is not an unexpected consequence. The overall effects of induction on overall plant fitness and costs associated with the induced state are discussed later in this volume by Heil (Chapter 9). In 1961, Frank Ross published the first of two papers on induced or, as he called it, acquired resistance of tobacco to tobacco mosaic virus (TMV). Using tobacco plants with the N gene for resistance to TMV, Ross (1961a) showed that the tissues immediately surrounding the TMV induced local lesions were highly resistant to infection by TMV and tobacco necrosis virus. In a companion paper, Ross (1961b) showed that infection of N gene tobacco with

TMV resulted in systemic increases in resistance to TMV. The systemic response was also induced by other local lesion viruses. Over the next few years, induced resistance in tobacco against fungi and bacterial pathogens was described, thus helping to illustrate the non-specific nature of this form of resistance (reviewed in Sequeira, 1983; Hammerschmidt & Kuć, 1995). In the late 1970s, Kuć and associates confirmed Cruickshank & Mandryk's observation that infection of tobacco with *P. tabacina* would induce resistance against this pathogen. These studies led to an extensive stream of publications from Kuć and colleagues on induced resistance to *P. tabacina* (see Tuzun & Kuć, 1989 and Hammerschmidt & Kuć, 1995 for a more thorough overview).

Cucumber plants have also proven to be an excellent model system for induced resistance studies. In 1975, Kuć *et al.* found that droplet inoculation of one leaf of anthracnose susceptible cucumber with the cucumber anthracnose fungus *Colletotrichum orbiculare* induced systemic resistance to the same pathogen. Similar to the case with *P. tabacina* on tobacco, a virulent isolate of a necrotic lesion inducing pathogen was capable of inducing systemic resistance. At about the same time, Hammerschmidt *et al.* (1976) reported that local resistance could be induced in cucumber with pathogens that were incompatible on this host. Subsequent work demonstrated that induced resistance in cucumber could be induced against and by a wide range of necrotic lesion inducing pathogens, as well as the hypersensitive response induced by bacteria pathogenic on hosts other than cucumber and provided good evidence that induced resistance could last for weeks (Hammerschmidt & Yang-Cashman, 1995). The biological spectrum was further expanded by Kuć and co-workers, who also showed that induced resistance was effective not only in multiple cultivars of the host, but also in other species and genera within a plant family (Kuć, 1982).

Systemic resistance implies the transmission of a systemic signal. The cucumber induced resistance model provided much of the early evidence for the presence and source of such a signal through grafting, petiole girdling and timing studies (reviewed in Hammerschmidt & Yang-Cashman, 1995). These experiments established a framework by which investigations on the nature of the systemic signal could be undertaken (e.g. Malamy *et al.*, 1990; Métraux *et al.*, 1990).

Green bean (*Phaseolus vulgaris*) played an important role in the development of our understanding of this phenomenon. Spray inoculation of etiolated bean hypocotyls with an incompatible race of the anthracnose pathogen, *Colletotrichum lindemuthianum*, resulted in the local induction of resistance against compatible races of the same pathogen (Rahe *et al.*, 1969). Skipp & Deverall (1973) expanded on these observations and showed that local resistance could be induced in leaves and the interior of seed pods as well as hypocotyls. Elliston *et al.* (1971) demonstrated that resistance could be induced in hypocotyls at a distance from the point of induction. Systemic resistance was demonstrated by Sutton (1979) and by Cloud & Deverall (1987), who induced resistance in upper leaves of bean plants by inoculation of lower leaves with droplets of *C. lindemuthianum* inoculum. More details on induced resistance in this plant family can be found in the review by Deverall & Dann (1995).

Arabidopsis thaliana has proven to be an invaluable tool in the study of plant–pathogen interactions, and induced resistance is no exception. Uknes *et al.* (1992) were first to demonstrate biologically induced resistance in Arabidopsis by inducing resistance to turnip crinkle virus (TCV) and *Pseudomonas syringae* pv. *tomato* by prior inoculation of the plants with necrosis inducing TCV. Cameron *et al.* (1994) expanded on this observation by showing that pre-inoculation of Arabidopsis leaves with an avirulent isolate of *P. syringae* pv. *tomato* induced resistance to infection for virulent isolates of the same pathogen and to *Pseudomonas maculicola*. Much of the current work illustrating the value of Arabidopsis as a tool for unravelling the biochemical, genetic and molecular basis of induced resistance is illustrated in Chapters 3, 4 and 6.

1.4 It's all about interactions

Induced resistance results from the interaction of a plant with a suitable inducing agent. The inducers, as discussed in Chapters 2 and 8, can be very diverse. However, in all cases, the interaction with the inducing agent or elicitor results in the expression of defences and in the priming of healthy tissues to quickly respond to infection (see Chapter 6). As discussed later in this book by Pieterse & Van Loon, interactions among and between various signalling pathways activated in the plant result in the final state of resistance but also illustrate that interactions occur within the plant as well. For us to fully understand the complexity of these interactions, it is essential to have a greater understanding of which genes are essential for induced resistance and how these genes are regulated (see Chapter 3). Because induced resistance is effective against a broad spectrum of pathogens, it is not surprising that some forms of induced resistance are effective against insects and that interactions with insect herbivores can induce similar types of defences as do pathogens (see Chapter 5). Within the induced plant, there are competing interactions. As Pieterse & Van Loon discuss in their chapter, crosstalk between induced resistance signalling pathways may help determine the type of resistance that is induced, but this may result in unexpected consequences (see Heil's Chapter 9). Interactions also occur within the plant to determine where resources should be allocated as the plant must 'decide' if it is better to enhance resistance or to allocate resources to growth and development. Thus, as discussed in Heil's chapter, inducing resistance may result in a fitness cost to the plant.

Interactions that affect induced resistance go beyond those that are within the plant. The environment can have a profound effect on whole plant physiology, and these environmental factors likely impact the induction and expression of induced resistance as described by Newton & Pons in this book (Chapter 7). Certainly, plants grown in natural and agricultural systems have different physiological characteristics from those from a growth chamber or greenhouse where most induced resistance work has been performed, and understanding the effects of the natural environment on induced resistance is critical.

The final, but no less important, interaction is with growers and those who are interested in implementing induced resistance as part of disease management programmes. As detailed throughout the book, there are many known inducers that present myriad means of delivering induced resistance. A key feature is to be able to apply these inducers using technologies already used by growers, and thus the ability to apply inducers as topical treatments is important for their acceptance (see Chapter 10). The interest of the private sector in the development of resistance inducing products has increased and is providing the tools to determine just how well induced resistance will perform in the field and the commercial marketplace (see Chapter 12). Induced resistance may not provide 100% control or control all pathogens. Thus, as discussed in Chapter 11 by Reglinski *et al.*, integration into practices that the grower can and will use is perhaps the most important interaction: that of the human application of this technology, as a grower will not use something that is neither effective nor reliable.

Our knowledge of induced resistance has come a very long way in the last 40 years or so. The chapters that follow reflect this progress and provide information and ideas needed to push forward both our understanding of mechanisms and how to apply this most fascinating form of disease resistance.

1.5 Acknowledgements

I would like to thank the Michigan Agricultural Experiment Station and the USDA for support of my work. More importantly, I would like to thank Professor Joe Kuć who introduced me to the wonders of induced resistance well over 30 years ago and gave me the opportunity to work in his programme on this phenomenon.

1.6 References

Allen PJ, 1959. Physiology and biochemistry of defense. In: Horsfall JG, Dimond AE, eds. *Plant Pathology, an Advanced Treatise* Vol. 1. New York: Academic Press, 435–467.

Cameron RK, Dixon R, Lamb CJ, 1994. Biologically induced systemic acquired resistance in *Arabidopsis thaliana. The Plant Journal* **5,** 715–725.

Chester KS, 1933. The problem of acquired physiological immunity in plants. *Quarterly Review of Biology* **8,** 129–154, 275–324.

Cloud AME, Deverall BJ, 1987. Induction and expression of systemic resistance to the anthracnose disease in bean. *Plant Pathology* **36,** 551–557.

Cohen Y, 2002. *β*-Aminobutyric acid-induced resistance against pathogens. *Plant Disease* **86,** 448–457.

Conrath U, Pieterse CMJ, Mauch-Mani B, 2002. Priming in plant–pathogen interactions. *Trends in Plant Science* **7,** 210–216.

Cruickshank IAM, Mandryk M, 1960. The effect of stem infection of tobacco with *Peronospora tabacina* on foliage reaction to blue mold. *Journal of Australian Institute of Agricultural Research* **26,** 369–372.

da Rocha AB, Hammerschmidt R, 2005. History and perspectives on the use of disease resistance inducers in horticultural crops. *Horttechnology* **15,** 518–529.

Deverall BJ, Dann EK, 1995. Induced resistance in legumes. In: Hammerschmidt R, Kuć J, eds. *Induced Resistance to Disease in Plants.* Dordrecht: Kluwer Academic Publishers, 1–30.

Elliston J, Kuć J, Williams EB, 1971. Induced resistance to bean anthracnose at a distance from the site of the inducing interaction. *Phytopathology* **61,** 1110–1112.

Fulton RW, 1986. Practices and precautions in the use of cross protection for plant-virus disease control. *Annual Review of Phytopathology* **24,** 67–81.

Gäumann EA, 1950. *Principles of Plant Infection; a Text-Book of General Plant Pathology for Biologists, Agriculturists, Foresters and Plant Breeders.* New York: Hafner.

Hammerschmidt R, 1999. Induced disease resistance: How do induced plants stop pathogens? *Physiological and Molecular Plant Pathology* **55,** 77–84.

Hammerschmidt R, Acres S, Kuć J, 1976. Protection of cucumber against *Colletotrichum lagenarium* and *Cladosporium cucumerinum. Phytopathology* **66,** 790–793.

Hammerschmidt R, Kuć J, 1995. *Induced Resistance to Disease in Plants.* Dordrecht: Kluwer Academic Publishers.

Hammerschmidt R, Nicholson RL, 1999. A survey of plant defense responses to pathogens. In: Agrawal A, Tuzun S, eds. *Induced Plant Defenses Against Pathogens and Herbivores.* St Paul, MN: APS Press, 55–71.

Hammerschmidt R, Yang-Cashman P, 1995. Induced resistance in cucurbits. In: Hammerschmidt R, Kuć J, eds. *Induced Resistance to Disease in Plants.* Dordrecht: Kluwer Academic Publishers, 63–85.

Heath MC, 2000. Nonhost resistance and nonspecific host defense. *Current Opinion in Plant Biology* **3,** 315–319.

Hijwegen T, 1963. Lignification, a possible mechanism of active resistance against pathogens. *European Journal of Plant Pathology* **69,** 314–317.

Kessmann H, Staub T, Hofmann C, Maetzke T, Herzog J, Ward E, Uknes S, Ryals J, 1994. Induction of systemic acquired disease resistance in plants by chemicals. *Annual Review of Phytopathology* **32,** 439–459.

Kuć J, 1957. A biochemical study of the resistance of potato tuber tissue to attack by various fungi. *Phytopathology* **47,** 676–680.
Kuć J, 1982. Induced immunity to plant disease. *BioScience* **32,** 854–860.
Kuć J, Barnes E, Daftsios A, Williams EB, 1959. The effect of amino acids on susceptibility of apple varieties to scab. *Phytopathology* **49,** 313–315.
Kuć J, Shockley G, Kearney K, 1975. Protection of cucumber against *Colletotrichum lagenarium* by *Colletotrichum lagenarium. Physiological Plant Pathology* **7,** 195–199.
Malamy J, Carr JP, Klessig DF, Raskin I, 1990. Salicylic acid – a likely endogenous signal in the resistance response of tobacco to tobacco mosaic virus. *Science* **250,** 1002–1004.
Matta A, 1971. Microbial penetration and immunization of noncongenial host plants. *Annual Review of Phytopathology* **9,** 387–410.
Métraux JP, Signer H, Ryals J, Ward E, Wyss-Benz M, Gaudin J, Raschdorf K, Schmid E, Blum W, Inverardi B, 1990. Increase in salicylic acid at the onset of systemic acquired resistance in cucumber. *Science* **250,** 1004–1006.
Müller KO, 1959. Hypersensitivity. In: Horsfall JG, Dimond AE, eds. *Plant Pathology: an Advanced Treatise* Vol. 1. New York: Academic Press, 459–519.
Pennazio S, Roggero P, Conti M, 2001. A history of plant virology. Cross protection. *Microbiologica* **24,** 99–114.
Price, WC, 1940. Acquired immunity from plant virus diseases. *Quarterly Review of Biology* 15, 338–361.
Rahe JE, Kuć J, Chuang C, Williams EB, 1969. Induced resistance in *Phaseolus vulgaris* to bean anthracnose. *Phytopathology* **59,** 1641–1645.
Ross AF, 1961a. Localized acquired resistance to plant virus infection in hypersensitive hosts. *Virology* **14,** 329–339.
Ross AF, 1961b. Systemic acquired resistance induced by localized virus infection in plants. *Virology* **14,** 340–358.
Sequeira L, 1983. Mechanisms of induced resistance in plants. *Annual Review of Microbiology* **37,** 51–79.
Skipp RA, Deverall BJ, 1973. Studies on cross-protection in the anthracnose disease of bean. *Physiological Plant Pathology* **3,** 299–314.
Sticher LB, Mauch-Mani B, Métraux JP, 1997. Systemic acquired resistance. *Annual Review of Phytopathology* **35,** 235–270.
Sutton DC, 1979. Systemic cross protection in bean against *Colletotrichum lindemuthianum. Australasian Plant Pathology* **8,** 4–5.
Tuzun S, Kuć J, 1989. Induced systemic resistance to blue mold of tobacco. In: McKeen WE, ed. *Blue Mold of Tobacco.* St. Paul, MN: APS Press, 177–200.
Uknes S, Mauch-Mani B, Moyer M, Potter S, Williams S, Dincher S, Chandler D, Slusarenko A, Ward E, Ryals J, 1992. Acquired resistance in Arabidopsis. *The Plant Cell* **4,** 645–656.
Vallad GE, Goodman RM, 2004. Systemic acquired resistance and induced systemic resistance in conventional agriculture. *Crop Science* **44,** 1920–1934.
Van Loon LC, 1997. Induced resistance in plants and the role of pathogenesis-related proteins. *European Journal of Plant Pathology* **103,** 753–765.
Van Loon LC, Bakker PAHM, Pieterse CMJ, 1998. Systemic resistance induced by rhizosphere bacteria. *Annual Review of Phytopathology* **36,** 453–483.
White, RF, 1979. Acetylsalicylic acid (aspirin) induces resistance to tobacco mosaic virus in tobacco. *Virology* **99,** 410–412.

Chapter 2
Agents that can elicit induced resistance

Gary Lyon
Scottish Crop Research Institute, Dundee, UK

2.1 Introduction

Induced resistance is a non-specific form of disease resistance in plants acting against a wide range of pathogens, and as such one would expect it to be activated by a range of non-specific inducers (elicitors). This is very much the case. Elicitors are characteristically non-specific in that they induce a general resistance effective against a range of pathogens and work in a taxonomically diverse range of plants. Some have systemic activity, inducing resistance some distance away from the site of application, while others induce resistance locally at the site of application.

It is neither feasible nor desirable to cite every publication on induced resistance. In this chapter, I outline the range of compounds that have been shown to induce some resistance-related mechanisms and which are thereby able to reduce the level of infection by subsequent pathogen challenge. It is this requirement of reducing pathogen infection which is important in this chapter, and compounds that have only been described as inducing some specific components of resistance cascades, e.g. syringolides (Ji *et al.*, 1997) and those cited in Kessler & Baldwin (2002) and Montesano *et al.* (2003), are therefore not included if there is no evidence of them increasing resistance to a subsequent inoculation with a pathogen. It is important to know whether there are a small number of compounds (or families of compounds) that are able to induce resistance or whether a relatively large number are able to do so. By implication, this will indicate whether there are multiple pathways or a rather limited number of pathways that can be activated.

Our knowledge concerning compounds that can stimulate disease resistance in plants is still somewhat ad hoc. Some are well characterized pure compounds, while other publications describe poorly characterized material or mixtures of compounds. What is clear is that there is a diverse range of chemically distinct compounds that are generally non-specific in their ability to induce resistance though with a proviso that some compounds are more effective at inducing resistance in some plant taxa than others. Some authors draw the distinction between an induced resistance where the pathogenesis-related protein PR1 is produced and those compounds that induce resistance without accumulation of PR1 and are suggesting that some compounds are 'priming' the plant, which is then able to respond more quickly to subsequent infection (Conrath *et al.*, 2002).

This chapter aims to outline the many different types of resistance elicitor. The effects of topical induction of resistance elicitors is dealt with in Chapter 10, while the effect of micro-organisms such as plant growth promoting rhizobacteria (PGPR) on induced resistance is described in Chapter 8.

2.2 Compounds inducing resistance

2.2.1 *Fungal, bacterial and PGPR products*

Plant pathogenic micro-organisms possess a battery of mechanisms through which they are able to infect and colonize plant tissues, and in addition, plants are able to recognize specific components of microbes to identify 'non-self'. For example, PGPR suppress infection by pathogens through a number of mechanisms which act directly against the pathogen, e.g. by producing antibiotics, siderophores (which compete for iron), and glucanases and chitinases which lyse the cells (Van Loon *et al.*, 1998). There is evidence that some of these types of compounds act through the indirect route of stimulating the plant's resistance mechanisms. Systemic resistance induced by rhizosphere bacteria (ISR) has been reviewed extensively by Van Loon *et al.* (1998) (see also Chapter 8), and that literature will not be covered here. What is relevant, though, are those components of micro-organisms that are capable of inducing resistance. There are also publications describing the use of crude or partially purified microbial extracts to induce resistance e.g. extracts of *Penicillium chrysogenum* (Thuerig *et al.*, 2006); these are not discussed in detail unless there is a clear indication of the chemical entity involved.

2.2.1.1 Antibiotics

A number of root colonizing strains of fluorescent *Pseudomonas* spp. are able to suppress disease in plants. The extent of such disease control depends on factors such as root colonization, induction of systemic resistance and the production of antimicrobial antibiotics (Haas & Keel, 2003). Many of these antibiotics have been well studied, and there are suggestions that some may have a dual role and may also be involved with induction of resistance. Iavicoli *et al.* (2003) tested a number of mutants of *Ps. fluorescens* and found that those with reduced 2,4-diacetylphloroglucinol (DAPG) production were less able to induce resistance to *Peronospora parasitica* on Arabidopsis. However, antibiotics such as pyoluteorin and DAPG from *Pseudomonas* spp. have also been reported to be phytotoxic at high concentrations (Maurhofer *et al.*, 1995), thus probably precluding their value in any practical application.

2.2.1.2 Chitin

Chitin (β-1,4 linked *N*-acetylglucosamine) is a common component of fungal cell walls, and various sized fragments (*N*-acetylchitooligosaccharides) have been shown to induce defence responses in a wide range of plant species, including barley, melon, parsley, rice, soybean, tomato and wheat (Zhang *et al.*, 2002 and references therein). Fragment size (chain length) has an effect on elicitor activity with very short chains (degree of polymerization 2–3) being less effective at inducing host responses than slightly longer chains (dp 7–8). A number of reports suggest that chitin is a stronger inducer of host responses than chitosan (see references in Shibuya & Minami, 2001).

Further information on chitosan (a deacetylated form of chitin) is provided in Section 2.3.1.

2.2.1.3 Ergosterol

The sterol ergosterol has been shown to elicit some resistance related responses including rapid alkalinization of the growth medium in tomato cell cultures (Granado *et al.*, 1995) and induction of H_2O_2 production in cucumber hypocotyls (Kauss & Jeblick, 1996). However, as yet, there is no evidence that ergosterol can induce resistance to subsequent infection by a plant pathogen.

2.2.1.4 Glucans from fungi

One of the first complex carbohydrates shown to be a resistance elicitor was a hepta-β-glucopyranoside isolated from mycelium of *Phytophthora megasperma* (Sharp *et al.*, 1984a, b). The work by Sharp *et al.* was ground-breaking as it clearly showed the effect that subtle structural changes in glucans have on phytoalexin elicitor activity. Purification of complex carbohydrates such as these is technically difficult on a large scale, and much subsequent work has used only partially purified glucans. Thus, partially purified oligosaccharide elicitors (Bio-Gel P-2 gel permeation void of acid hydrolysed mycelium) from *P. megasperma* f. sp. *glycinea* were later shown to increase resistance of tobacco leaves to several taxonomically different groups of viruses by between 50 and 100%, as assessed by symptom production and virus accumulation (Kopp *et al.*, 1989). Interestingly, no protection against virus infection was observed when the same elicitor fraction was tested on bean or turnip. Chemical characterization of complex carbohydrates in such preparations is difficult, and it is not surprising that there are few reports of well characterized naturally occurring complex carbohydrates as resistance elicitors. Nevertheless, it is likely that such complex carbohydrates are likely to be initiating resistance-related responses in many other microbial–plant interactions.

Yeast derived elicitors, though frequently not fully characterized, have been widely used to induce various resistance-related responses such as the upregulation of genes associated with resistance, and have been shown to be particularly effective against powdery mildew, for example on barley (Reglinski *et al.*, 1994a, b). Some activity of yeast-derived elicitors has also been reported on other crops such as lettuce against *Botrytis cinerea* and *Rhizoctonia solani* (Reglinski *et al.*, 1995).

Recently, Bru *et al.* (2006) showed that a modified cyclodextrin (heptakis(2,6-di-*O*-methyl)-βCD) can induce a number of resistance related responses in grapevine cell cultures. This result offers exciting opportunities as such compounds are well defined and can be chemically modified to produce a wide range of related structures and are available commercially. A cyclic 1,3–1,6-linked β-glucan secreted from the symbiotic nitrogen-fixing bacterium *Bradyrhizobium japonicum* has been reported to suppress phytoalexin accumulation in soybean induced by a fungal β-glucan elicitor (Mithöfer *et al.*, 1996). In contrast, a cyclic 1,2-β-glucan from *Rhizobium meliloti*, which does not nodulate on soybean, was inactive both as an elicitor and a suppressor of host responses. This seems to indicate that potential symbionts are successful, in part, because they are capable of suppressing host defence responses.

2.2.1.5 LPS

Lipopolysaccharides (LPS) and lipooligosaccharides from the outer surface of Gram negative bacteria are known to induce a number of disease resistance components (Dow *et al.*, 2000; Erbs & Newman, 2003). For example, LPS from Gram-negative bacteria induces a rapid burst of nitric oxide and induces defence-related genes in Arabidopsis (Zeidler *et al.*, 2004). LPS from a number of non-pathogenic bacteria have been shown to induce resistance to infection including, for instance, resistance of carnation to Fusarium (Van Peer & Schippers, 1992) and of *Nicotiana tabacum* to *Phytophthora nicotianae* (Coventry & Dubery, 2001). The elicitor activity of LPS is possibly due to the lipid A core region, as this component was also shown to be effective at inducing nitric oxide (Zeidler *et al.*, 2004). In addition, Silipo *et al.* (2005) showed that the lipid A and core oligosaccharides derived from the lipooligosaccharide from *Xanthomonas campestris* pv. *campestris* were able to induce PR1 and PR2 in Arabidopsis and prevent the hypersensitive response (HR) induced by avirulent bacteria. Newman *et al.* (2000) reported that LPS from plant pathogens and enteric bacteria did not induce necrosis and that no cell death occurred in protected tissue. This protective effect of LPS has been described as localized induced resistance (LIR) (cited in Newman *et al.*, 2000).

The spent growth medium and purified exopolysaccharides (EPS) from the Gram-negative bacterium *Pantoea agglomerans* have been shown to prime suspension-cultured wheat cells (Ortmann & Moerschbacher, 2006). Culture filtrate of *P. agglomerans* sprayed on to wheat leaves suppressed infection by *Puccinia recondita* f. sp. *tritici* (Kempf & Wolf, 1989).

2.2.1.6 Proteins and peptides

A number of proteins with enzymatic activity have been shown to induce some resistance related responses. For example, a xylanase from *Trichoderma viride* induced ethylene (Fuchs *et al.*, 1989) and PR proteins in tobacco (Lotan & Fluhr, 1990), and a xylanase from *Phytophthora parasitica* (Farmer & Helgeson, 1987) induced ethylene and phytoalexin accumulation in tobacco. Such results suggest that plant cell wall derived xylans will possess biological activity and be able to elicit some plant responses associated with resistance. This of course is not the same as inducing resistance to disease but does suggest that wall derived complex carbohydrates need to be tested. Other proteins and peptides have been isolated from *Trichoderma virens* including a possible serine proteinase (Hanson & Howell, 2004).

Trichoderma spp. commonly produce a variety of compounds, including peptaibols, that induce localized or systemic resistance responses in plants (Harman *et al.*, 2004). Peptaibols are polypeptides typically between 15 and 20 residues long, with a high proportion of non-standard amino acids, and the chain has an alkyl N terminus (usually acetyl) and a hydroxy-amino acid at the C terminus. Peptaibols frequently have antimicrobial activity and are isolated from fungi such as *Trichoderma* and *Emericellopsis*. A database of peptaibols is available at http://www.cryst.bbk.ac.uk/peptaibol (Whitmore *et al.*, 2003). Induction of systemic resistance by *Trichoderma* spp. has been reviewed by Hoitink *et al.* (2006).

There are other proteins that, when applied to plants, can enhance responses normally associated with enhanced resistance. For example, elicitins are small proteins secreted by

Phytophthora and *Pythium* spp. that cause necrosis but can also induce resistance in tobacco to, for example, *P. parasitica* or phytoplasma (Lherminier *et al.*, 2003) and to viruses (Cordelier *et al.*, 2003). Similarly, Séjalon-Delmas *et al.* (1997) purified a 34 kDa protein (GP 34) from mycelium of *P. parasitica* var. *nicotianae* that, when applied to tobacco roots, enhanced lipoxygenase activity and hydroxyproline-rich glycoprotein accumulation. Similarly, PB90, a 90 kDa protein of unknown function secreted by *P. boehmeriae*, induces an HR response in tobacco and induces local and systemic resistance to *P. nicotianae* and TMV (Zhang *et al.*, 2004). The increased resistance, but not the HR, was shown to involve a salicylic acid-dependent pathway. Cell wall proteins have been isolated from *Pythium oligandrum* that, when applied to roots of sugar beet, reduced the severity of damping off caused by *R. solani* and when applied to wheat reduced the number of spikelets infected with *Fusarium graminearum* (Takenaka *et al.*, 2003).

Interestingly, the N terminus of the bacterial elongation factor Tu from Gram-negative bacteria such as *Escherichia coli* has been shown to induce resistance in Arabidopsis and other Brassicaceae (Kunze *et al.*, 2004). Arabidopsis recognizes the N terminus of Tu, and an N-acetylated peptide matching the first 18 amino acids, termed elf18, is elicitor-active against subsequent inoculation with pathogenic bacteria. A shorter peptide, named elf12, containing the first 12 amino acids, is inactive as an elicitor. Flagellin is the main protein component of the bacterial flagellum and has been described as a general elicitor triggering some similar responses in plants and animals (Gómez-Gómez & Boller, 2002). A synthetic peptide (flg22) consisting of 22 amino acids corresponding to the N-terminal domain of flagellin, when used to treat Arabidopsis plants increased resistance to *P. syringae* pv. *tomato* DC3000 decreasing bacterial numbers by approximately 100 fold two days after inoculation (Zipfel *et al.*, 2004).

There are several commercial products based on microbial proteins. These include ELMGuard (ArborSciences, Canada), for controlling Dutch Elm disease, which is based on a proteinaceous elicitor from *Ophiostoma ulmi*, and Messenger™ (Eden Bioscience Corp., Bothell, WA; http://www.edenbio.com/), which is based on the 44 kDa protein harpin obtained from *Erwinia amylovora* (Wei *et al.*, 1992). Messenger™ was registered with the EPA, USA in April 2000 and is a wettable dry granule containing 3% of the harpin protein HarpinEA. It has broad activity on a wide spectrum of crops. Proact™ is described as the next generation of foliar applied Harp-N-Tek™ products from Eden Bioscience and is currently sold for use on cotton, corn and rice. In addition, Eden Bioscience also market N-Hibit™ for application to cotton seed to induce resistance against nematodes.

We therefore see a number of proteins, belonging to limited protein families, which are capable of eliciting resistance. Some of these proteins seem to act in a fairly direct manner (e.g. flagellin) while others with enzyme activity may be releasing elicitor active products.

2.2.1.7 Salicylic acid

Some plant growth promoting rhizobacteria produce extracellular salicylic acid which may be responsible, in part, for their ISR activity (Van Loon *et al.*, 1998). See also section 2.2.2.6.

2.2.1.8 Sphingolipids

Sphingolipids occur widely in membranes in eukaryotic cells and have a multitude of functions (Shah, 2005). One group of sphingolipids, the cerebrosides, described as non-race-specific elicitors (Umemura *et al.*, 2004), have been isolated from a range of fungal pathogens including *Fusarium oxysporum*, *Pythium* sp. and *Botrytis* sp., and have been shown to be effective inducers of a hypersensitive response and SAR (Keller *et al.*, 1996; Picard *et al.*, 2000; Baillieul *et al.*, 2003; Cordelier *et al.*, 2003). Treatment of *Lactuca sativa* (lettuce), *Lycopersicon esculentum* (tomato), *Cucumis melo* (melon) and *Ipomoea batatas* (sweet potato) with cerebroside B induced resistance to infection by *F. oxysporum* (Umemura *et al.*, 2004).

2.2.1.9 Volatile organic compounds

Ryu *et al.* (2004), studying the induction of resistance by PGPR, analysed various volatiles emitted by bacteria, demonstrated that 2,3-butanediol induced resistance to soft rot caused by *E. carotovora* subsp. *carotovora* in Arabidopsis and showed that it was associated with ethylene signalling and was independent of salicylic acid and jasmonic acid signalling pathways.

2.2.2 Plant extracts and plant products

2.2.2.1 Brassinolide

Brassinolide is a naturally occurring plant growth regulator and is perhaps the most active of the brassinosteroid family of plant hormones. Brassinolide is an effective elicitor on both monocots and dicots, and has been shown to enhance resistance to a range of pathogens in both tobacco and rice (Nakashita *et al.*, 2003). Brassinolide does not induce acidic or basic PR genes and does not require salicylic acid biosynthesis, suggesting it is acting through a different mechanism than SAR responses which are dependent on salicylic acid. Interestingly, brassinosteroids have also been shown to enhance resistance to cold stress (Hotta *et al.*, 1998).

2.2.2.2 Jasmonates and related compounds

Induction of jasmonates in plants is often associated with a wound response. Walters *et al.* (2006) showed that mechanical wounding of the first leaves of broad bean (*Vicia faba*) led to a reduction in rust (*Uromyces fabae*) infection in the wounded leaf as well as the unwounded second leaf. The increase in resistance was accompanied by an accumulation of jasmonic acid and two trihydroxy-oxylipins. The important role of jasmonates in intracellular signalling associated with resistance to pests and pathogens is well documented (Turner *et al.*, 2002), and there are many publications describing the topical application of methyl jasmonate to plants and the subsequent induction of resistance to a range of pathogens (references cited in Pozo *et al.*, 2005).

2.2.2.3 Oligogalacturonides

Oligogalacturonides (OGAs) obtained through pectic enzyme degradation or acid hydrolysis of pectic polysaccharides from plant cell walls have been shown to elicit a number of defence-related plant responses (see references in Shibuya & Minami, 2001) although they have not yet been shown to induce resistance to subsequent infection by a pathogen. Interestingly, there is a strong synergistic interaction between OGAs and the hepta-β-glucan isolated from fungal mycelium (Davis *et al.*, 1986). While some researchers have suggested application of more than one elicitor to plants to increase disease control on the basis of triggering different receptors, and hence different components of a resistance response, there has been no systematic study of elicitors to look for synergism.

2.2.2.4 Oxalate

Doubrava *et al.* (1988) extracted oxalate from spinach and rhubarb leaves and showed it was able to induce systemic resistance to *Colletotrichum lagenarium* in cucumber. By the nature of the inducer, such induction is likely to be due to a non-specific effect on the plant rather than through a specific oxalate receptor.

2.2.2.5 Plant extracts

A number of plant extracts have been shown to possess elicitor activity including extracts of *Hedera helix* (Baysal *et al.*, 2002). A commercial product Milsana (KHH BioScience Inc, Raleigh, NC) contains an ethanolic extract of *Reynoutria sachalinensis* and has shown activity against a range of pathogens on many crops, though the active ingredient has not been published. Lysaplant (previously known as Elorisan) produced by Bugico, Switzerland is described as a biostimulant and contains extracts from a number of different plant species (listed in Thompson, 2004). Lysaplant has been successfully used to control a number of diseases, particularly on trees, though it is not clear from publications how much of this control is due to induced resistance and whether some may be a direct action.

Von Rad *et al.* (2005) looked at gene expression in Arabidopsis treated with several commercially available elicitors including 'Neudo Vital', which is an ethanolic plant extract produced by W. Neudorff GmbH KG, Emmerthal, Germany, 'Bio-S', which is an extract of several plant species and is produced by Gebrüder Schätte KG, Bad Waldsee, Germany, and 'PRORADIX', which is an ethanolic extract of *Pseudomonas fluorescens* ssp. *proradix* and is produced by Sourcon Padena GmbH & Co, KG, Tübingen, Germany.

2.2.2.6 Salicylic acid

Salicylic acid (SA) is perhaps one of the first discovered compounds to induce resistance and is often associated with accumulation of pathogenesis-related (PR) proteins such as PR1. A number of compounds which are structurally related to SA have also been shown to possess the ability to induce resistance, and in fact BTH (benzothiadiazole) (see section 2.2.5.2), perhaps the first commercial produced resistance elicitor, is structurally related. SA is covered in some detail in Chapter 10.

2.2.2.7 Spermine

The polyamine spermine was shown to induce resistance in leaves of *Nicotiana tabacum* to infection by tobacco mosaic virus (TMV) via an SA-independent pathway (Yamakawa *et al.*, 1998).

2.2.2.8 Volatile organic compounds

Plants release a range of volatile organic compounds in response to wounding or herbivores, and some of these compounds have been reported to induce resistance responses under laboratory conditions. For example, (*E*)-2-hexenal which is released from wounded plants, can also induce the expression of defence-related genes in intact plants (Bate & Rothstein, 1998). In addition, Kessler *et al.* (2006) showed that clipped sagebrush released a number of volatiles, including methyl jasmonate, methacrolein and terpenoids. In laboratory and field experiments, these volatiles were said to prime the response of *Nicotiana attenuata* rather than directly inducing resistance.

The potential for complex interactions involving plant volatiles is demonstrated by the biological activity of *cis*-jasmone. Birkett *et al.* (2000) showed that *cis*-jasmone, which is a component of plant volatiles that can be induced by physical damage, when applied to intact bean plants induced the production of volatiles such as the monoterpene (*E*)-β-ocimene, which affects plant defence by stimulating the activity of parasitic insects. Similarly, *cis*-jasmone, applied to wheat plants in laboratory and field studies stimulated resistance to the grain aphid *Sitobion avenae* (Bruce *et al.*, 2003).

2.2.2.9 Ethylene

Ethylene has been widely described as having a role in signalling in response to biotic and abiotic stress, but its effect on disease resistance when applied externally is more variable. For example, the timing of applications appears to be critical, and it can sometimes increase resistance if applied before inoculation but seems to increase susceptibility if applied after inoculation (Van Loon *et al.*, 2006).

2.2.3 Carbohydrates

2.2.3.1 Chitosan

Chitosan has been reported to induce many defence-related responses in plants (see references in Cabrera *et al.*, 2006 and Chapter 10) and to possibly have a dual mode of action by directly affecting fungal growth.

Chitosan is derived from chitin, which is commercially extracted from shells of crustaceans such as crab and shrimp, and as it is widespread in nature, it is regarded as having a low potential for toxicity. Different forms of chitosan are available, depending on how they are produced. Chitosan obtained by alkaline deacetylation of chitin results in a product which is 20–30% acetylated with the acetyl groups uniformly distributed along the polymer. In contrast, in chitosan with a similar degree of acetylation which is obtained from fungi, the acetyl groups are clustered in groups.

Elexa™ is a commercial formulation containing 4% chitosan derived from crab shells (Sharathchandra *et al.*, 2004) which was developed and marketed by GlycoGenesys Inc. (Boston, MA) (http://www.glycogenesys.com/) and has been reported to protect a wide range of crops against many pathogens (Bhaskara Reddy *et al.*, 1999; Agostini *et al.*, 2003; Sharathchandra *et al.*, 2004). It was then marketed as Elexa by SafeScience (Salem, UT), but Elexa™ is now marketed as Elexa™ 4 Plant Defense Booster (Elexa™ 4 PDB) by Plant Defense Boosters Inc. (Syracuse, NY) and has been described as the first pesticide to be registered by the US Environmental Protection Agency as a 'Plant Defense Booster' (http://www.plantdefenseboosters.com/). Elexa™ 4 PDB is also said to have broad applicability on a wide range of crops (http://www.plantdefenseboosters.com/pdf_files/pdb7-12b.pdf). Another chitosan-based product, AgroChit, is marketed by Bioprogress as 3% chitosan solution in lactic acid and is sold as a plant growth regulator and inducer of disease resistance in potato plants (http://en.bioprogress.ru/production/production13.php).

2.2.3.2 Saccharin

Saccharin is a metabolite of probenazole (Uchiyama *et al.*, 1973), a compound known to induce resistance in rice (see section 2.2.5.1), and was first shown to be an inducer of SAR by Siegrist *et al.* (1998), who showed that it induced resistance to fungal and viral diseases when applied to cucumber, tobacco and bean plants. Subsequently, saccharin was shown to induce resistance to the rust *U. fabae* on *V. faba* and to the powdery mildew fungus *Blumeria graminis* f. sp. *hordei* on barley (Boyle & Walters, 2005, 2006).

2.2.3.3 Seaweed glucans

Laminarin is a water-soluble β-1,3 glucan from the brown alga *Laminaria digitata* and has an average degree of polymerization of 25 glucosyl residues and up to three single β-1,6 glucose branches (Read *et al.*, 1996). Though laminarin induces a number of defence responses in plants, it is only capable of inducing a low level of resistance to infection by pathogens. However, the level of induced resistance is much greater if the glucan is sulfated (e.g. to produce laminarin sulfate PS3 which has a degree of sulfation of 2.4) (Ménard *et al.*, 2004). Ménard *et al.* also showed that a minimum glucan chain length is essential for biological activity and that the sulfate residue is essential and could not be replaced by other anionic groups. The sulfated glucan PS3 induces only localized resistance and not systemic resistance (Ménard *et al.*, 2005).

2.2.4 Others

2.2.4.1 BABA

The non-protein amino acid DL-3-aminobutyric acid (BABA) has been shown to induce broad spectrum disease resistance in a wide range of crops (see Chapter 10 for details) and is effective, with few side effects, when applied as a soil drench. The structurally related compounds 2- and 3-aminobutyric acid isomers seem to be ineffective as resistance inducers (Ovadia *et al.*, 2000)

Interestingly, BABA is one of the few compounds that is effective as an elicitor in solanaceous plants (Cohen, 1994a, b; Cohen *et al.*, 1994; Siegrist *et al.*, 2000; Andreu *et al.*, 2006).

The potential specificity of some resistance elicitors was demonstrated by Ton & Mauch-Mani (2004), who showed that BABA induced resistance in Arabidopsis against *Alternaria brassicicola* and *Plectosphaerella cucumerina*, unlike BTH which had no effect against these two pathogens.

There are fewer reports of elicitors being tested for controlling insects, but Hodge *et al.* (2005) showed that BABA induced resistance in legumes to attack by the pea aphid *Acyrthosiphon pisum* when applied as a soil drench.

Interestingly, BABA-treated Arabidopsis plants were also sensitized to react faster and more efficiently to various abiotic stresses such as high temperature, high salinity and drought stress, suggesting there are some common molecular components responsive to biotic and abiotic stress (Jakab *et al.*, 2005).

Because of its broad spectrum of activity on a wide range of crops, BABA has many characteristics of an elicitor which could have practical applications.

A BABA–Cu complex (Makhteshim-Agan, Israel) has been tested on grapevines where it was very effective in controlling downy mildew (*Plasmopara viticola*) (Reuveni *et al.*, 2001) and may perhaps offer a useful way forward in combining the concept of induced resistance with direct antimicrobial activity.

2.2.4.2 Lipids/fatty acids

Systemic resistance to *Phytophthora infestans* is induced in potato by unsaturated fatty acids such as arachidonic, eicosapentaenoic, linoleic, linolenic and oleic acids (Cohen *et al.*, 1991). Arachidonic and eicosapentaenoic acids were particularly effective at inducing resistance, though they also caused some necrosis, while linoleic and oleic acids did not induce necrotic spots. This group of elicitors is particularly important for potato and tomato, as some of the other groups of elicitors (particularly glucans) seem to be rather ineffective on these plants.

2.2.4.3 Nitric oxide

Nitric oxide (NO) is an important signalling molecule (Delledonne, 2005) that is involved in the establishment of SAR (for example) in tobacco, and NO-releasing compounds such as nitrosoglutathione (GSNO) induce systemic resistance against TMV in tobacco (Song & Goodman, 2001). NO is induced by LPS; see section 2.2.1.5.

The effects of NO on plants are varied and extensive including some which are detrimental such as cytotoxicity (Romero-Puertas *et al.*, 2004). Thus, it seems that although it may possess some ability to initiate induced resistance, it is unlikely to become a molecule that would have practical applications for disease control.

2.2.5 Commercially available products

2.2.5.1 Fungicides

A number of fungicides have been shown to have a dual mode of action, i.e. direct antifungal activity as well as activating a (low) level of induced resistance. For example, in glasshouse experiments, plant gene expression patterns induced by fenpropimorph were similar, though less intense, to those induced by BTH, and azoxystrobin also induced some defence-related genes (Pasquer *et al.*, 2005). Interestingly, these genes were already expressed at a high level in field experiments and did not show any further increase in response to fungicide or BTH treatment.

Using NahG and *nim1* (non-inducible immunity) mutants of Arabidopsis, Molina *et al.* (1998) showed that the fungicides metalaxyl, fosetyl and $Cu(OH)_2$ were much less effective in controlling *P. parasitica* than on wild type plants, suggesting that in part they controlled pathogens through an induction of host defence responses. In the plant, Fosetyl-Al is converted into a phosphite ion, which itself is known to induce resistance when applied externally to plants. A number of commercial products containing phosphite as an active component are available.

The synthetic compound probenazole (Oryzemate®) has been used for many years to control rice blast caused by *Magnaporthe grisea* and has long been known to have a dual mode of action, by having a weak direct antifungal activity as well as stimulating the host's resistance mechanisms. Probenazole (PBZ) and its active metabolite 1,2-benzisothiazole-1,1-dioxide (BIT) have been shown to induce SAR in Arabidopsis and tobacco upstream of salicylic acid, thereby acting through salicylic acid accumulation (Nakashita *et al.*, 2002b). Probenazole breaks down in plants to produce the related compound saccharin (Uchiyama *et al.*, 1973) (see section 2.2.3.2).

2,2-Dichloro-3,3-dimethylocyclopropane carboxylic acid (DDCC) (WL28325) application to rice caused an accumulation of momilactone phytoalexins which coincided with inhibition of hyphal growth of *Pyricularia oryzae* (syn *M. grisea*) (Cartwright *et al.*, 1980). A related fungicide, carpropamid ((1*RS*,3*SR*)2,2-dichloro-*N*-[1-(4-chlorophenyl) ethyl-1-ethyl-3-methyl cyclopropanecarboxamide]) is also reported to have a dual activity acting, in part, as a resistance activator and also by inhibiting melanin biosynthesis (Oostendorp *et al.*, 2001).

A strobilurin fungicide, Pyraclostrobin (BASF F500), also possesses some elicitor activity and enhances the resistance of tobacco against TMV and wildfire (*P. syringae* pv. *tabaci*) possibly by priming the plants prior to subsequent attack (Herms *et al.*, 2002).

A new product, proquinazid (6-iodo-2-propoxy-3-propylquinazolin-4(3H)-one), from DuPont and recommended for controlling mildew in cereals and grapes has been reported to induce or switch on the crop's defence mechanisms to mildew as well as having a direct antifungal activity (Abram, 2005). Proquinazid is sold in Poland as Talius®.

A number of other products are available which have been variously described by their manufacturers as enhancing disease resistance. However, many of these may be better described as nutritional supplements; categorical evidence that they are stimulating resistance mechanisms is still lacking, and for that reason they have not been listed here.

2.2.5.2 Synthetic resistance inducers

Benzothiadiazole or BTH (benzo(1,2,3)thiadiazole-7-carbothioic acid *S*-methyl ester: CGA 245704) has been shown to possess activity in a wide range of plant species against a wide range of pathogens (Friedrich *et al.*, 1996; Lawton *et al.*, 1996; Benhamou & Bélanger, 1998; Tally *et al.*, 1999) and is perhaps the best known synthetic elicitor and was developed commercially as Bion® by Novartis (now marketed by Syngenta) (Kessmann *et al.*, 1996). It is perhaps particularly effective against mildew in wheat (Görlach *et al.*, 1996) and could be effective for up to 10 weeks (Ruess *et al.*, 1996). Interestingly, BTH also induced resistance in tomato plants to the whitefly *Bemisia tabaci* (Nombela *et al.*, 2005) and in sunflower (*Helianthus annuus*) to the parasitic weed broomrape (*Orobanche cumana*) (Buschmann *et al.*, 2005).

Busam *et al.* (1997) suggested that 2,6-dichloroisonicotinic acid (INA) and BTH act more selectively than salicylic acid and that the molecular responses to BTH and SA are not the same. For instance, Heidel & Baldwin (2004) showed that there were some differences in the defence-related genes induced by BTH compared with SA. Defence genes induced in

N. attenuata by BTH but not by SA included *α-DOX*, 5-epi-aristolchene synthase, proteinase inhibitor, *WRKY2*, *WRKY3*, xyloglucan *endo*-transglycosylase, germin, *PR-3* and *PAL*. Similarly, Bovie *et al.* (2004) showed a very strong induction of PR8 mRNA in cucumber in response to BTH, while the response to salicylic acid was virtually negligible.

N-cyanomethyl-2-chloroisonicotinamide (NCI) has been shown to induce a broad range of disease resistance in tobacco and rice without stimulating salicylic acid biosynthesis (Nakashita *et al.*, 2002a) and appears to activate SAR by acting at a point between salicylic acid and NPR1 (Yasuda *et al.*, 2003). A structurally related compound N-phenylsulfonyl-2-chloroisonicotinamide is also effective as an elicitor and has also been reported to protect rice against *M. grisea* (Yoshida *et al.*, 1990). Oxycom™ (Redox Chemicals Inc., Burley, ID) consists of two components, viz. component A, which is 5% v/v stabilized solution of peracetic acid containing 10–12% acetic acid and 20–22% hydrogen peroxide, and component B which is a mixture of plant nutrients (Kim *et al.*, 2001). Thus, its activity is based in part on the production of 'active oxygen species' and in part on a 'weak' antifungal activity. It has been reported to induce host responses associated with disease resistance and shows some systemic activity. It has been tested under field conditions on a number of crops against a range of pathogens (Kim *et al.*, 2001).

ReZist (initially manufactured by Stoller Enterprises, Inc. (Houston, TX) but later marketed by Micromix International Ltd (Nottingham, UK) has activity against citrus scab (*Elsinoë fawcettii*) on rough lemon (*Citrus jambhiri*) seedlings (Agostini *et al.*, 2003). Although ReZist is often described as an inducer of SAR, it is not clear how much of its activity is due to a direct effect on the pathogen and how much is due to an induced host response. ReZist is reported to contain 1.75% copper, 1.75% manganese and 1.75% zinc with polyamines and natural plant extracts.

2.2.5.3 Others

Vitamin B_1 (thiamine) induces SAR when applied to rice, Arabidopsis and vegetable crops, and increased resistance to fungal, bacterial and viral infections (Ahn *et al.*, 2005).

Menadione sodium bisulfite, a water-soluble addition compound of menadione (vitamin K_3), which was first studied as a plant growth regulator, induces resistance in banana to Panama disease caused by the wilt disease *F. oxysporum* f. sp. *cubense* (Borges *et al.*, 2004) and in oilseed rape against *Brassica napus* (Borges *et al.*, 2003). The use of menadione and its derivatives to induce resistance in plants has been patented (Borges-Pérez & Fernández-Falcón, 1995, 1996).

Riboflavin, or vitamin B_2, applied to Arabidopsis induces systemic resistance to *P. parasitica* and *P. syringae* pv. *tomato*, and resistance to TMV and *Alternaria alternata* in tobacco (Dong & Beer, 2000).

Three synthetic amides of adipic acid have been shown to induce resistance in pepper plants to subsequent infection with *Alternaria solani* (Flors *et al.*, 2003). Although their mode of action is unknown, current evidence suggests that they are acting through an induction of resistance rather than having a direct antimicrobial action, and interestingly no phytotoxic effects have yet been observed, thus making them potentially of some commercial value.

Cholic acid, a bile acid in animals, when applied to rice leaves not only induced the accumulation of phytoalexins, a hypersensitive response and PR proteins, but also increased resistance to subsequent infection by *M. grisea* (Koga *et al.*, 2006). Interestingly, while a fungal cerebroside isolated from *M. grisea* induced both the phytocassane and momilactone phytoalexins, the cholic acid induced mainly phytocassanes, suggesting a level of specificity in the induction process.

UV-C has been shown to induce a number of resistance-related responses including induction of chitinase, β-1,3-glucanase and phenylalanine ammonia-lyase in peach fruit (El Ghaouth *et al.*, 2003).

Reignault *et al.* (2004) tested the effect of chitosan, Iodus 40®, Milsana®, salicylyl heptanoate, trehalose, and pectic oligosaccharides on wheat to control powdery mildew and showed a reduction in the level of infection with Milsana®, salicylyl heptanoate and trehalose. Iodus 40® is a product of Goëmar (Saint-Malo, France; www.goemar.com) and is recommended by them for the control of diseases on wheat.

Silicon increases resistance of plants to pathogenic fungi possibly through an interaction with defence responses and silicic acid was postulated to play a possible role in both local and systemic resistance (Fauteux *et al.*, 2005) (see also Chapter 10).

2.3 Conclusions

A wide range of structurally diverse compounds have been shown to have the ability to induce resistance, and a very limited screen of ad hoc compounds by Fought & Kuć (1996) clearly demonstrated that many commonly occurring compounds do not elicit resistance. There is little information in the literature on the results of wide scale screening of compounds for elicitor activity, though undoubtedly some will have been carried out by agrochemical companies. Relatively minor changes in chemical structure can influence elicitor ability, thus providing ample opportunity to discover new compounds which have a greater effectiveness than currently used compounds. Though many compounds are generally non-specific and induce resistance in a wide range of crop species against a diverse range of plant pathogens, differences in efficacy do exist. Thus, it is right to expect that some elicitors will be more effective than others and that opportunities exist to have additive effects if different elicitors are applied at the same time.

Though some of the PR proteins such as PR1 are considered to be an almost diagnostic indicator of a compound's ability to induce resistance, other responses do vary between different groups of elicitors. Where the composition of an elicitor formulation is complex, for example with plant or microbial extracts, then host responses at the gene transcription level can also be complex and can vary between different elicitors (von Rad *et al.*, 2005).

However, if we are to discover really new elicitors which are robust and effective, and have few side effects, there needs to be confidence in the biotechnology industry that such screening efforts will be rewarded with a successful commercial product.

2.3.1 REDOX regulation

To adequately understand how this wide variety of compounds can induce a common response, or at least a small number of similar responses, one needs to discover a common mechanism through which they function. For example, SAR is often associated with an oxidative burst and a subsequent accumulation of SAR genes including genes for PR proteins. Thus, one recent suggestion is that elicitors may operate through redox-related changes

(Pavet *et al.*, 2005). This involves conformational changes to NPR1 from an inactive oligomer to an active monomer (Mou *et al.*, 2003) which accumulates in the nucleus, and changes involving cysteine residues in TGA1 and TGA4 transcription factors (Fobert & Després, 2005). Interestingly, redox-related changes have been shown to be affected by a number of abiotic stresses (Pastori & Foyer, 2002) including light (Mateo *et al.*, 2004) and humidity (Zhou *et al.*, 2004) in Arabidopsis. Thus, to fully understand the process of induced resistance induction, one needs to understand how all these resistance elicitors could act through common regulatory pathways. Similarly, in terms of looking for an over-arching explanation of how some compounds can induce SAR, Sticher & Métraux (2000) showed that inhibitors of N-glycosylation such as tunicamycin or amphomycin induced SAR in cucumber. This does not seem to be a primary trigger for induction of resistance but does seem to be an essential step in the process.

2.3.2 Factors affecting efficacy

The effectiveness of various elicitors is affected by plant taxonomy, and there are many examples where an elicitor is more effective on some plant species than on others. The molecular basis for these differences is not understood and perhaps highlights a potential problem when setting up a screening system for elicitors.

There is already evidence that some elicitors are at least family- or genus-specific, and there has been evidence that efficacy can be affected markedly by adjuvants. Edwards *et al.* (2005) showed that herbicide safeners could act in both a chemical- and species-specific manner, thus demonstrating the enormous potential when considering elicitor formulation.

2.3.3 Synergism

Molina *et al.* (1998) showed that application of BTH in combination with the fungicides metalaxyl, fosetyl and $Cu(OH)_2$ resulted in a synergistic effect on pathogen resistance in wild type Arabidopsis plants and an additive effect in NahG and the BTH-unresponsive *nim1* mutants. The extent to which different types of elicitor can be used together to give additive or even synergistic effects has not been fully explored.

2.3.4 Assays

Although cell cultures have been successfully tested for their use to screen for resistance inducers (Siegrist *et al.*, 1998), most studies have relied on the topical application of compounds to whole plants. Setting up assays to detect elicitor activity may not always be simple and straightforward, and there are many anecdotal stories to suggest that elicitor screening using standard screens for fungicides is not appropriate for resistance inducers. In addition, Koga *et al.* (1998) suggested that such assays had to be optimized with regard to environmental conditions and recommended 22°C, high light (30,000 lux) and 80% humidity to maximize the response of rice leaves to elicitor-active extracts of *M. grisea*. Such conditions may simply be close to the optimum for healthy growth of the rice plants but do indicate the importance of environmental conditions. Koga *et al.* (1998) also showed an effect of leaf age with younger leaves producing a lower response than older leaves. That humidity can affect resistance responses has also been demonstrated by Zhou *et al.* (2004) who showed that the enhanced disease resistance phenotype of an Arabidopsis *ssi4* mutant, which exhibits spontaneous lesion formation, was suppressed by high (95%) relative humidity and hypothesized that a humidity sensitive factor may be present in the *ssi4* signalling pathway.

A number of elicitor-active compounds have been isolated from fungi as pathologists have sought to understand molecular events associated with infection processes. The search for naturally occurring, and in particular synthetic, compounds does not yet appear to have been exhaustive, and many more families of natural and synthetic elicitor-active compounds undoubtedly await discovery.

To provide the right commercial environment (see Chapter 12), we therefore need to understand why we have inconsistency with some of our existing compounds and to fully understand how best to use these products within an integrated crop protection strategy. When these shortfalls are resolved, we will undoubtedly continue to discover many new and diverse groups of compounds able to induce resistance.

2.4 Acknowledgements

I am grateful to The Scottish Executive Environment and Rural Affairs Department (SEERAD) for continued support and funding.

2.5 References

Abram M. 2005. Ammunition for mildew. *Farmers Weekly* 16 December 2005, 59.

Agostini JP, Bushong PM, Timmer LW, 2003. Greenhouse evaluation of products that induce host resistance for control of scab, melanose and Alternaria brown spot of citrus. *Plant Disease* **87,** 69–74.

Ahn IP, Kim S, Lee YH, 2005. Vitamin B1 functions as an activator of plant disease resistance. *Plant Physiology* **138,** 1505–1515.

Andreu AB, Guevara MG, Wolski EA, Daleo GR, Caldiz DO, 2006. Enhancement of natural disease resistance in potatoes by chemicals. *Pest Management Science* **62,** 162–170.

Baillieul F, de Ruffray P, Kauffmann S, 2003. Molecular cloning and biological activity of α-, β-, and γ-megaspermin, three elicitins secreted by *Phytophthora megasperma* H20. *Plant Physiology* **131,** 155–166.

Bate NJ, Rothstein SJ, 1998. C_6-volatiles derived from the lipoxygenase pathway induce a subset of defense-related genes. *The Plant Journal* **16,** 561–569.

Baysal O, Laux P, Zeller W, 2002. Further studies on the induced resistance (IR) effect of plant extract from *Hedera helix* against fire blight *(Erwinia amylovora). Acta Horticulturae* **590,** 273–277.

Benhamou N, Bélanger RR, 1998. Induction of systemic resistance to *Pythium* damping-off in cucumber plants by benzothiadiazole: ultrastructure and cytochemistry of the host response. *The Plant Journal* **14,** 13–21.

Bhaskara Reddy MV, Arul J, Angers P, Couture L, 1999. Chitosan treatment of wheat seeds induces resistance to *Fusarium graminearum* and improves seed quality. *Journal of Agricultural and Food Chemistry* **47,** 1208–1216.

Birkett MA, Campbell CAM, Chamberlain K, Guerrieri E, Hick AJ, Martin JL, Matthes M, Napier JA, Pettersson J, Pickett JA, Poppy GM, Pow EM, Pye BJ, Smart LE, Wadhams GH, Wadhams LJ, Woodcock CM, 2000. New roles for cis-jasmone as an insect semiochemical and in plant defense. *Proceedings of the National Academy of Sciences, USA* **97,** 9329–9334.

Borges AA, Borges-Pérez M, Fernández-Falcón MJ, 2004. Induced resistance to Fusarial wilt of banana by menadione sodium bisulphite treatments. *Crop Protection* **23,** 1245–1247.

Borges AA, Cools HJ, Lucas JA, 2003. Menadione sodium bisulphite: a novel plant defence activator which enhances local and systemic resistance to infection by *Leptosphaeria maculans* in oilseed rape. *Plant Pathology* **52,** 429–436.

Borges-Pérez M, Fernández-Falcón MJ, 1995. Compositions for inducing resistance to tracheomycosis in plants. Patent WO 95/0372.

Borges-Pérez M, Fernández-Falcón MJ, 1996. Utilization of compositions which contain menadione for the stimulation of plant metabolism in order to induce their resistance to pathogen and pest and/or accelerate their blooming. Patent WO/9628026.

Bovie C, Ongena M, Thonart P, Dommes J, 2004. Cloning and expression analysis of cDNAs corresponding to genes activated in cucumber showing systemic acquired resistance after BTH treatment. *BMC Plant Biology* **4,** 15.

Boyle C, Walters D, 2005. Induction of systemic protection against rust infection in broad bean by saccharin: effects on plant growth and development. *New Phytologist* **167,** 607–612.

Boyle C, Walters DR, 2006. Saccharin-induced protection against powdery mildew in barley: effects on growth and phenylpropanoid metabolism. *Plant Pathology* **55,** 82–91.

Bru R, Sellés S, Casado-Vela J, Belchí-Navarro S, Pedreño MA, 2006. Modified cyclodextrins are chemically defined glucan inducers of defense responses in grapevine cell cultures. *Journal of Agricultural and Food Chemistry* **54,** 65–71.

Bruce TJA, Martin JL, Pickett JA, Pye BJ, Smart LE, Wadhams LJ, 2003. cis-Jasmone treatment induces resistance in wheat plants against the grain aphid, *Sitobion avenae* (Fabricius) (Homoptera: Aphididae). *Pest Management Science* **59,** 1031–1036.

Busam G, Kassemeyer H-H, Matern U, 1997. Differential expression of chitinases in *Vitis vinifera* L. responding to systemic acquired resistance activators or fungal challenge. *Plant Physiology* **115,** 1029–1038.

Buschmann H, Fan Z-W, Sauerborn J, 2005. Effect of resistance-inducing agents on sunflower (*Helianthus annuus* L.) and its infestation with the parasitic weed *Orobanche cumana* Wallr. *Journal of Plant Diseases and Protection* **112,** 386–397.

Cabrera JC, Messiaen J, Cambier P, Van Cutsem P, 2006. Size, acetylation and concentration of chitooligosaccharide elicitors determine the switch from defence involving PAL activation to cell death and water peroxide production in Arabidopsis cell suspensions. *Physiologia Plantarum* **127,** 44–56.

Cartwright DW, Langcake P, Ride JP, 1980. Phytoalexin production in rice and its enhancement by a dichlorocyclopropane fungicide. *Physiological Plant Pathology* **17,** 259–267.

Cohen Y, 1994a. Local and systemic control of *Phytophthora infestans* in tomato plants by DL-3-amino-*n*-butanoic acids. *Phytopathology* **84,** 55–59.

Cohen Y, 1994b. 3-Aminobutyric acid induces systemic resistance against *Peronospora tabacina. Physiological and Molecular Plant Pathology* **44,** 273–288.

Cohen Y, Gisi U, Mosinger E, 1991. Systemic resistance of potato plants against *Phytophthora infestans* induced by unsaturated fatty acids. *Physiological and Molecular Plant Pathology* **38,** 255–263.

Cohen Y, Niderman T, Mössinger E, Fluhr R, 1994. β-Aminobutyric acid induces the accumulation of pathogenesis-related proteins in tomato (*Lycopersicon esculentum* L.) plants and resistance to late blight infection caused by *Phytophthora infestans*. *Plant Physiology* **104,** 59–66.

Conrath U, Corné MJP, Mauch-Mani B, 2002. Priming in plant–pathogen interactions. *Trends in Plant Science* **7,** 210–216.

Cordelier S, de Ruffray P, Fritig B, Kauffmann S, 2003. Biological and molecular comparison between localized and systemic acquired resistance induced in tobacco by a *Phytophthora megasperma* glycoprotein elicitin. *Plant Molecular Biology* **51,** 109–118.

Coventry HS, Dubery IA, 2001. Lipopolysaccharides from *Burkholderia cepacia* contribute to an enhanced defensive capacity and the induction of pathogenesis-related proteins in *Nicotiana tabacum. Physiological and Molecular Plant Pathology* **58,** 149–158.

Davis K, Darvill A, Albersheim P, 1986. Several biotic and abiotic elicitors act synergistically in the induction of phytoalexin accumulation in soybean. *Plant Molecular Biology* **6,** 23–32.

Delledonne M, 2005. NO news is good news for plants. *Current Opinion in Plant Biology* **8,** 390–396.

Dong H, Beer SV, 2000. Riboflavin induces resistance in plants by activating a novel signal transduction pathway. *Phytopathology* **90,** 801–811.

Doubrava NS, Dean RA, Kuć J, 1988. Induction of systemic resistance to anthracnose caused by *Colletotrichum lagenarium* in cucumber by oxalate and extracts from spinach and rhubarb leaves. *Physiological and Molecular Plant Pathology* **33,** 69–79.

Dow M, Newman M-A, von Roepenack E, 2000. The induction and modulation of plant defense responses by bacterial lipopolysaccharides. *Annual Review of Phytopathology* **38,** 241–261.

Edwards R, Del Buono D, Fordham M, Skipsey M, Brazier M, Dixon DP, Cummings I, 2005. Differential induction of glutathione transferases and glucosyltransferases in wheat, maize and *Arabidopsis thaliana* by herbicide safeners. *Zeitschrift für Naturforschung* **60,** 307–316.

El Ghaouth A, Wilson CL, Callahan AM, 2003. Induction of chitinase, β-1,3-glucanase, and phenylalanine ammonia lyase in peach fruit by UV-C treatment. *Phytopathology* **93,** 349–355.

Erbs G, Newman M-A, 2003. The role of lipopolysaccharides in induction of plant defence responses. *Molecular Plant Pathology* **4,** 421–425.

Farmer EE, Helgeson JP, 1987. An extracellular protein from *Phytophthora parasitica* var. *nicotianae* is associated with stress metabolite accumulation in tobacco callus. *Plant Physiology* **85,** 733–740.

Fauteux F, Rémus-Borel W, Menzies JG, Bélanger RR, 2005. Silicon and plant disease resistance against pathogenic fungi. *FEMS Microbiology Letters* **249,** 1–6.

Flors V, Miralles C, González-Bosch C, Carda M, García-Agustín P, 2003. Three novel synthetic amides of adipic acid protect *Capsicum annuum* plants against the necrotrophic pathogen *Alternaria solani*. *Physiological and Molecular Plant Pathology* **63,** 151–158.

Fobert PR, Després C, 2005. Redox control of systemic acquired resistance. *Current Opinion in Plant Biology* **8,** 378–382.

Fought L, Kuć JA, 1996. Lack of specificity in plant extracts and chemicals as inducers of systemic resistance in cucumber plants to anthracnose. *Journal of Phytopathology* **144,** 1–6.

Friedrich L, Lawton K, Ruess W, Masner P, Specker N, Rella MG, Meier B, Dincher S, Staub T, Uknes S, Métraux J-P, Kessmann H, Ryals J, 1996. A benzothiadiazole derivative induces systemic acquired resistance in tobacco. *The Plant Journal* **10,** 61–70.

Fuchs Y, Saxena A, Gamble HR, Anderson JD, 1989. Ethylene biosynthesis-inducing protein from cellulysin is an endoxylanase. *Plant Physiology* **89,** 138–143.

Gómez-Gómez L, Boller T, 2002. Flagellin perception: a paradigm for innate immunity. *Trends in Plant Science* **7,** 251–256.

Görlach J, Voirath S, Knauf-Beiter G, Hengy G, Beckhove U, Kogel K-H, Oostendorp M, Staub T, Ward E, Kessmann H, Ryals J, 1996. Benzothiadiazole, a novel class of inducers of systemic acquired resistance, activates gene expression and disease resistance in wheat. *The Plant Cell* **8,** 629–643.

Granado J, Felix G, Boller T, 1995. Perception of fungal sterols in plants. *Plant Physiology* **107,** 485–490.

Haas D, Keel C, 2003. Regulation of antibiotic production in root-colonizing *Pseudomonas* spp. and relevance for biological control of plant disease. *Annual Review of Phytopathology* **41,** 117–153.

Hanson LE, Howell CR, 2004. Elicitors of plant defense responses from biological control strains of *Trichoderma virens*. *Phytopathology* **94,** 171–176.

Harman GE, Howell CR, Viterbo A, Chet I, Lorito M, 2004. *Trichoderma* species – opportunistic, avirulent plant symbionts. *Nature Reviews in Microbiology* **2,** 43–56.

Heidel AJ, Baldwin IT, 2004. Microarray analysis of salicylic acid- and jasmonic acid-signalling in responses of *Nicotiana attenuata* to attack by insects from multiple feeding guilds. *Plant, Cell and Environment* **27,** 1362–1373.

Herms S, Seehaus K, Koehle H, Conrath U, 2002. A strobilurin fungicide enhances the resistance of tobacco against tobacco mosaic virus and *Pseudomonas syringae* pv. *tabaci*. *Plant Physiology* **130,** 120–127.

Hodge S, Thompson GA, Powell G, 2005. Application of DL-β-aminobutyric acid (BABA) as a root drench to legumes inhibits the growth and reproduction of the pea aphid *Acyrthosiphon pisum* (Hemiptera: Aphididae). *Bulletin of Entomological Research* **95,** 449–455.

Hoitink HAJ, Madden LV, Dorrance AE, 2006. Systemic resistance induced by *Trichoderma* spp.; interactions between the host, the pathogen, the biocontrol agent and soil organic matter quality. *Phytopathology* **96,** 186–189.

Hotta Y, Tanaka T, Bingshan L, Takeuchi Y, Konnai M, 1998. Improvement of cold resistance in rice seedlings by 5-aminolevulinic acid. *Journal of Pesticide Science* **23,** 29–33.

Iavicoli A, Boutet E, Buchala A, Métraux J-P, 2003. Induced systemic resistance in *Arabidopsis thaliana* in response to root inoculation with *Pseudomonas fluorescens* CHA0. *Molecular Plant–Microbe Interactions* **16,** 851–858.

Jakab G, Ton J, Flors V, Zimmerli L, Métraux J-P, Mauch-Mani B, 2005. Enhancing Arabidopsis salt and drought stress tolerance by chemical priming for its abscisic acid responses. *Plant Physiology* **139,** 267–274.

Ji C, Okinaka Y, Takeuchi Y, Tsurushima T, Buzzell RI, Sims JJ, Midland SL, Slaymaker D, Yoshikawa M, Yamaoka N, Keen NT, 1997. Specific binding of the syringolide elicitors to a soluble protein fraction from soybean leaves. *The Plant Cell* **9,** 1425–1433.

Kauss H, Jeblick W, 1996. Influence of salicylic acid on the induction of competence for H_2O_2 elicitation. *Plant Physiology* **111,** 755–763.

Keller H, Blein J-P, Bonnet P, Ricci P, 1996. Physiological and molecular characteristics of elicitin-induced systemic acquired resistance in tobacco. *Plant Physiology* **110,** 365–376.

Kempf H-J, Wolf G, 1989. *Erwinia herbicola* as a biocontrol agent of *Fusarium culmorum* and *Puccinia recondita* f. sp. *tritici* on wheat. *Phytopathology* **79,** 990–994.
Kessler A, Baldwin IT, 2002. Plant responses to insect herbivory: the emerging molecular analysis. *Annual Review of Plant Biology* **53,** 299–328.
Kessler A, Halitschke R, Diezel C, Baldwin IT, 2006. Priming of plant defense responses in nature by airborne signalling between *Artemisia tridentata* and *Nicotiana attenuata*. *Oecologia* **148,** 280–292.
Kessmann H, Oostendorp M, Ruess W, Staub T, Kunz W, Ryals J, 1996. Systemic activated resistance – a new technology for plant disease control. *Pesticide Outlook* **7,** 10–13.
Kim YC, Blee KA, Robins J, Anderson AJ, 2001. Oxycom™ under field and laboratory conditions increases resistance responses in plants. *European Journal of Plant Pathology* **107,** 129–136.
Koga J, Kubota H, Gomi S, Umemura K, Ohnishi M, Kono T, 2006. Cholic acid, a bile acid elicitor of hypersensitive cell death, pathogenesis-related protein, and phytoalexins accumulation in rice. *Plant Physiology* **140,** 1475–1483.
Koga J, Oshima K, Ogawa N, Ogasawara N, Shimura M, 1998. A new bioassay for measuring elicitor activity in rice leaves. *Annals of the Phytopathology Society of Japan* **64,** 97–101.
Kopp M, Rouster J, Fritig B, Darvill A, Albersheim P, 1989. Host-pathogen interactions XXXII. A fungal glucan preparation protects Nicotianae against infection by viruses. *Plant Physiology* **90,** 208–216.
Kunze G, Zipfel C, Robatzek S, Niehaus K, Boller T, Felix G, 2004. The N-terminus of bacterial elongation factor Tu elicits innate immunity in Arabidopsis plants. *The Plant Cell* **16,** 3496–3507.
Lawton KA, Friedrich L, Hunt M, Weymann K, Delaney T, Kessmann H, Staub T, Ryals J, 1996. Benzothiadiazole induces disease resistance in *Arabidopsis* by activation of the systemic acquired resistance signal transduction pathway. *The Plant Journal* **10,** 71–82.
Lherminier J, Benhamou N, Larrue J, Milat M-L, Boudon-Padieu E, Nicole M, Blein J-P, 2003. Cytological characterization of elicitin-induced protection in tobacco plants infected by *Phytophthora parasitica* or phytoplasma. *Phytopathology* **93,** 1308–1319.
Lotan T, Fluhr R, 1990. Xylanase, a novel elicitor of pathogenesis-related proteins in tobacco, uses a non-ethylene pathway for induction. *Plant Physiology* **93,** 811–817.
Mateo A, Mühlenbock P, Rustérucci C, Chang CC-C, Miszalski Z, Karpinska B, Parker JE, Mullineaux PM, Karpinski S, 2004. LESION SIMULATING DISEASE 1 is required for acclimation to conditions that promote excess excitation energy. *Plant Physiology* **136,** 2828–2830.
Maurhofer M, Keel C, Haas D, Défago G, 1995. Influence of plant species on disease suppression by *Pseudomonas fluorescens* strain CHA0 with enhanced antibiotic production. *Plant Pathology* **44,** 40–50.
Ménard R, Alban S, de Ruffray P, Jamois F, Franz G, Fritig B, Yvin J-C, Kauffmann S, 2004. β-1,3 glucan sulphate, but not β-1,3 glucan, induces the salicylic acid signaling pathway in tobacco and Arabidopsis. *The Plant Cell* **16,** 3020–3032.
Ménard R, de Ruffray M, Fritig B, Yvin J-C, Kauffmann S, 2005. Defense and resistance-inducing activities in tobacco of the sulfated β-1,3 glucan PS3 and its synergistic activities with the unsulfated molecule. *Plant and Cell Physiology* **46,** 1964–1972.
Mithöfer A, Bhagwat AA, Feger M, Ebel J, 1996. Suppression of fungal β-glucan-induced plant defence in soybean (*Glycine max* L.) by cyclic 1,3–1,6-β-glucans from the symbionts *Bradyrhizobium japonicum*. *Planta* **199,** 270–275.
Molina A, Hunt MD, Ryals JA, 1998. Impaired fungicide activity in plants blocked in disease resistance signal transduction. *The Plant Cell* **10,** 1903–1914.
Montesano M, Brader G, Palva ET, 2003. Pathogen derived elicitors: searching for receptors in plants. *Molecular Plant Pathology* **4,** 73–79.
Mou Z, Fan W, Dong X, 2003. Inducers of plant systemic acquired resistance regulate NPR1 function through redox changes. *Cell* **113,** 935–944.
Nakashita H, Yasuda M, Nishioka M, Hasegawa S, Arai Y, Uramoto M, Yoshida S, Yamaguchi I, 2002a. Chloroisonicotinamide derivative induces a broad range of disease resistance in rice and tobacco. *Plant and Cell Physiology* **43,** 823–831.
Nakashita H, Yasuda M, Nitta T, Asami T, Fujioka S, Arai Y, Sekimata K, Takatsuto S, Yamaguchi I, Yoshida S, 2003. Brassinosteroid functions in a broad range of disease resistance in tobacco and rice. *The Plant Journal* **33,** 887–898.
Nakashita H, Yoshioka K, Yasuda M, Nitta T, Arai Y, Yoshida S, Yamaguchi I, 2002b. Probenazole induces systemic acquired resistance in tobacco through salicylic acid accumulation. *Physiological and Molecular Plant Pathology* **61,** 197–203.
Newman M-A, von Roepenack E, Daniels M, Dow M, 2000. Lipopolysaccharides and plant responses to phytopathogenic bacteria. *Molecular Plant Pathology* **1,** 25–31.

Nombela G, Pascual S, Aviles M, Guillard E, Muñiz M, 2005. Benzothiadiazole induces local resistance to *Bemisia tabaci* (Hemiptera: Aleyrodidae) in tomato plants. *Journal of Economic Entomology* **98,** 2266–2271.

Oostendorp M, Kunz W, Dietrich B, Staub T, 2001. Induced disease resistance in plants by chemicals. *European Journal of Plant Pathology* **107,** 19–28.

Ortmann I, Moerschbacher BM, 2006. Spent growth medium of *Pantoea agglomerans* primes wheat suspension cells for augmented accumulation of hydrogen peroxide and enhanced peroxidase activity upon elicitation. *Planta* **224,** 963–970.

Ovadia A, Biton R, Cohen Y, 2000. Induced resistance to downy mildew and Fusarium wilt in cucurbits. *Acta Horticulturae (ISHS)* **510**, 55–60.

Pasquer F, Isidore E, Zarn J, Keller B, 2005. Specific patterns of changes in wheat gene expression after treatment with three antifungal compounds. *Plant Molecular Biology* **57,** 693–707.

Pastori GM, Foyer CH, 2002. Common components, networks, and pathways of cross-tolerance to stress. The central role of 'redox' and abscisic acid-mediated controls. *Plant Physiology* **129,** 460–468.

Pavet V, Olmos E, Kiddle G, Mowla S, Kumar S, Antoniw J, Alvarez ME, Foyer CH, 2005. Ascorbic acid deficiency activates cell death and disease resistance responses in Arabidopsis. *Plant Physiology* **139,** 1291–1303.

Picard K, Ponchet M, Blein J-P, Rey P, Tirilly Y, Benhamou N, 2000. Oligandrin. A proteinaceous molecule produced by the mycoparasite *Pythium oligandrum* induces resistance to *Phytophthora parasitica* infection in tomato plants. *Plant Physiology* **124,** 379–395.

Pozo MJ, Van Loon LC, Pieterse CMJ, 2005. Jasmonates – signals in plant–microbe interactions. *Journal of Plant Growth Regulation* **23,** 211–222.

Read SM, Currie G, Bacic A, 1996. Analysis of the structural heterogeneity of laminarin by electrospray-ionisation-mass spectrometry. *Carbohydrate Research* **281,** 187–201.

Reglinski T, Lyon GD, Newton AC, 1994a. Induction of resistance mechanisms in barley by yeast-derived elicitors. *Annals of Applied Biology* **124,** 509–517.

Reglinski T, Lyon GD, Newton AC, 1995. The control of *Botrytis cinerea* and *Rhizoctonia solani* on lettuce using elicitors extracted from yeast cell walls. *Journal of Plant Diseases and Protection* **102,** 257–266.

Reglinski T, Newton AC, Lyon GD, 1994b. Assessment of the ability of yeast-derived elicitors to control powdery mildew in the field. *Journal of Plant Diseases and Protection* **101,** 1–10.

Reignault P, Randoux B, Renard D, Sanssen J, Durand R, 2004. Defence responses activated by resistance inducers and natural compounds during the wheat/powdery mildew compatible interaction. Workshop on Resistance of cereals to biotic stress. IHAR Radzików, Poland. November 28–December 1, 2004. p 47–51.

Reuveni M, Zahavi T, Cohen Y, 2001. Controlling downy mildew (*Plasmopara viticola*) in field-grown grapevine with β-aminobutyric acid (BABA). *Phytoparasitica* **29,** 125–133.

Romero-Puertas MC, Perazzolli M, Zago ED, Delledonne M, 2004. Nitric oxide signalling functions in plant–pathogen interactions. *Cellular Microbiology* **6,** 795–803.

Ruess W, Mueller K, Knauf-Beiter G, Staub T, 1996. Plant activator CGA-245704: an innovative approach for disease control in cereals and tobacco. In: Brighton Crop Protection Conference – Pests and Diseases p 53–60.

Ryu C-M, Farag MA, Hu C-H, Reddy MS, Kloepper JW, Paré PW, 2004. Bacterial volatiles induce systemic resistance in Arabidopsis. *Plant Physiology* **134,** 1017–1026.

Séjalon-Delmas N, Villalba Mateos F, Botín A, Rickauer M, Dargent R, Esquerré-Tugayé MT, 1997. Purification, elicitor activity, and cell wall localization of a glycoprotein from *Phytophthora parasitica* var. *nicotianae*, a fungal pathogen of tobacco. *Phytopathology* **87,** 899–909.

Shah J, 2005. Lipids, lipases, and lipid-modifying enzymes in plant disease resistance. *Annual Review of Phytopathology* **43,** 229–260.

Sharathchandra RG, Niranjan Raj S, Shetty NP, Amruthesh KN, Shekar Shetty H, 2004. A chitosan formulation Elexa™ induces mildew disease resistance and growth promotion in pearl millet. *Crop Protection* **23,** 881–888.

Sharp JK, McNeil M, Albersheim P, 1984a. The primary structures of one elicitor-active and seven elicitor-inactive hexa (β-D-glucopyranosyl)-D-glucitols isolated from the mycelial walls of *Phytophthora megasperma* f. sp. *glycinea*. *The Journal of Biological Chemistry* **259,** 11321–11336.

Sharp JK, Valent B, Albersheim P, 1984b. Purification and partial characterization of a β-glucan fragment that elicits phytoalexin accumulation in soybean. *The Journal of Biological Chemistry* **259,** 11312–11320.

Shibuya N, Minami E, 2001. Oligosaccharide signalling for defence responses in plant. *Physiological and Molecular Plant Pathology* **59,** 223–233.

Siegrist J, Mühlenbeck S, Buchenauer H, 1998. Cultured parsley cells, a model system for the rapid testing of abiotic and natural substances as inducers of systemic acquired resistance. *Physiological and Molecular Plant Pathology* **53,** 223–238.

Siegrist J, Orober M, Buchenauer H, 2000. β-aminobutyric acid-mediated enhancement of resistance in tobacco mosaic virus depends on the accumulation of salicylic acid. *Physiological and Molecular Plant Pathology* **56,** 95–106.

Silipo A, Molinaro A, Sturiale L, Dow JM, Erbs G, Lanzetta R, Newman M-A, Parrilli M, 2005. The elicitation of plant innate immunity by lipooligosaccharide of *Xanthomonas campestris*. *Journal of Biological Chemistry* **280,** 33660–33668.

Song F, Goodman RM, 2001. Activity of nitric oxide is dependent on, but is partially required for function of, salicylic acid in the signaling pathway in tobacco systemic acquired resistance. *Molecular Plant–Microbe Interactions* **14,** 1458–1462.

Sticher L, Métraux J-P, 2000. Inhibitors of N-linked glycosylation induce systemic acquired resistance in cucumber. *Physiological and Molecular Plant Pathology* **56,** 245–252.

Takenaka S, Nishio Z, Nakamura Y, 2003. Induction of defense reactions in sugar beet and wheat by treatment with cell wall protein fractions from the mycoparasite *Pythium oligandrum*. *Phytopathology* **93,** 1228–1232.

Tally A, Oostendorp M, Lawton K, Staub T, Bassi B, 1999. Commercial development of elicitors of induced resistance to pathogens. In: Agrawal AA, Tuzun S, Bent E, eds. *Induced plant defenses against pathogens and herbivores*. St. Paul, MN: APS Press, pp 357–369.

Thompson BE, 2004. Five years of Irish trials on biostimulants – the conversion of a skeptic. *USDA Forest Service Proceedings* RMRS-P-33 72–79.

Thuerig B, Binder A, Boller T, Guyer U, Jiménez S, Rentsch C, Tamm L, 2006. An aqueous extract of the dry mycelium of *Penicillium chrysogenum* induces resistance in several crops under controlled and field conditions. *European Journal of Plant Pathology* **114,** 185–197.

Ton J, Mauch-Mani B, 2004. β-amino-butyric acid-induced resistance against necrotrophic pathogens is based on ABA-dependent priming for callose. *The Plant Journal* **38,** 119–130.

Turner JG, Ellis C, Devoto A, 2002. The jasmonate signal pathway. *The Plant Cell* S153–S164.

Uchiyama M, Abe H, Sato R, Shimura M, Watanabe T, 1973. Fate of 3-allyloxy-1,2-benzisothiazole 1, 1-dioxide (oryzemate®) in rice plants. *Agricultural and Biological Chemistry* **37,** 737–745.

Umemura K, Tanino S, Nagatsuka T, Koga J, Iwata M, Nagashima K, Amemiya Y, 2004. Cerebroside elicitor confers resistance to Fusarium disease in various plant species. *Phytopathology* **94,** 813–819.

Van Loon LC, Bakker PAHM, Pieterse CMJ, 1998. Systemic resistance induced by Rhizosphere bacteria. *Annual Review of Phytopathology* **36,** 453–483.

Van Loon LC, Geraats BPJ, Linthorst HJM, 2006. Ethylene as a modulator of disease resistance in plants. *Trends in Plant Science* **11,** 184–191.

Van Peer R, Schippers B, 1992. Lipopolysaccharides of plant-growth promoting *Pseudomonas* sp. strain WCS417r induce resistance in carnation to Fusarium wilt. *Netherlands Journal of Plant Pathology* **98,** 129–139.

von Rad U, Mueller MJ, Durner J, 2005. Evaluation of natural and synthetic stimulants of plant immunity by microarray technology. *New Phytologist* **165,** 191–202.

Walters DR, Cowley T, Weber H, 2006. Rapid accumulation of trihydroxy oxylipins and resistance to the bean rust pathogen *Uromyces fabae* following wounding in *Vicia faba*. *Annals of Botany* **97,** 779–784.

Wei ZM, Laby RJ, Zumoff CH, Bauer DW, He SY, Collmer A, Beer SV, 1992. Harpin, elicitor of the hypersensitive response produced by the plant pathogen *Erwinia amylovora*. *Science* **257,** 85–88.

Whitmore L, Chugh JK, Snook CF, Wallace BA, 2003. The peptaibols database: a sequence and structures resource. *Journal of Peptide Science* **9,** 663–665.

Yamakawa H, Kamada H, Satoh M, Ohashi Y, 1998. Spermine is a salicylate-dependent endogenous inducer for both tobacco acidic pathogenesis-related proteins and resistance against tobacco mosaic virus infection. *Plant Physiology* **118,** 1213–1222.

Yasuda M, Nakashita H, Hasegawa S, Nishioka M, Arai Y, Uramoto M, Yamaguchi I, Yoshida S, 2003. *N*-cyanomethyl-2-chloroisonicotinamide induces systemic acquired resistance in Arabidopsis without salicylic acid accumulation. *Bioscience, Biotechnology, and Biochemistry* **67,** 322–328.

Yoshida H, Konishi K, Nakagawa T, Sekido S, Yamaguchi I, 1990. Characteristics of *N*-phenylsulfonyl-2-chloroisonicotinamide as an anti-rice blast agent. *Journal of Pesticide Science* **15,** 199–203.

Zeidler D, Zähringer U, Gerber I, Dubery I, Hartung T, Bors W, Hutzler P, Durner J, 2004. Innate immunity in *Arabidopsis thaliana*: lipopolysaccharides activate nitric oxide synthase (NOS) and induce defense genes. *Proceedings of the National Academy of Sciences USA*, **101,** 15811–15816.

Zhang B, Ramonell K, Somerville S, Stacey G, 2002. Characterization of early, chitin-induced gene expression in Arabidopsis. *Molecular Plant–Microbe Interactions* **15,** 963–970.

Zhang Z-G, Wang Y-C, Li J, Ji R, Shen G, Wang S-C, Zhou X, Zheng X-B, 2004. The role of SA in the hypersensitive response and systemic acquired resistance induced by elicitor PB90 from *Phytophthora boehmeriae*. *Physiological and Molecular Plant Pathology* **65,** 31–38.

Zhou F, Menke FLH, Yoshioka K, Moder W, Shirano Y, Klessig DF, 2004. High humidity suppresses *ssi4*-mediated cell death and disease resistance upstream of MAP kinase activation, H_2O_2 production and defense gene expression. *The Plant Journal* **39,** 920–932.

Zipfel C, Robatzek S, Navarro L, Oakeley EJ, Jones JDG, Felix G, Boller T, 2004. Bacterial disease resistance in Arabidopsis through flagellin perception. *Nature* **428,** 764–767.

Chapter 3
Genomics in induced resistance

Kemal Kazan[1] *and Peer M. Schenk*[2]

[1]Commonwealth Scientific and Industrial Research Organization Plant Industry, St Lucia, Australia
[2]School of Integrative Biology, University of Queensland, St Lucia, Australia

'No man is an island,
Entire of itself.
Each is a piece of the continent,
A part of the main . . .'

John Donne (1573–1631)

3.1 Introduction

Genomics, the study of genomes using large-scale and high throughput methods, has recently led to a paradigm shift in every aspect of plant biology, providing a wealth of information about the intricate nature of living systems at an unprecedented rate. The application of genomic technologies to the study of plant defence in the recent past has similarly provided revolutionary new insights into how plants defend themselves from pathogen attack. Perhaps somewhat expectedly and largely owing to the suitability to genomic studies, the research on the model plant Arabidopsis has paved the way in plant genomics research. The Arabidopsis genome is the first plant genome ever sequenced, and this represents a landmark event in modern plant biology. Arabidopsis is suitable for both reverse and forward genetic approaches. Random mutagenesis in Arabidopsis has so far produced many useful mutant phenotypes for many key traits, including plant defence responses. Importantly, approximately 75% of all known genes in Arabidopsis are disrupted by at least one T-DNA insertion, and ambitious efforts are currently under way to reveal the functions of all its ~25,000 genes by the year 2010. Some of the first large-scale gene expression studies aimed at analysing plant defence responses were conducted on this species. These initial expression profiling studies followed by functional analyses of genes altered in expression have revealed unexpectedly complex interactions between plant signalling pathways. As a result, we are now slowly moving beyond the limitations of understanding single processes such as examining the behaviour of an individual transcription factor in plant defence. Instead, we are asking increasingly more how a single component of a particular signalling pathway interacts with other components of the same or different signalling pathways and how this interaction contributes to the behaviour of the plant as a whole. Indeed, an improved understanding of how plant genomes actually work will lead to a systems biological approach, integrating input from multiple sources and disciplines.

This chapter will examine the use of genomics in helping to understand the molecular basis of induced resistance. This will include the discovery of genes associated with or involved in induced resistance. It will also deal with the development of suitable promoters for use in molecular genetic studies. The chapter will also cover various genome–phenome analysis tools available for helping researchers to understand induced resistance and to aid exploitation for molecular plant improvement. Examples of such tools will include DNA microarrays and quantitative real-time RT-PCR for gene discovery and expression profiling, high throughput functional studies and assays (e.g. transient expression systems), as well as mutant plants and transgenic plants with modified gene expression (e.g. gene-silencing or over-expression) for phenotypic analyses in order to elucidate gene function and the complicated network that involves crosstalk between multiple signalling pathways in induced resistance (see Chapter 4 for further detail on signalling in induced resistance).

Owing to the reasons explained above, most of the discussion in this Chapter is devoted to the advances made in the model plant Arabidopsis. It should also be emphasized, however, that the studies are currently under way for the application of the genomic revolution into crop plants which undoubtedly would benefit most from this research. For example, it is almost certain that the completion of the genome sequencing of the economically important monocot model rice and the development of tools for understanding gene function will further boost genomics research in economically important monocot species such as cereals.

3.2 Transcriptome analyses for discovery of genes involved in induced resistance

The gene content of an organism actively transcribed in a given tissue and time is called the 'transcriptome'. The transcriptome provides a dynamic link between the genome and the protein content (or proteome) of the same organism and the phenotype. Transcriptomics studies the behaviour of the transcriptome in a genome-wide scale using extremely powerful techniques such as DNA sequencing, DNA microarrays and real-time quantitative reverse transcriptase PCR (RT-Q-PCR). As mentioned above, while the sequencing of the Arabidopsis genome has influenced all aspects of plant biology, the most profound impact was on the study of gene expression itself. In fact, the availability of the whole genome sequence in Arabidopsis has coincided with the development of new technologies suitable for large-scale gene expression analyses. As a result, it is now clear that induced resistance (IR) most likely results in a coordinated action of many genes with diverse functions. Although it may be somewhat naïve to assume that all genes induced or repressed in response to pathogen challenge would have direct roles in induced resistance, pathogen responsive genes are certainly good candidates for further functional studies. In the following sections, some of the genomic technologies extensively used for large-scale identification of genes potentially involved in induced resistance as well as some of the novel insights revealed are discussed briefly.

3.2.1 EST sequencing

Historically, large-scale sequencing of anonymous cDNA clones for the identification of genes expressed under certain conditions has been one of the first steps in the new genomics era and has progressed in parallel to the significant advances made in DNA sequencing technologies. The aim of the EST (Expressed Sequence Tag) projects is to develop collections of cDNA clones with a broad representation of genes active during defence responses. Comparison of the ESTs with sequence databases reveals preliminary evidence whether these ESTs have any sequence similarity to the previously identified genes with defensive functions. Table 3.1 shows various examples of EST sequencing projects undertaken for the identification of new genes potentially involved in induced resistance. In one of the recent examples, Jantasuriyarat *et al.* (2005) monitored the transcriptional changes in rice at early stages of the infection by the rice blast pathogen *Magnaporthe grisea*. A large collection of 68,920 EST sequences was generated from cDNA libraries derived from pathogen-challenged and control (unchallenged)-leaf tissues, representing a total of 13,570 unique sequences. From the sequence analysis, a large number of genes that were highly induced or suppressed in resistant and susceptible conditions were identified. As expected, comparison of the *M. grisea*-challenged libraries with the mock-inoculated control library revealed an increase in the percentage of genes in the functional categories of defence and signal transduction mechanisms (Jantasuriyarat *et al.*, 2005). Undoubtedly, the availability of well developed EST collections has significantly accelerated the development of large-scale gene expression profiling (e.g. microarray experiments) as explained below.

3.2.2 cDNA microarrays/DNA chips

The development of DNA microarray/chip technology for large-scale gene expression analyses has been the real power behind the significant advances made in functional genomics of plant disease resistance (see Table 3.2 for examples). A typical DNA microarray used for expression profiling contains between hundreds and up to hundreds of thousands of cDNA probes arrayed on a solid surface. The DNA probes used for this purpose could be either PCR amplified cDNA fragments (cDNA microarrays) or *in silico* synthesized oligonucleotides (GeneChips) with complementary sequences to target sequences. Recently, several laboratories have developed in-house facilities to fabricate DNA microarrays by arraying random cDNA clones on a glass slide and hybridizing these with the cDNAs derived from pathogen inoculated and control RNAs samples (Campbell *et al.*, 2003). Proprietary GeneChip microarrays have also been developed by Affymetrix (a commercial company) for Arabidopsis, wheat, barley, soybean, rice, grape vine, sugar cane, etc.

Table 3.1 Examples of the EST-sequencing projects for discovery of host genes involved in induced resistance responses.

Host species	Pathogen used	Reference
Soybean (*Glycine max* L.)	*Phytophthora sojae*	Qutob *et al.* (2000)
Rice (*Oryza sativa* L.)	*Magnaporthe grisea*	Kim *et al.* (2001); Ebbole *et al.* (2004); Jantasuriyarat *et al.* (2005)
Wheat (*Triticum aestivum* L.)	*Fusarium graminearum*	Kruger *et al.* (2002)
Vetch (*Lathyrus sativus* L.)	*Mycosphaerella pinodes*	Skiba *et al.* (2005)
Chickpea (*Cicer arietinum* L.)	*Ascochyta rabiei*	Coram & Pang (2005a, b)

Table 3.2 Examples of transcriptome and proteome analyses for discovery of genes involved in induced resistance.

Species	Pathogen/elicitor	Reference
Defence-related transcriptome studies (e.g. microarrays – cDNA AFLP and RT-Q-PCR analyses)		
Arabidopsis thaliana	*Alternaria brassicicola*	Schenk *et al.* (2000, 2003), Narusaka *et al.* (2003), van Wees *et al.* (2003), McGrath *et al.* (2005)
Arabidopsis thaliana	*Pseudomonas syringae*	Maleck *et al.* (2000), Chen *et al.* (2002), Cheong *et al.* (2002), Scheideler *et al.* (2002), Tao *et al.* (2003), Glombitza *et al.* (2004), Verhagen *et al.* (2004)
Arabidopsis thaliana	*Phytophthora infestans*	Huitema *et al.* (2003)
Arabidopsis thaliana	*Blumeria graminis* f. sp. *hordei*	Zimmerli *et al.* (2004)
Arabidopsis thaliana	*Erysiphe cichoracearum*	Zimmerli *et al.* (2004)
Arabidopsis thaliana	Tobacco mosaic virus	Golem & Culver (2003), Whitham *et al.* (2003)
Arabidopsis thaliana	Cucumber mosaic virus	Marathe *et al.* (2004)
Arabidopsis thaliana	Flagellin 22 peptide	Navarro *et al.* (2004)
Arabidopsis thaliana	Chitin	Ramonell *et al.* (2002, 2005), Zhang *et al.* (2002)
Arabidopsis thaliana	Harpin	Krause & Durner (2004)
Arabidopsis thaliana	Nitric oxide	Polverari *et al.* (2003)
Arabidopsis thaliana	Lipopolysaccharides	Zeidler *et al.* (2004)
Arabidopsis thaliana	Cell death	Swidzinski *et al.* (2002)
Arabidopsis thaliana	Jasmonate	Schenk *et al.* (2000), Sasaki *et al.* (2001), Chen *et al.* (2002), Glazebrook *et al.* (2003), Glombitza *et al.* (2004), Devoto *et al.* (2005), McGrath *et al.* (2005)
Arabidopsis thaliana	Salicylate and salicylate analogues	Maleck *et al.* (2000), Schenk *et al.* (2000), Chen *et al.* (2002), Glombitza *et al.* (2004), Wang *et al.* (2005b)
Arabidopsis thaliana	Ethylene	Schenk *et al.* (2000), De Paepe *et al.* (2004), Eckey *et al.* (2004), Glombitza *et al.* (2004)
Arabidopsis thaliana	Rhizobacterium	Cartieaux *et al.* (2003)
Barley (*Hordeum vulgare*)	*Blumeria graminis* (powdery mildew)	Caldo *et al.* (2004), Zierold *et al.* (2005)
Rice (*Oryza sativa*)	*Magnaporthe grisea* (rice blast)	Lu *et al.* (2004)
Rice (*Oryza sativa*)	Flagellin	Fujiwara *et al.* (2004)
Rice (*Oryza sativa*)	*N*-Acetylchitooligosaccharide	Akimoto-Tomiyama *et al.* (2003)
Rice (*Oryza sativa*)	Rice yellow mottle virus	Ventelon-Debout *et al.* (2003)
Tomato (*Lycopersicon esculentum*)	*Xanthomonas campestris* pv. *vesicatoria*	Gibly *et al.* (2004)
Tomato (*Lycopersicon esculentum*)	*Pseudomonas syringae* and jasmonate	Zhao *et al.* (2003)
Potato (*Solanum tuberosum* L.)	*Phytophthora infestans*	Restrepo *et al.* (2005)
Sugarcane (*Saccharum officinarum*)	Methyl jasmonate	Bower *et al.* (2005)

Table 3.2 (*Continued*)

Species	Pathogen/elicitor	Reference
Wheat (*Triticum aestivum*)	Powdery mildew	Bruggmann *et al.* (2005)
Cassava (*Manihot esculenta*)	*Xanthomonas axonopodis* pv. *manihotis*	Lopez *et al.* (2005)
Cassava (*Manihot esculenta*)	*Pseudomonas syringae*	Kemp *et al.* (2005)
Soybean (*Glycine max*)	*Phytophthora sojae*	Moy *et al.* (2004)
Soybean (*Glycine max*)	*Pseudomonas syringae*	Zou *et al.* (2005)
Cotton (*Gossypium herbaceum*)	*Fusarium oxysporum* f. sp. *vasinfectum*	Dowd *et al.* (2004)
Sorghum (*Sorghum vulgare*)	Salicylate and jasmonate	Salzman *et al.* (2005)
Tobacco (*Nicotiana benthamiana*)	Enveloped viruses	Senthil *et al.* (2005)
Peanut (*Arachis hypogaea*)	*Cercosporidium personatum* (leaf spot disease)	Luo *et al.* (2005)
Chickpea (*Cicer arietinum*)	*Ascochyta rabiei* (ascochyta blight)	Coram & Pang (2005a, b)
Proteome studies		
Arabidopsis thaliana	Fungal elicitors	Ndimba *et al.* (2003)
Arabidopsis thaliana	Salicylate	Oh *et al.* (2005)
Arabidopsis thaliana	*Fusarium moniliforme*	Ndimba *et al.* (2003)
Arabidopsis thaliana	SA-treated cell cultures	Gruhler *et al.* (2005)
Tobacco (trichomes)	Unchallenged	Amme *et al.* (2005)
Rice (*Oryza sativa*)	*Magnaporthe grisea*	Kim *et al.* (2003)
Rice (*Oryza sativa*)	Rice yellow mottle virus	Ventelon-Debout *et al.* (2004)
Medicago truncatula	Cell suspension cultures	Lei *et al.* (2005)
Wheat (*Triticum aestivum*)	*Fusarium graminearum* (head blight or scab)	Wang *et al.* (2005c)

(see Table 3.3 for the website), and this list is still growing. The Affymetrix Arabidopsis GeneChip contains nearly 24,000 gene sequences of the approximately 25,000 predicted genes. Undoubtedly, the GeneChip array is constantly being improved (see, for instance, Allemeersch *et al.*, 2005 for CATMA array technology), supersedes the traditional cDNA microarrays in specificity and reproducibility and may well be the future of the array technology.

As indicated above, sequenced cDNA clones (i.e. ESTs), if available, could be the ideal material to be arrayed on microarray slides for use in expression analysis. However, the availability of well characterized EST collections is not a prerequisite for construction of cDNA arrays. PCR amplified anonymous clones from pathogen infected cDNA libraries can also be arrayed and hybridized with probes derived from infected and uninfected

Table 3.3 Examples of plant genomic resources/databases on the World Wide Web (refer to Rensink & Buell (2005) for additional information on plant microarray databases).

Database	Content
GENEVESTIGATOR https://www.genevestigator.ethz.ch/	Arabidopsis Microarray Database and Analysis Toolbox; Zimmermann *et al.* (2004)
Affymetrix http://www.affymetrix.com/index.affx	Arabidopsis genome array that can monitor up to 24,000 gene sequences
TAIR (The Arabidopsis Information Resource) http://arabidopsis.org/; and NACS (Nottingham Arabidopsis Stock Centre) http://arabidopsis.info/home.html	Anything you want to know about Arabidopsis; Arabidopsis information and germplasm resource
SIGNAL (SALK Institute Genomic Analysis Laboratory) http://signal.salk.edu/tabout.html	A searchable database containing the insertion site information and the availability of the corresponding mutant lines; Alonso *et al.* (2003)
TIGR (The Institute of Genomic Research) http://www.tigr.org/	A non-for-profit centre dedicated to deciphering and analysing genomes
NCBI (National Center for Biotechnology Information) http://www.ncbi.nlm.nih.gov/	A rich resource for genomic information from all living organisms
RMOS (The Rice Microarray Opening Site) http://cdna01.dna.affrc.go.jp/RMOS/main_en.html	Rice microarray database
Rice PIPELINE http://cDNA01.dna.affrc.go.jp/PIPE/	A compilation of rice genomics data such as genome sequences, full length cDNAs, gene expression profiles, mutant lines, *cis* elements from various databases; Yazaki *et al.* (2004)
Oryzabase http://www.shigen.nig.ac.jp/rice/oryzabase/	A genomic database for rice; Yamazaki & Jaiswal (2005)
PGIR (the Plant Genome Initiative at Rutgers) http://pgir.rutgers.edu/	A useful resource for rice, maize and sorghum genome sequencing
PLEXdb http://barleybase.org/plexdb/html/modules.php?name=PD_general&page=links.php	A community resource for plant–pathogen microarrays
ArrayExpress http://www.ebi.ac.uk/arrayexpress/	A public repository for microarray data; Parkinson *et al.* (2005)
Plant Genome http://www.plantgenome.uga.edu/links.htm	The Plant Genome Mapping Laboratory (A web site maintained at the University of Georgia with extensive links to a number of other databases)
BarleyBase http://www.barleybase.org/	The United States Department of Agriculture (USDA)-funded public repository for plant microarray data; Shen *et al.* (2005)
DRASTIC http://www.drastic.org.uk/	A database resource for analysis of signal transduction in cells; Button *et al.* (2006)

(control-mock) tissues. This approach has been used by Campbell *et al.* (2003) for rapid identification of pathogen responsive genes in Arabidopsis. These authors have first constructed a suppression subtractive hybridization cDNA library to minimize the number of clones that presumably do not show differential expression between inoculated and un-inoculated samples. The anonymous clones from this library were then arrayed on glass microarray slides and hybridized with the probes derived from *Alternaria brassicicola*-inoculated and mock-inoculated control tissues. Clones that showed significant induction/repression have been sequenced to reveal the identities of induced/repressed genes. As a result, a novel ABC transporter gene significantly induced by salicylic acid (SA) and infection by various fungal pathogens has been identified. Table 3.2 shows several examples of defence related microarray studies undertaken for rapid identification of genes involved in induced disease resistance responses.

Although microarray technology is now within the reach of many laboratories, the overall cost involved in these analyses can still be somewhat prohibitive. Therefore, searching the publicly available microarray databases which contain expression data from large numbers of genes whose expression is altered during infection by pathogens or treatment with signalling molecules can be a rapid and cost effective way to identify the expression patterns of individual genes of interest. Information about some of these databases is given in Table 3.3. It is expected that in the near future, the availability of such microarray databases will expand significantly, providing a useful resource for plant scientists.

3.2.3 Genome-wide real-time quantitative RT-PCR (RT-Q-PCR)

The RT-Q-PCR technique provides a quantitative measurement of the expression profile of a gene by monitoring the fluorescence emitted during each PCR cycle. The exponential phase of the fluorescence signal, where the first significant increase in the amount of PCR product correlates to the initial amount of target template, increases in direct proportion to the amount of PCR product. Therefore, the higher the starting copy number of the nucleic acid target, the sooner a significant increase in fluorescence is observed. Although RT-Q-PCR is often utilized for validating and extending the results of microarray experiments (Schenk *et al.*, 2003), until recently the use of this technique for large-scale gene expression analyses has been limited. Recently, Czechowski *et al.* (2004) developed a genome-wide RT-Q-PCR-based resource for quantitative measurements of transcripts of 1465 Arabidopsis TF (transcription factor) genes in root and shoot tissues. The same resource was also used for identification of Arabidopsis genes induced by methyl jasmonate (MeJA) and *A. brassicicola* (McGrath *et al.*, 2005). These analyses have identified 134 transcription factors belonging to the AP2/ERF, MYB, WRKY and NAC TF families that showed a significant change in expression. Functional analysis of the selected AP2/ERFs belonging to the activator and repressor type AP2/ERFs revealed that over-expression of a positive regulator and inactivation of negative regulator both resulted in increased resistance to the Fusarium wilt pathogen *Fusarium oxysporum* (McGrath *et al.*, 2005). This study suggests that plant defence responses are tightly controlled by transcriptional activation of multiple repressors and activators.

3.2.4 cDNA AFLPs

cDNA-AFLP (amplified fragment length polymorphism) is a fragment-based technique for genome-wide expression analysis of genes expressed under certain conditions. In this approach, unique transcript tags derived from the 3′ end region of expressed genes are PCR amplified and displayed on acrylamide gels. Selective amplification of subsets of transcript tags allows one to fractionate the initial pool of tags and to detect low abundant messengers (Brugmans *et al.*, 2002). Using this method, Kemp *et al.* (2005) identified 78 transcript derived fragments (TDFs) showing differential expression during a hypersensitive response of cassava (*Manihot esculenta*) leaves induced by *Pseudomonas syringae*. Although the confirmation of the function of the proteins encoded by these altered transcripts requires further analyses, many genes found were putative homologues of known defence-related genes (Kemp *et al.*, 2005).

3.2.5 Novel insights into induced resistance revealed by transcriptome analysis

As mentioned above, the application of gene expression profiling into the analysis of plant defence has revealed several major insights into how plants defend themselves from pathogen attack. First of all, these methods have allowed the identification of new genes associated with plant defence. For instance, microarray analysis of gene expression in Arabidopsis identified the pathogen- and the jasmonic acid (JA)-inducible AtMYC2 gene (Schenk *et al.*, 2000). Functional analysis of AtMYC2 has revealed that this gene is a negative regulator of JA/ethylene (ET) responsive defence gene expression. In fact, the *myc2* mutant shows increased expression from a number of defence genes such as *PDF1.2*, *PR1*, *CHIB* and *PR4* and also tolerance to a number of pathogens, including *F. oxysporum* and *Botrytis cinerea* (Anderson *et al.*, 2004; Lorenzo *et al.*, 2004). Identification of AtMYC2 as a negative regulator of defence also suggested the existence of an antagonistic interaction between abscisic acid (ABA) and JA/ET pathways (Figure 3.1) because AtMYC2 was previously identified as a positive regulator of ABA and drought signalling pathways (Abe *et al.*, 2003).

Second, global gene expression profiling during plant defence has allowed identification of new physiological processes involved in induced defence responses. For instance, inoculation of Arabidopsis plants with the bacterial pathogen *P. syringae* resulted in

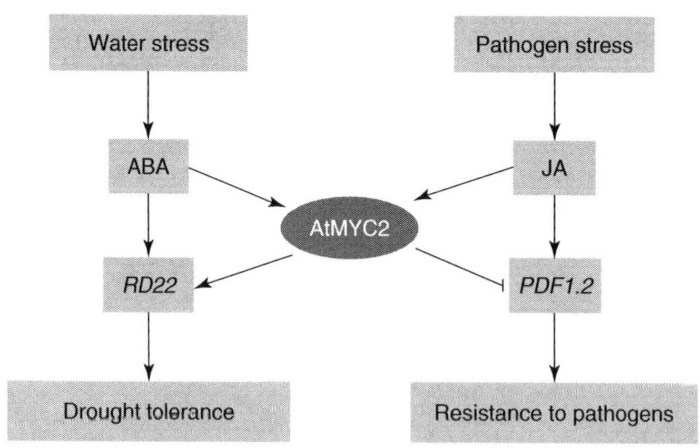

Figure 3.1 Antagonistic interaction between ABA and JA/ET pathways modulated by the transcription factor AtMYC2. Transcriptome analysis identified the jasmonate- and pathogen-responsive AtMYC2 as negatively regulating the expression of *PDF1.2*, a JA/ET responsive defence gene, and resistance to several pathogens. In contrast, AtMYC2 positively regulates the expression of *RD22*, a drought and ABA inducible gene.

an expression change from housekeeping to defence metabolism, indicating an increased demand for energy and biosynthetic capacity in plants fighting off a pathogen attack (Scheideler *et al.*, 2002). Similarly, Schenk *et al.* (2003) observed activation of genes involved in the β-oxidation pathway in Arabidopsis plants inoculated with the incompatible fungal pathogen *A. brassicicola*. This pathway is involved in fatty acid metabolism, and its activation during plant defence may be significant for the synthesis of plant defence signalling molecules such as JA.

Third, global gene expression profiling has allowed determination of the extent by which pathogen signalling pathways overlap with those involved in defence hormone signalling (Schenk *et al.*, 2000; Kazan *et al.*, 2001; Chapman *et al.*, 2002). Microarray analyses conducted by Schenk *et al.* (2000) have shown both synergistic and antagonistic interactions between SA and JA signalling pathways in regulating expression from many genes, while Salzman *et al.* (2005) have found that a common set of genes is induced by both SA and JA in the monocot plant sorghum. The complex signalling interactions between different defence signalling pathways are obviously critical in fine tuning the overall plant response to pathogen challenge, and synergistic and/or antagonistic interactions seem to be a common feature of many signalling pathways.

Harnessing the power of transcriptome analysis has also led to the explorations of the molecular events underpinning pathogen compatibility and incompatibility. For instance, Zimmerli *et al.* (2004), using microarray expression profiling, compared the global gene expression patterns of Arabidopsis inoculated with the non-host barley powdery mildew to those inoculated with a virulent, host powdery mildew, *Erysiphe cichoracearum*. In these experiments, although the Arabidopsis transcriptional responses to host and non-host inoculations overlapped substantially, an earlier and stronger activation or repression of gene expression was observed after inoculation with the non-host powdery mildew. Similarly, expression analyses of Arabidopsis plants inoculated with the non-host pathogen *Phytophthora infestans* revealed a significant overlap between Arabidopsis non-host response and known defence responses triggered by defence signalling compounds. Particularly, the non-host response to *P. infestans* was clearly associated with the activation of the jasmonate pathway (Huitema *et al.*, 2003), suggesting that manipulation of the JA pathway may provide increased disease resistance.

More recently, microarray analyses have been extended to the analysis of plant responses triggered during the *Rhizobacteria*-induced systemic resistance (ISR) which primes the plants to mount a stronger and more effective defence response upon challenge with a virulent pathogen. Interestingly, these analyses showed that the genes expressed during ISR differed from those expressed during pathogen-induced systemic acquired resistance (SAR). Although the ISR-inducing bacteria *Pseudomonas fluorescens* elicited a substantial change in the expression of 97 genes in root treatments, none of the approximately 8,000 genes tested showed a consistent change in expression systemically in the leaves. As expected, a large number of genes showed a stronger expression pattern in ISR-expressing leaves after challenge by the compatible bacteria *P. syringae*, suggesting that these genes were primed to respond faster or more strongly upon pathogen attack (Verhagen *et al.*, 2004).

Another significant contribution of expression profiling to plant defence has been in the area of genome-wide identification of genes affected in defence-related mutants or plant lines over-expressing transcriptional activators involved in induced resistance. For instance,

Lorenzo *et al.* (2003), using Arabidopsis lines over-expressing the transcriptional activator ETHYLENE RESPONSE FACTOR1 (ERF1), have identified a large number of JA-ET responsive genes whose expression is enhanced by *ERF1* over-expression. This indicated that ERF1 acts downstream of the intersection between ET and JA pathways and is a key element in the integration of both signals for the regulation of defence response genes.

Microarray analyses of Arabidopsis mutants have also revealed essential information about the genes whose expression is affected by the mutated signalling component. In one recent study, Devoto *et al.* (2005) studied the gene expression patterns in response to jasmonate and wounding in wild-type and the *coi1* mutant by microarray analysis. The results of this study showed that COI1, an F-box protein functioning in JA signalling, is required for expression of approximately 84% of 212 genes induced by JA, and for expression of approximately 44% of 153 genes induced by wounding. One unexpected finding in this study was that an intact *COI1* gene was also required for JA-dependent repression of 53% of 104 genes and for repression of approximately 46% of 83 genes whose expression was suppressed by wounding, providing further insights into the role of COI1 as a regulator of wound- and JA signalling.

One of the assumptions behind the large-scale gene expression profiling is that functionally associated genes tend to be co-expressed. This indicates that they could also be co-regulated. Since co-regulation is usually governed by transcription factors via their specific binding elements, putative regulators can be identified from promoter sets of (co-expressed) genes by screening for over-represented nucleotide patterns. Using this logic, the W-box sequence was found to be the major sequence element in the promoters of genes co-regulated with *PR1* (called the *PR1* regulon) (Maleck *et al.*, 2000).

Finally, global analysis of gene expression has identified additional functions for the defence associated genes that have been already studied in some detail. For instance, using gene expression profiling in Arabidopsis, Wang *et al.* (2005a) have found that in addition to controlling the expression of *PR* genes, NPR1, a regulatory protein involved in SAR, directly controls the expression of the protein secretory pathway genes. Up-regulation of genes involved in protein folding and secretion (e.g. BiP2, DAD1 and SEC61) was essential for SAR because mutations in these genes compromised the plant's ability to efficiently secrete PR1 after treatment with benzothiadiazole (BTH, a SA analogue; see Chapters 2, 11 and 13 for details) (Wang *et al.*, 2005a). The mutations in these genes also resulted in increased susceptibility of the mutants against the bacterial pathogen *P. syringae* pv. *maculicola* ES4326 (Wang *et al.*, 2005a).

3.3 Proteome analyses and induced resistance

Although transcriptome analyses have so far revealed many novel insights into the induced resistance responses, a number of defence-associated genes, particularly those with regulatory functions (e.g. kinases) may not be responsive to defence-related signals at the transcriptional level, and thus such genes cannot be reliably identified by transcriptome analyses. One complementary approach to transcriptome-based defence gene discovery would be the identification of pathogen-induced proteins using proteome analysis. In this method, proteins isolated from challenged/treated tissue are separated on two-dimensional sodium dodecyl sulfate–polyacrylamide gel electrophoresis (2D SDS–PAGE) and visualized using a staining procedure such as Coomassie blue or silver stain. Proteins of interest are then excised from the gel and digested with a protease (e.g. trypsin) to produce peptides which are then identified by using matrix assisted laser desorption ionization (MALDI) mass spectrometry. So far, a few preliminary proteome analyses conducted in rice have successfully identified some of the known PR proteins (see Table 3.2 for examples) that accumulate abundantly in response to JA treatment or inoculation by the pathogenic fungus *M. grisea* (Kim *et al.*, 2001, 2003). In Arabidopsis, Oh *et al.* (2005) analysed the secreted proteins in response to SA treatment. This analysis identified a novel secreted protein encoding a lipase (named GLIP1 for GDSL LIPASE1) with antimicrobial activity. Arabidopsis lines containing a T-DNA insertion in this gene showed significantly increased susceptibility to the necrotrophic fungal pathogen *A. brassicicola*, demonstrating the utility of proteomics approaches in identifying genes involved in induced resistance (Oh *et al.*, 2005).

Although the induced resistance response is an important component of the plant's defence against microbial invaders, some of the defences are preformed or constitutively expressed, contributing to the basal defence responses (see also Chapter 6). For instance, vegetative parts of a plant can secrete antimicrobial proteins to the plant surface for interaction with potential invaders. These proteins can be collected by simply washing the leaf surface with water and subjecting the extract to 2-D gel analysis. Leaf water washes of tobacco leaves identified highly hydrophobic, basic proteins termed phylloplanins that inhibited spore germination and leaf infection by the oomycete pathogen *Peronospora tabacina* (Shepherd *et al.*, 2005). The hairy appendages, called trichomes, found on the surface of most plants also play a significant role in delivering such compounds to the plant surface. Indeed, proteomic analysis of tobacco trichomes showed that among the proteins specifically enriched in trichomes, the components of stress defence responses were strongly represented (Amme *et al.*, 2005).

Application of a high throughput yeast two-hybrid screening system to study possible interactions between signalling components involved in induced resistance can be a promising proteomic tool (Fang *et al.*, 2002; Immink & Angenent, 2002; Kersten *et al.*, 2002), particularly when used in conjunction with other techniques. However, one of the obvious limitations of this technique is the fact that physical interactions are typically established in *in vitro* environments, in the absence of many co-factors and post-translational modifications.

In summary, apart from a few preliminary studies conducted on rice and Arabidopsis, proteome-based gene identification approaches have not yet fulfilled their promise for discovery of new defence genes. One major drawback associated with the analysis of the proteome is that many proteins, especially those involved in signalling processes, are below the threshold of detection (Thurston *et al.*, 2005). Those readers wishing to obtain more information about proteome analyses in plants should refer to the recent extensive reviews by Hirano *et al.* (2004), Laugesen *et al.* (2004), Newton *et al.* (2004) and Rose *et al.* (2004).

3.4 Metabolome analysis and induced resistance

Genomics and proteomics cover the analysis of the entire set of genes and proteins, respectively, while metabolomics has been defined as the quantitative measurement of all low molecular weight metabolites in a given cell or tissue. Accumulation of plant secondary metabolites with roles in induced resistance often occurs in plants subjected to signal molecules (reviewed by Zhao *et al.*, 2005). The plant secondary metabolites with potential roles in induced resistance responses may include glucosinolates, alkaloids, plant hormones (e.g. SA, JA, ET, ABA and nitric oxide [NO]) and phytoalexins (e.g. stilbene synthase, camelexin, rishitin and saponin). Monitoring a whole set of secondary metabolites by various high resolution spectrometry techniques, such as gas chromatography-mass spectrometry (GC-MS), liquid chromatography-mass spectrometry (LC-MS), gas chromatography/electron ionization-time of flight-mass spectrometry (GC/EI-TOF-MS), electron ionization-tandem mass spectrometry (EI-MS/MS), and nuclear magnetic resonance (NMR) spectrometry, during defence responses can be a powerful way to determine the changes occurring in plant metabolism to counter pathogen attack.

Historically, the potential role(s) of roots in induced resistance has been somewhat ambiguous mainly due to difficulties encountered in studying plant roots. Recently, metabolome analyses have revealed potentially crucial roles that roots can play in plant defence. For instance, it was found that plant roots secrete a number of metabolites in response to elicitation by salicylic acid, jasmonic acid, chitosan and two fungal cell wall elicitors. Such metabolites include butanoic acid, *trans*-cinnamic acid, *o*-coumaric acid, *p*-coumaric acid, ferulic acid, *p*-hydroxybenzamide, methyl *p*-hydroxybenzoate, 3-indolepropanoic acid, syringic acid, and vanillic acid. Most of these secreted secondary metabolites identified by high performance liquid chromatography exhibit a wide range of antimicrobial activity against both soil-borne bacteria and fungi at the concentration detected in the root exudates (Walker *et al.*, 2003).

Another study in Arabidopsis further strengthens the roles of root-derived antimicrobial metabolites to confer tissue-specific resistance to a wide range of bacterial pathogens that show sensitivity to these metabolites (Bais *et al.*, 2005). A *P. syringae* strain that is both partly resistant to these compounds and successfully inhibits the synthesis/exudation of these metabolites was able to cause disease on plants that were otherwise resistant to the infection by metabolite sensitive strains of the bacteria (Bais *et al.*, 2005).

Interestingly, evidence from metabolome analyses also indicated new defensive roles for a plant metabolite that was initially associated with abiotic stress responses. It has long been known that proline accumulates in plant cells in response to abiotic stresses. Recent research reveals a link between proline accumulation and incompatible plant–pathogen interactions (Fabro *et al.*, 2004). Inoculation of Arabidopsis leaves with avirulent strains of the bacterial pathogen *P. syringae* pv. *tomato* triggers proline accumulation in Arabidopsis leaf tissues while proline levels remain unchanged after infection with isogenic disease-causing virulent bacteria (Fabro *et al.*, 2004). Interestingly, proline accumulation after challenge with avirulent bacteria was also dependent on the plant's SA levels. Arabidopsis plants deficient in SA biosynthesis (e.g. *nahG* plants and *eds5* mutants) were also compromised in proline accumulation in response to avirulent bacteria (Fabro *et al.*, 2004). The actual physiological role of proline in modulating plant defence responses warrants further investigation. However, it is likely that the protective mode of action of proline may be linked to its antioxidant effects. Indeed, treatment of Arabidopsis plants with ROS (reactive oxygen species) activates expression of genes encoding rate limiting steps in proline biosynthesis as well as increases in tissue proline contents (Fabro *et al.*, 2004).

3.5 Forward genetic approaches for discovery of genes involved in induced resistance

It is the forward genetic approach that has been traditionally used for defence gene discovery as part of the analysis of many plant traits. This approach starts with a disease phenotype in a mutant individual and then identifies the gene(s) or mutation(s) that controls or causes this phenotype. Because Arabidopsis is rather amenable for generating mutants by application of chemical and physical mutagens affected in plant defence responses and subsequently cloning of the mutated genes, so far most such mutants have been generated in this plant species. Some of the Arabidopsis mutants were isolated by virtue of their increased susceptibility phenotype to virulent or avirulent pathogens. The genes altered in such mutants possibly encode positive regulators of plant disease resistance. Some of the examples of this class of mutants may include *npr1*, *eds5*, *pad4* and *sid2*. Among the genes mutated in these mutants, EDS5 and SID2 are involved in SA biosynthesis, while NPR1 is a positive regulator of the SA signalling pathway. In contrast, the Arabidopsis mutants (e.g. *mpk4*, *jin1* and *cpr5*) affected in the genes encoding negative regulators of defence, show increased resistance to pathogen attack. *JIN1* (also known as *AtMYC2*) encodes a negative regulator of diverse classes of defence genes such as *PR1*, *PDF1.2*, *CHIB* (basic chitinase) and *PR4* while MPK4 negatively regulates SA biosynthesis and is required for jasmonate responsive expression of the *PDF1.2* gene (Petersen *et al.*, 2000). A detailed account of Arabidopsis mutants compromised in plant defence gene induction and disease resistance was recently given by Thatcher *et al.* (2005). Here, only a small subset of relatively well characterized mutants identified through forward genetics (i.e. mutant screening) with an altered disease phenotype to fungal or bacterial pathogens are presented (Table 3.4).

In other screenings, mutants that show increased resistance to pathogens have also been identified. For instance, mutations in the *PMR6* (*POWDERY MILDEW RESISTANT6*) gene encoding a pectate lyase enzyme cause increased resistance to the powdery mildew pathogen *Erysiphe cichoracearum* without any alterations in the expression of known defence genes *PR1* and *PDF1.2*. It is postulated therefore that *pmr6*-mediated resistance is a novel form of resistance based on the loss of a gene required during a compatible interaction (Vogel *et al.*, 2002).

Another mutant that shows increased resistance to *E. cichoracearum* is *pmr4*. Surprisingly, the cloning of the gene mutated in the *pmr4* mutant revealed that this locus encodes a callose synthase involved in the production of callose (a β-1,3-glucan) deposited in the cell wall following pathogen attack (Nishimura *et al.*, 2003). This result certainly questions the role of callose deposition in contributing to disease resistance.

In addition to Arabidopsis mutants mentioned in Table 3.4, the barley *mlo* mutant has been subjected to extensive studies. Mutations in the *MLO* gene of barley result in enhanced resistance to the biotrophic barley pathogen powdery mildew (*Blumeria graminis* f. sp. *hordei*), due to increased accumulation of H_2O_2 in epidermal cells and activation of a cell death programme (Piffanelli *et al.*, 2002). This suggests that the wild-type MLO is, in fact, a negative regulator of cell death and biotrophic pathogen resistance. However, the wild-type *MLO* gene is required to enhance resistance to necrotrophic pathogens by suppressing the oxidative burst mediated cell death. In line with this notion, the *mlo* plants show enhanced susceptibility to the hemibiotrophic rice blast fungus *M. grisea* (Jarosch *et al.*, 2003).

Table 3.4 Examples of some of the better characterized Arabidopsis mutants compromised in induced resistance responses.

Mutant/gene	Disease and molecular phenotype of the mutant	Reference
JA-ET pathway mutants		
bos1/MYB transcription factor	Increased susceptibility to necrotrophic pathogens, i.e. *Botrytis cinerea*	Mengiste *et al.* (2003); Veronese *et al.* (2004)
coi1/F-box protein	Increased pathogen susceptibility	Xie *et al.* (1998)
cev/cellose synthetase	Constitutive defence gene expression/ increased resistance to pathogens	Ellis *et al.* (2002)
ein2/membrane protein	Increased susceptibility to *Botrytis cinerea* and lack of JA-ET responsive defence gene expression in response to JA and ethylene	Thomma *et al.* (1999)
jin1/basic helix loop helix transcription factor	Increased resistance to necrotrophic pathogen	Lorenzo *et al.* (2004)
jar1/adenylate forming enzyme	Reduced expression from JA-responsive genes	Staswick *et al.* (1998)
fad3, *fad7*, *fad8*/fatty acid desaturase	Increased susceptibility to necrotrophic pathogens/reduced JA responsive defence gene expression	Vijayan *et al.* (1998)
SA pathway mutants		
npr1/ankyrin repeat protein	Reduced expression from SA-responsive genes/increased susceptibility to *Pseudomonas syringae*	Cao *et al.* (1994)
eds5/MATE transporter		
eds1/lipase-like protein	Increased susceptibility to biotrophic pathogens/reduced SA-dependent defence gene expression	Parker *et al.* (1996)
pad4/lipase like protein	Increased susceptibility to biotrophic pathogens/reduced SA-dependent defence gene expression	Jirage *et al.* (1999)
sid2/isochorismate synthase	Increased susceptibility to biotrophic pathogens/reduced SA-responsive defence gene expression	Wildermuth *et al.* (2001)
cpr5/transmembrane domain protein	Increased resistance to biotrophic pathogens/increased SA-responsive gene expression	Bowling *et al.* (1994)
mpk4/map kinase	Increased resistance to *Pseudomonas syringae*. Increased SA responsive defence gene expression/reduced JA-responsive gene expression.	Petersen *et al.* (2000)

3.6 Reverse genetic approaches

A complementary approach used in functional analysis of genes is inactivation of the function of the candidate genes and testing the effect of inactivation on the disease resistance phenotype. This approach is commonly known as 'reverse genetic'. One potential drawback associated with the gene inactivation studies, however, is that due to functional degeneracy (presence of elements with similar functions) of the genome, not all knock-outs or gene silenced lines show an observable disease phenotype. Nevertheless, reverse genetic approaches offer a significant resource for functional genomics studies. In the following sections, various reverse genetic approaches used for studying gene function will be briefly discussed.

3.6.1 Insertional inactivation

Insertional inactivation of genes by T-DNA or transposon insertions harnesses the power of transformation technologies. To identify the functional roles of all Arabidopsis genes, multiple laboratories have produced gene knock-out lines using T-DNA or transposon insertions. One of the most extensive gene-knock out projects has been conducted in the SALK Institute Genomic Analysis Laboratory (SIGNAL) (Alonso *et al*., 2003). Currently, 75% of all Arabidopsis genes contain at least one T-DNA insertion. During preparation of this chapter, efforts were under way to identify at least two homozygous T-DNA insertion lines for all 25,000 Arabidopsis genes (http://signal.salk.edu/gabout.html) by 2010, an estimated dateline for identification of the function of every Arabidopsis gene. However, it is evident that this enormous task cannot only be achieved by simply generating gene knock-outs. As mentioned above, not all knock-outs produce an easily distinguishable phenotype, so additional studies such as gene over-expression experiments, etc. may be required to complement the data from knock-out lines.

Nevertheless, there are numerous examples of exploring the power of insertional mutagenesis for functional genomic studies at least in Arabidopsis, and similar resources are being developed in other plants such as tomato, rice and maize (Table 3.5). For instance, the response of Arabidopsis seedlings to chitin treatment, a major component of the fungal cell wall, was analysed by Ramonell *et al.* (2002, 2005) to identify chitin-responsive plant genes. Subsequently, T-DNA insertion mutants for nine chito-oligomer responsive genes were analysed, and it was found that three of the mutants (two disease-resistance-like protein and one E3 ligase mutant) were more susceptible to the fungal pathogen, powdery mildew, than wild type as measured by conidiophore production in the infected leaves. This obviously confirms the role of chitin as an important elicitor of plant defences.

In another example, the genome-wide RT-Q-PCR analyses undertaken by McGrath *et al.* (2005) have identified a MJ and pathogen inducible transcriptional repressor called AtERF4. Insertional inactivation of this gene caused increased *PDF1.2* expression and improved resistance towards the Fusarium wilt pathogen *F. oxysporum.*

Unfortunately, the generation of T-DNA or transposon insertion lines using transformation technologies may not be practical for many crop plants due to the limitations imposed by the unavailability of efficient transformation technologies. Recently, to accelerate the functional genomic efforts, a non-transgenic method for reverse genetics called

Table 3.5 Examples of large scale insertional mutagenesis studies undertaken to identify gene function.

Species	Resource type	References/source
Arabidopsis thaliana	88,122 T-DNA insertions covering ~74% of the all predicted genes	Alonso *et al.* (2003); The Salk Institute Genomic Analysis Laboratory (SIGNAL) (http://signal.salk.edu/); available through Arabidopsis Biological and Resource Center (http://www.Arabidopsis.org) or Nottingham Arabidopsis Stock Centre-NACS (http://arabidopsis.info/home.html)
Arabidopsis thaliana	52,964 T-DNA Insertion Lines	Sessions *et al.* (2002); the Syngenta Arabidopsis Insertion Library (SAIL) (http://www.arabidopsis.org/abrc/sail.jsp)
Arabidopsis thaliana	TILLING	University of Washington; user fee paid service for screening point mutations in the genes of interest (http://tilling.fhcrc.org:9366/)
Arabidopsis thaliana	Enhancer Trap Lines	University of Pennsylvania, (http://enhancertraps.bio.upenn.edu/). Available through Arabidopsis Biological and Resource Center (http://www.Arabidopsis.org) or Nottingham Arabidopsis Stock Centre-NACS (http://arabidopsis.info/home.html)
Arabidopsis thaliana	1125 transposon tagged lines	Ito *et al.* (2002); available through RIKEN BRC (http://rarge.gsc.riken.jp/)
Arabidopsis thaliana	58,000 T-DNA insertion lines	Strizhov *et al.* (2003); GABI-kat generation of Flanking Sequence Tags (FSTs) from T-DNA mutagenized plants. Available through Max Planck Institute of Plant Breeding Research http://www.gabi-kat.de/ or Arabidopsis Biological and Resource Center (http://www.Arabidopsis.org)
Maize	Transposon tagged lines	McCarty *et al.* (2005); http://iniformmu.org; Cowperthwaite *et al.* (2002); May *et al.* (2003)
Rice	~29,000 T-DNA enhancer trap lines	Sallaud *et al.* (2004)
Rice	5200 T-DNA tagged lines	Sha *et al.* (2004)
Rice	Transposon tagged lines	Kolesnik *et al.* (2004); Upadhyaya *et al.* (2002)
Rice	T-DNA insertion lines	Sallaud *et al.* (2003, 2004)
Tomato	Activation tagging/silencing	http://www.onderzoekinformatie.nl/en/oi/nod/onderzoek/OND1300668/ http://www.north-south.nl/index.php/item/194 http://www.sgn.cornell.edu/help/about/tomato_project/index.pl

Targeting Induced Local Lesions IN Genomes (TILLING) has been developed (reviewed by Slade & Knauf, 2005). In this method, first, point mutations (i.e. single base pair changes) are induced in plants by treating seeds with a chemical mutagen. Second, DNA is extracted from M2 individuals, and seeds are stored for future use. Third, for the TILLING assay, dye-labelled PCR primers are designed based on available sequence information to amplify a single gene of interest using pooled DNA from several individuals. These PCR products are denatured and re-annealed to allow the formation of mismatched base pairs which are recognized and cleaved by the use of an endonuclease CelI. The CelI treated DNA fragments are then analysed on the gel to identify where the mutation resides. Although, using this method, mutations in genes involved in induced resistance have yet to be demonstrated, mutant plants identified by TILLING approach are not subjected to the same regulatory approval requirements and thus offer a unique advantage over those generated by transgenic technologies.

3.6.2 Insertional activation

Although T-DNA insertions mostly inactivate the genes when inserted into the coding regions, some insertions into coding as well as other regions have the potential to activate gene expression. This is called 'activation tagging', which generates 'gain-of-function' mutations instead of loss of function mutations often generated by the insertional inactivation. Results from several studies indicated that gain-of-function mutants produced by activation-tagging T-DNAs may have different spectra of mutants that have never been isolated as conventional loss-of-function mutations. For instance, transformation of Arabidopsis plants with T-DNA carrying cauliflower mosaic virus 35S enhancers and subsequent screening for increased resistance to *P. syringae* identified the *cdr1* (*constitutive disease resistance 1*) mutant with constitutively active SAR responses. *CDR1* encodes an extracellular aspartic protease, and it was proposed that CDR1 mediates a peptide signal system involved in the activation of inducible resistance mechanisms (Xia *et al.*, 2004). Similarly, screening T-DNA tagged lines of Arabidopsis for mutants specifically compromised in SAR identified the *defective in induced resistance 1-1* (*dir1-1*) mutant which exhibits wild-type local resistance to avirulent and virulent *P. syringae*, but that pathogenesis-related gene expression is abolished in uninoculated distant leaves, and *dir1-1* fails to develop SAR to virulent *P. syringae* or *Peronospora parasitica* (Maldonado *et al.*, 2002).

Inclusion of reporter genes such as *PR1* promoter-driven reporter gene constructs into the T-DNA construct used in mutagenesis facilitates high throughput screening for mutants that show increased reporter gene expression from the pathogen and SA inducible *PR1* promoter. Screening of lines via high throughput luciferase imaging identified an Arabidopsis mutant that exhibited enhanced *PR1* gene expression, designated *activated disease resistance 1* (*adr1*). This line showed constitutive expression of a number of key defence marker genes and accumulated SA. Furthermore, *adr1* plants exhibited resistance against the biotrophic pathogens *P. parasitica* and *E. cichoracearum* (Grant *et al.*, 2003).

3.6.3 Post-transcriptional gene silencing

RNAi has recently emerged as an alternative to interruption of gene expression and thus as a useful functional genomics tool (Kusaba, 2004). In this method, a double-stranded (dsRNA) construct containing sense and antisense portions of the gene to be targeted, separated by an intron, is delivered into plant cells by a variety of means (e.g. stable transformation or transient expression mediated by microprojectile bombardment, tissue infiltration by Agrobacterium or expression from the genome of a virus). Obviously, stable transformation offers a heritable and stable reduction of gene expression, but this is limited to plants for which suitable transformation methods are available (see Table 3.6 for a comparative analysis of insertional gene inactivation and post-transcriptional gene silencing methods used in reverse genetic studies). The gene silencing process is initiated by dsRNA that is recognized by a member of the RNase III family enzyme called Dicer, and digested into 21 nt small interfering RNA (siRNA) duplexes. These small RNA species guide further destruction of the mRNA.

Douchkov *et al.* (2005) developed a method for high throughput, transient induced gene silencing (TIGS) by RNAi in barley epidermal cells that is based on biolistic transgene delivery. This method was shown to be useful to test the roles of genes in resistance or susceptibility to the powdery mildew fungus *B. graminis* f. sp. *hordei*. Libraries of RNAi constructs can be built up by new, cost-efficient methods that allow cloning of any blunt-ended DNA fragments without the need for adaptor sequences (Schenk *et al.*, 2004) or that combine highly efficient ligation and recombination (Gateway cloning system; http://www.invitrogen.com).

Table 3.6 Comparison of the advantages and disadvantages of the two main methods used in reverse genetic studies.

Insertional inactivation (e.g. T-DNA of transposon insertions)	**Transcriptional gene silencing (e.g. RNAi, VIGS)**
Advantages	
1. Specific to a single gene only	1. Multiple ways of delivery (see text for details)
2. Large collections of T-DNA lines in Arabidopsis and transposon insertion lines for Arabidopsis, maize, rice, tomato and potato are already available (see Table 3.5 for details)	2. Can be directed to a specific gene or multiple related genes can be silenced with a single construct
3. Application is not dependent on the availability of sequence information from the genes to be inactivated	3. Variable levels of reduction in gene expression helps better characterization of the phenotypic effects
	4. Target genes can be silenced by using inducible expression of the silencing construct in a specific developmental stage or tissue type
	5. Suitable for large-scale functional genomics studies
Disadvantages	
1. Applicability to high throughput reverse-genetic analysis is dependent on the availability of relatively efficient transformation methods	1. Sequence information is required for the genes to be silenced
2. Large numbers of insertions required for moderate genome coverage	2. In the absence of whole genome sequence, the extent of similar genes being silenced with a single gene construct cannot be fully predicted
3. Cannot be targeted to specific genes	3. Reduced but not absolute silencing can cause only subtle changes in phenotype
4. Multiple insertions in some lines could be problematic	4. No distinct phenotypes observed for the functionally redundant genes
5. Homozygous lines are required to observe the phenotypes resulting from recessive mutations	
6. No distinct phenotypes observed for the functionally redundant genes	

By using this method, a role of the t-SNARE protein HvSNAP34 in three types of durable, race-nonspecific resistance was observed (Douchkov *et al.*, 2005).

For plant species for which highly developed T-DNA and/or transposon insertion lines are not available, silencing gene expression following its transcription into mRNA offers a valuable promise into examining gene function. RNA silencing has been successfully applied for identification of a function of a tobacco gene (*NpPDR1*) encoding an ABC (ATP-Binding Cassette) transporter protein (Stukkens *et al.*, 2005). Transgenic tobacco plants in which *NpPDR1* expression was prevented by RNA interference showed reduced resistance to *B. cinerea*, suggesting a direct role for this gene in pathogen resistance (Stukkens *et al.*, 2005). Similarly, in soybean, RNAi-mediated silencing of a gene encoding an isoflavone synthase resulted in enhanced susceptibility to the oomycete pathogen *Phytophthora sojae* (Subramanian *et al.*, 2005), suggesting a role for isoflavones in induced defence responses.

RNAi-based silencing of *NIMIN1*, a tobacco gene suspected of encoding a negative regulator in the SA pathway, has indeed shown that transgenic plants with reduced *NIMIN1* mRNA levels showed hyperactivation of *PR1* gene expression after SA treatment (Weigel *et al.*, 2005). Further confirming the role of this gene as a negative regulator, transgenic plants constitutively expressing high amounts of *NIMIN1* showed reduced SA-mediated *PR* gene induction and increased susceptibility to *P. syringae* (Weigel *et al.*, 2005).

In Arabidopsis, RNAi offers a useful resource particularly for functional analysis of genes for which no T-DNA or transposon insertion lines are currently available. For instance, a loss-of-function approach was used to demonstrate that the Arabidopsis mitogen activated protein kinase (MAPK) MPK6 plays a role in induced resistance. MPK6-silenced Arabidopsis showed compromised resistances to an avirulent strain of *P. parasitica* and avirulent and virulent strains of *P. syringae*, suggesting that MPK6 plays a role in both resistance gene mediated and basal resistance (Menke *et al.*, 2004).

Virus induced gene silencing (VIGS) has recently emerged as a useful tool for the analysis of gene function (see Burch-Smith *et al.* 2004 for a recent review). This technique is particularly suitable for plants where the effects of single gene knockouts cannot be immediately observed due to functional redundancy or difficulties encountered in stable transformation of species under study. In VIGS, plants are infected with viruses (or viral transcripts) that are engineered to carry sequences derived from plant gene transcripts. The presence of such transcripts activates a sequence-specific RNA degradation system in the host that, using the mechanism explained above, leads to the destruction of both viral- and homologous host mRNA sequences. VIGS has been successfully used for silencing of genes involved in induced resistance studies especially in dicotyledonous plants such as tobacco. More recently, VIGS silencing has been adapted for use in wheat (*Triticum aestivum* L.; Scofield *et al.*, 2005) and barley (*Hordeum vulgare* L.; Holzberg *et al.*, 2002) using barley stripe mosaic virus (BSMV). Infection of wheat plants with BSMV constructs carrying a 150 bp fragment of the rust resistance gene *Lr21* caused successful silencing of the *Lr21* gene, and this made the gene silenced plants susceptible to leaf rust infection (Scofield *et al.*, 2005).

Recently, VIGS was also adapted to high throughput screening for genes that alter disease resistance phenotypes. In this method, random sequences from a cDNA library are first placed under a constitutive promoter and then cloned into the virus genome. The binary vector carrying the virus construct is then introduced into *Agrobacterium tumefaciens* for transfer into plants by vacuum or syringe infiltration. The virus systemically spreads from the local inoculation point to the upper parts of the plant, and during this process the expression from the endogenous plant gene is suppressed. If the endogenous plant gene being silenced is involved in disease resistance, inoculation of VIGS plants could result in increased susceptibility to the pathogen. Using this system, the role of protein phosphate 2A catalytic subunits as negative regulators of plant defence responses has been uncovered (He *et al.*, 2004) as protein phosphatase 2A-silenced plants showed increased defence gene expression, an accelerated HR response and increased resistance to the bacterial pathogen *P. syringae* and the fungal pathogen, *Cladosporium fulvum*. More recently, a VIGS approach has been used for functional characterization of genes associated with powdery mildew resistance in barley (Hein *et al.*, 2005).

3.7 Manipulation of master switches for activation of induced resistance

Reverse and forward genetic approaches have so far revealed that the genes encoding master switches are promising candidates for engineering disease resistance. A master switch is a regulatory protein or in some cases a transcription factor acting relatively upstream in the signalling cascade. They regulate gene expression by activating transcription factors which, in turn, directly bind to the promoter region of target genes. In some instances, a master switch can be constitutively present in the cell or can be activated by protein–protein interactions and/or phosphorylation/dephosphorylation.

Protein kinase and MAP kinase signalling cascades constitute an integral part of the plant defence signalling pathways mainly for their roles of relaying of pathogen signals to downstream components by protein phosphorylation and dephosphorylation. Mitogen activated kinase kinase kinase (MAPKKK) acts upstream from MAPKK and MAPK pathways and functions in activating downstream MAPKs in response to a stimulus (reviewed by Pedley & Martin, 2005). In Arabidopsis, inactivation of *EDR1*, encoding a MAPKKK, showed that EDR1 is a negative regulator of plant defence as this resulted in increased resistance to the powdery mildew pathogen, *E. cichoracearum* (Frye *et al.*, 2001).

In Arabidopsis, the roles of three MAPKs (MAPK3, MAPK4 and MPK6) in plant defence have been relatively well studied. Mutational analyses have indicated that MAPK4 is required for JA-responsive expression of *PDF1.2* and *THI2.1* defence genes (Petersen *et al.*, 2000). Similarly, silencing of *MAPK6* has compromised resistance of Arabidopsis to the biotrophic pathogen *P. parasitica*, and virulent and avirulent strains of bacterial pathogen *P. syringae*. So far, no studies have examined the effect of stable over-expression of these kinases in Arabidopsis. However, transient over-expression studies have revealed that MAPK3, MAPK4 and MAPK6 all act downstream of the flagellin receptor FLS2 and upstream of the WRKY22 and WRKY29 transcription genes (Asai *et al.*, 2002). Recent studies have also placed MAPK pathways downstream from the tobacco mosaic virus resistance gene *N* in tobacco and *Pto* bacterial resistance gene in tomato (Ekengren *et al.*, 2003). Owing to their position upstream in the signalling pathway, the possibility exists that constitutive over-expression of these genes can cause compromised plant growth and development.

In addition to MAP kinases, a number of other regulatory genes controlling downstream defence gene expression have been identified in Arabidopsis. The Arabidopsis genes encoding NPR1, NDR1, EDS1, PAD4, SGT1, COI1 and JAR1 proteins are just a few examples of this class of regulators (Eulgem, 2005). Most of these genes have been identified through forward genetic approaches. The *NPR1* gene of Arabidopsis is one of the better studied master regulators functioning at the crossroads of a few different signalling pathways such as SAR, ISR and R-gene mediated resistance, as well as regulating antagonistic crosstalk between SA and JA signalling pathways (Spoel *et al.*, 2003; Pieterse & Van Loon, 2004; see also Chapter 4). NPR1 activates SA-responsive defence gene expression through interaction with members of the TGA family of bZIP transcription factors (Fan & Dong, 2002). Manipulation of regulatory genes like these can potentially provide increased resistance in transgenic plants. Indeed, inducible over-expression of *NPR1* in Arabidopsis provides increased resistance to diverse pathogens, while constitutive over-expression of *NPR1* can lead to developmental abnormalities (Cao *et al.*, 1998).

While most master regulators so far have been identified by reverse/forward genetic approaches, transcriptome analyses also identified a number of TFs based on their response to pathogens. The Arabidopsis genome contains more than 1500 transcription factors belonging to different transcription factor gene families including AP2/ERFs, MYBs, bHLH, NACs and WRKYs, and specific transcription factors potentially involved in plant defence were identified using the strategy outlined in Figure 3.2 (Chen *et al.*, 2002; McGrath *et al.*, 2005), that is transcriptome analyses followed by functional characterization of individual genes using over-expression or gene knock-out techniques. Examples of defence related TFs are given in Table 3.7.

The regulatory genes mentioned in this section may act as either negative or positive regulators. Accordingly, over-expression of positive regulators or inactivation of negative regulators may activate defence responses and cause increased disease resistance. Some transcription factors act as positive regulators of one signalling pathway but as negative regulators of another. For instance, WRKY70 and AtMYC2 transcription factors inversely regulate SA and JA, and JA and ABA signalling pathways (Anderson *et al.*, 2004; Li *et al.*, 2004; see also Figure 3.1). Recently, excellent reviews have been published about transcription factors involved in regulating induced defence responses (Singh *et al.*, 2002; Eulgem, 2005).

3.8 Suitable promoters for defence gene expression

Ideally, switching on the production of a recombinant protein with defensive functions only under challenge by a pathogen necessitates the use of pathogen inducible promoters. This would reduce the physiological cost that may result when the transgene is expressed constitutively, even in the absence of pathogen attack; although constitutive expression of transgenes may still be preferred to maximize the transgene effect. A typical example is the constitutive expression of the *NPR1* gene which has detrimental effects (reduced plant size, spontaneous lesion development, etc.) on plant development, while inducible expression of this gene under a pathogen inducible promoter alleviated such undesirable effects (Cao *et al.*, 1998). Specific expression of defence genes in a tissue such as in the epidermal tissue could also be important to encounter the pathogen challenge (Altpeter *et al.*, 2005).

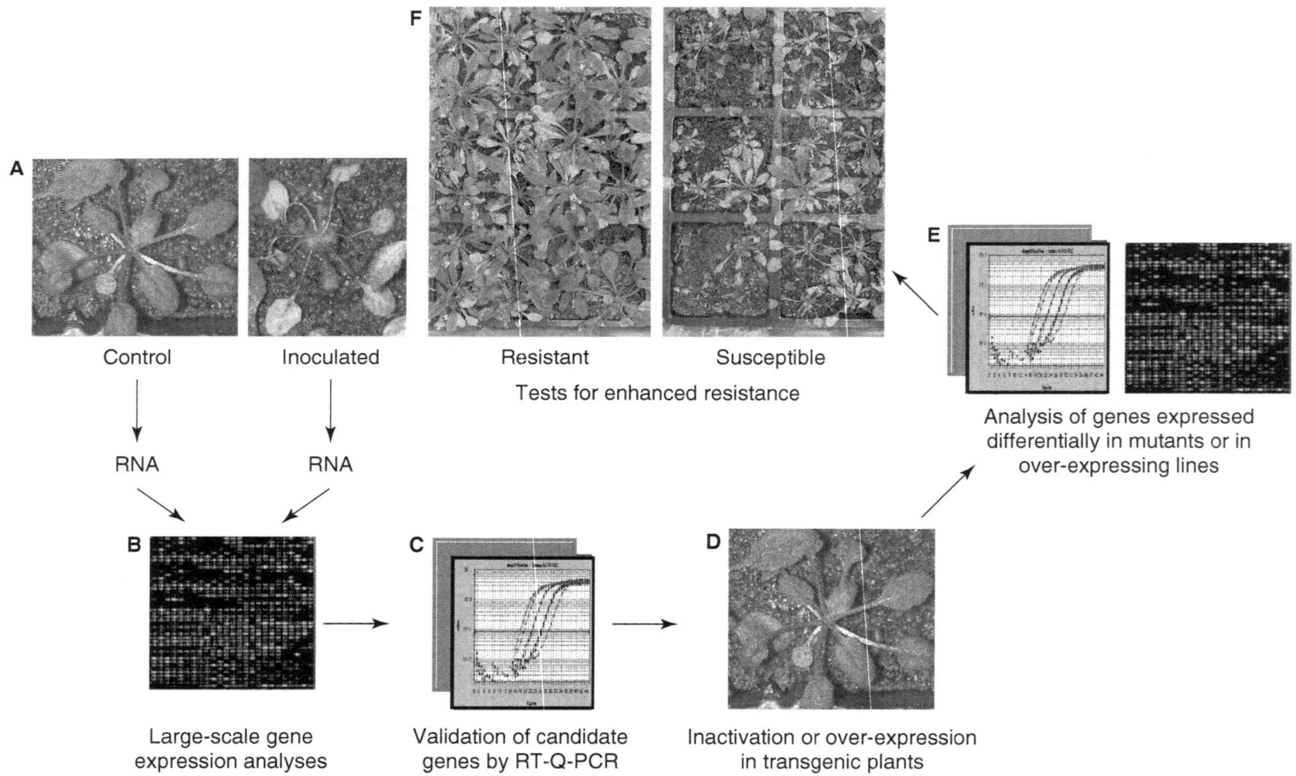

Figure 3.2 Strategy used for large-scale identification of genes potentially involved in induced resistance. (A) RNA samples isolated from pathogen inoculated and uninoculated (control) plants are analysed using large-scale gene expression analyses such as microarray analysis (B) to identify candidate genes. Inducibility of such genes during plant defence is subsequently confirmed by RT-Q-PCR or other independent gene expression analyses (C). Selected genes are either inactivated by T-DNA, transposon insertions or RNAi or over-expressed (D) to observe the effect of such manipulations on defence gene expression (E) and disease resistance (F).

Table 3.7 Examples of Arabidopsis transcription factors, protein kinases and other regulatory genes involved in plant defence responses.

Regulatory genes	Defensive function	Reference
Transcription factors/family		
AtMYC2/basic helix loop helix	Negative regulation of plant defence. The *myc2* mutant shows increased disease resistance to *Fusarium oxysporum, Botrytis cinerea* and *Plectosphaerella cucumerina*	Lorenzo *et al.* (2004), Anderson *et al.* (2004)
WRKY70/WRKY	Positive and negative regulator of SA and JA pathways, respectively. WRKY-70 over-expression increases resistance to virulent pathogens and results in constitutive expression of SA-responsive pathogenesis-related genes	Li *et al.* (2004)
AtWRKY18/WRKY	Potentiation of plant defence genes and increased resistance to *Pseudomonas syringae*	Chen & Chen (2002)
AtMYB30/MYB	Positive regulator of the hypersensitive cell death programme in plants in response to pathogen attack	Vailleau *et al.* (2002)
BOS1/MYB	Required for biotic and abiotic stress responses in Arabidopsis	Mengiste *et al.* (2003)
AtAF2/NAC	Negative regulation of plant defence. Over-expression shows reduced defence gene expression and increased susceptibility to *Fusarium oxysporum*	Delessert *et al.* (2005)
LOL1/C_2H_2 Zinc-finger	Positive regulator of cell death	Epple *et al.* (2003)
LSD1/C_2H_2 Zinc-finger	Negative regulator of cell death	Dietrich *et al.* (1997)
TGA2, TGA5, TGA6	Positive regulators of SA-dependent defence gene expression and pathogen resistance. Induction of PR gene expression and pathogen resistance by the SA analogue 2,6-dichloroisonicotinic acid (INA) was blocked in the *tga6-1, tga2-1, tga5-1* mutants	Zhang *et al.* (2003)
AtWHY1/WHIRLY	Required for disease resistance responses	Desveaux *et al.* (2004)
ERF1/APETELA2/Ethylene Response Factor-AP2/ERF	Positive regulator of plant defences. Over-expression provides increased pathogen resistance	Lorenzo *et al.* (2003), Berrocal-Lobo & Molina (2004)
AtERF2/AP2/ERF	Positive regulator of plant defences. Over-expression provides increased pathogen resistance	Brown *et al.* (2003), McGrath *et al.* (2005)
AtERF4/AP2/ERF TF	Negative regulator of defence responses. Over-expression provides increased pathogen susceptibility	McGrath *et al.* (2005)

(Continued)

Table 3.7 (*Continued*)

Regulatory genes	Defensive function	Reference
Protein kinases		
MPK6/Mitogen Activated (MAP) Kinase	Positive regulator of defence responses; *mpk6* silenced plants show increased susceptibility to *Pseudomonas syringae* and *Peronospora parasitica*	Menke *et al.* (2004)
MPK4/Mitogen Activated (MAP) Kinase	Negative regulation of SA biosynthesis; the *mpk4* mutant shows increased SA accumulation and increased resistance to the bacterial pathogen *Pseudomonas syringae*	Petersen *et al.* (2000)
EDR1/Mitogen Activated kinase kinase kinase (MAPKKK)	Negative regulation of defence responses; the *edr1* mutant shows increased resistance to the necrotrophic pathogen *Erysiphe cichoracearum*	Frye *et al.* (2001)
FLS2/Flagellin receptor/ Leucine rich repeat LRR_RLK	The *fls2* mutant shows increased susceptibility to *Pseudomonas syringae*	Zipfel *et al.* (2004)
ERECTA/receptor like kinase	Required for resistance to the necrotrophic fungus *Plectosphaerella cucumerina*	Llorente *et al.* (2005)
OXI1/Serine threonine kinase (STK)	Required for activation of MPK3 and MPK6 in response to reactive oxygen species and elicitors and basal resistance to *Peronospora parasitica*	Rentel *et al.* (2004)
MKS1/MPK4 substrate	Overexpression of MKS1 in wild-type plants is sufficient to activate SA-dependent resistance	Andreasson *et al.* (2005)
RFO/Wall-associated kinase like protein	Required for resistance to *Fusarium oxysporum* in Arabidopsis	Diener and Ausubel (2005)
Others		
NPR1/Ankyrin repeat protein	Required for activation of SA-responsive defence genes and resistance to *Pseudomonas syringae*	Cao *et al.* (1997)
Heterotrimeric G-protein	Required for resistance to the necrotrophic fungus *Plectosphaerella cucumerina*	Llorente *et al.* (2005)

Those readers who wish to obtain further information about various expression strategies used in genetic engineering of disease resistance should consult the recent review by Gurr & Rushton (2005).

3.9 Conclusions: a systems biological approach to induced plant defence?

So far, research on model plant species has provided a wealth of information about the genes involved in plant defence. It is expected that such information will help speed up the research in crop species. However, one of the current challenges in transferring this knowledge to crop species is the identification of crop genes that are functionally identical (orthologues) to the genes identified in the model plant species such as Arabidopsis. Bioinformatic approaches coupled with molecular mapping, cloning and functional complementation could make this task relatively easy. Indeed, a recent study identified potato homologues of many Arabidopsis genes functional in defence signalling (Pajerowska *et al.*, 2005). The fact that most genes in plants are members of large gene families can make this task even more challenging. In addition, characterized Arabidopsis mutants offer a useful resource in this respect. Crop genes that are suspected of being orthologous to Arabidopsis genes can be introduced by transformation into Arabidopsis mutant lines. If the introduced gene complements the defect (i.e. increased or reduced disease resistance), this is taken as strong evidence of similarity in gene function. Once their predicted function is confirmed, these genes can then be manipulated in crop species of interest.

Genomic analysis of induced defence has clearly established that the induced resistance response requires coordinate action of many genes and/or defence signalling pathways. The inherent complexity associated with individual defence signalling pathways can be further complicated by the extensive crosstalk that occurs among the multiple stress and/or defence signalling pathways (see Chapter 4). Therefore, the development of holistic approaches may be required to integrate the effects of multiple parameters on the plant system as a whole and to estimate the responses of plants exposed to stress. The methodology of systems biology requires the identification of individual components and their respective interactions. As outlined in Figure 3.3, this information is then integrated into a predictive model to explain the behaviour of the system as a whole. Hypotheses developed through this process can then be tested experimentally by disturbing the system (e.g. by use of knock-outs) and testing the effect of such disturbance on the whole plant. This may then lead to the refinement of the existing models as well as the development of new hypotheses.

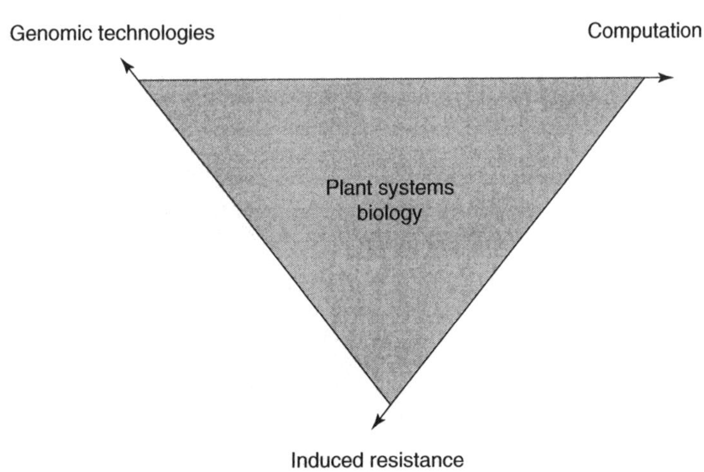

Figure 3.3 Use of systems biology to predict plant responses, e.g. induction of resistance via activation of complex signalling pathways. See text for additional details.

In conclusion, although the systems biology of plant stress and disease tolerance is still in its infancy, the gene regulatory networks that control stress responses are emerging with the aid of reverse genetics, large-scale gene expression, as well as other emerging disciplines such as metabolomics and proteomics, at least in the model plant species such as Arabidopsis and rice. Combinatorial use of these high throughput multiparallel analytical approaches (Hirai *et al*., 2005), as well as new data analyses and model building methods (Prusinkiewicz, 2004) will certainly help better understand the nature of induced plant stress and defence biology.

3.10 Acknowledgements

We apologise to our colleagues whose work on this subject could not be highlighted due to space limitations. We are grateful to Dr. John M. Manners, for his continuous encouragement and leadership over many years of plant defence-related genomic research. We thank B. Dombrecht, E. Campbell and K. McGrath, for the photographs used in Figure 3.2, and B. Dombrecht and K. McGrath, for critical manuscript reading. We also acknowledge valuable contributions made by many of our students to this exciting area of research.

3.11 References

Abe H, Urao T, Ito T, Seki M, Shinozaki K, Yamaguchi-Shinozaki K, 2003. Arabidopsis AtMYC2 (bHLH) and AtMYB2 (MYB) function as transcriptional activators in abscisic acid signaling. *The Plant Cell* **15,** 63–78.

Akimoto-Tomiyama C, Sakata K, Yazaki J, Nakamura K, Fujii F, Shimbo K, Yamamoto K, Sasaki T, Kishimoto N, Kikuchi S, Shibuya N, Minami E, 2003. Rice gene expression in response to *N*-acetylchitooligosaccharide elicitor: comprehensive analysis by DNA microarray with randomly selected ESTs. *Plant Molecular Biology* **52,** 537–551.

Allemeersch J, Durinck S, Vanderhaeghen R, Alard P, Maes R, Seeuws K, Bogaert T, Coddens K, Deschouwer K, Van Hummelen P, Vuylsteke M, Moreau Y, Kwekkeboom J, Wijfjes AH, May S, Beynon J, Hilson P, Kuiper MT, 2005. Benchmarking the CATMA microarray. A novel tool for Arabidopsis transcriptome analysis. *Plant Physiology* **137**, 588–601.

Alonso JM, Stepanova AN, Leisse TJ, Kim CJ, Chen H, Shinn P, Stevenson DK, Zimmerman J, Barajas P, Cheuk R, Gadrinab C, Heller C, Jeske A, Koesema E, Meyers CC, Parker H, Prednis L, Ansari Y, Choy N, Deen H, Geralt M, Hazari N, Hom E, Karnes M, Mulholland C, Ndubaku R, Schmidt I, Guzman P, Aguilar-Henonin L, Schmid M, Weigel D, Carter DE, Marchand T, Risseeuw E, Brogden D, Zeko A, Crosby WL, Berry CC, Ecker JR, 2003. Genome-wide insertional mutagenesis of *Arabidopsis thaliana*. *Science* **301,** 653–657.

Altpeter F, Varshney A, Abderhalden O, Douchkov D, Sautter C, Kumlehn J, Dudler R, Schweizer P, 2005. Stable expression of a defence-related gene in wheat epidermis under transcriptional control of a novel promoter confers pathogen resistance. *Plant Molecular Biology* **57,** 271–283.

Amme S, Rutten T, Melzer M, Sonsmann G, Vissers JP, Schlesier B, Mock HP, 2005. A proteome approach defines protective functions of tobacco leaf trichomes. *Proteomics* **5,** 2508–2518.

Anderson JP, Badruzsaufari E, Schenk PM, Manners JM, Desmond OJ, Ehlert C, Maclean DJ, Ebert PR, Kazan K, 2004. Antagonistic interaction between abscisic acid and jasmonate-ethylene signaling pathways modulates defence gene expression and disease resistance in Arabidopsis. *The Plant Cell* **16,** 3460–3479.

Andreasson E, Jenkins T, Brodersen P, Thorgrimsen S, Petersen NH, Zhu S, Qiu JL, Micheelsen P, Rocher A, Petersen M, Newman MA, Bjorn Nielsen H, Hirt H, Somssich I, Mattsson O, Mundy J, 2005. The MAP kinase substrate MKS1 is a regulator of plant defence responses. *EMBO Journal* **24,** 2579–2589.

Asai T, Tena G, Plotnikova J, Willmann MR, Chiu WL, Gomez-Gomez L, Boller T, Ausubel FM, Sheen J, 2002. MAP kinase signalling cascade in Arabidopsis innate immunity. *Nature* **415,** 977–983.

Bais HP, Prithiviraj B, Jha AK, Ausubel FM, Vivanco JM, 2005. Mediation of pathogen resistance by exudation of antimicrobials from roots. *Nature* **434,** 217–221.

Berrocal-Lobo M, Molina A, 2004. Ethylene response factor 1 mediates Arabidopsis resistance to the soilborne fungus *Fusarium oxysporum. Molecular Plant–Microbe Interactions* **17,** 763–770.

Bower NI, Casu RE, Maclean DJ, Reverter A, Chapman SC, Manners JM, 2005. Transcriptional response of sugarcane roots to methyl jasmonate. *Plant Science* **168,** 761–772.

Bowling SA, Guo A, Cao H, Gordon AS, Klessig DF, Dong X, 1994. A mutation in Arabidopsis that leads to constitutive expression of systemic acquired resistance. *The Plant Cell* **6,** 1845–1857.

Brown RL, Kazan K, McGrath KC, Maclean DJ, Manners JM, 2003. A role for the GCC-box in jasmonate-mediated activation of the *PDF1.2* gene of Arabidopsis. *Plant Physiology* **132,** 1020–1032.

Bruggmann R, Abderhalden O, Reymond P, Dudler R, 2005. Analysis of epidermis- and mesophyll-specific transcript accumulation in powdery mildew-inoculated wheat leaves. *Plant Molecular Biology* **58,** 247–267.

Brugmans B, Fernandez del Carmen A, Bachem CW, van Os H, van Eck HJ, Visser RG, 2002. A novel method for the construction of genome wide transcriptome maps. *The Plant Journal* **31,** 211–222.

Burch-Smith TM, Anderson JC, Martin GB, Dinesh-Kumar SP, 2004. Applications and advantages of virus-induced gene silencing for gene function studies in plants. *The Plant Journal* **39,** 734–746.

Button DK, Gartland KMA, Ball LD. Natanson L, Gartland JS, Lyon GD, 2006. DRASTIC – INSIGHTS: Querying information in a plant gene expression database. *Nucleic Acids Research* **34,** D712-D716.

Caldo RA, Nettleton D, Wise RP, 2004. Interaction-dependent gene expression in Mla-specified response to barley powdery mildew. *The Plant Cell* **16,** 2514–2528.

Campbell EJ, Schenk PM, Kazan K, Iris AMA, Penninckx IA, Anderson JP, Maclean DJ, Cammue BP, Ebert PR, Manners JM, 2003. Pathogen-responsive expression of a putative ATP-binding cassette transporter gene conferring resistance to the diterpenoid sclareol is regulated by multiple defense signaling pathways in Arabidopsis. *Plant Physiology* **133,** 1272–1284.

Cao H, Bowling SA, Gordon AS, Dong XN, 1994. Characterization of an Arabidopsis mutant that is non-responsive to inducers of systemic acquired resistance. *The Plant Cell* **6,** 1583–1592.

Cao H, Glazebrook J, Clarke JD, Volko S, Dong X, 1997. The Arabidopsis NPR1 gene that controls systemic acquired resistance encodes a novel protein containing ankyrin repeats. *Cell* **88,** 57–63.

Cao H, Li X, Dong X, 1998. Generation of broad-spectrum disease resistance by overexpression of an essential regulatory gene in systemic acquired resistance. *Proceedings of the National Academy of Sciences USA* **95,** 6531–6536.

Cartieaux F, Thibaud MC, Zimmerli L, Lessard P, Sarrobert C, David P, Gerbaud A, Robaglia C, Somerville S, Nussaume L, 2003. Transcriptome analysis of Arabidopsis colonized by a plant-growth promoting rhizobacterium reveals a general effect on disease resistance. *The Plant Journal* **36,** 177–188.

Chapman S, Schenk P, Kazan K, Manners J, 2002. Using biplots to interpret gene expression patterns in plants. *Bioinformatics* **18,** 202–204.

Chen C, Chen Z, 2002. Potentiation of developmentally regulated plant defence response by AtWRKY18, a pathogen-induced Arabidopsis transcription factor. *Plant Physiology* **129,** 706–716.

Chen WQ, Provart NJ, Glazebrook J, Katagiri F, Chang HS, Eulgem T, Mauch F, Luan S, Zou GZ, Whitham SA, Budworth PR, Tao Y, Xie ZY, Chen X, Lam S, Kreps JA, Harper JF, Si-Ammour A, Mauch-Mani B, Heinlein M, Kobayashi K, Hohn T, Dangl JL, Wang X, Zhu T, 2002. Expression profile matrix of Arabidopsis transcription factor genes suggests their putative functions in response to environmental stresses. *The Plant Cell* **14,** 559–574.

Cheong YH, Chang H-S, Gupta R, Wang X, Zhu T, Luan S, 2002. Transcriptional profiling reveals novel interactions between wounding, pathogen, abiotic stress, and hormonal responses in Arabidopsis. *Plant Physiology* **129,** 661–677.

Coram TE, Pang ECK, 2005a. Isolation and analysis of candidate ascochyta blight defence genes in chickpea. Part I. Generation and analysis of an expressed sequence tag (EST) library. *Physiological and Molecular Plant Pathology* **66,** 192–200.

Coram TE, Pang ECK, 2005b. Isolation and analysis of candidate ascochyta blight defence genes in chickpea. Part II. Microarray expression analysis of putative defence-related ESTs. *Physiological and Molecular Plant Pathology* **66,** 201–210.

Cowperthwaite M, Park W, Xu Z, Yan X, Maurais SC, Dooner HK, 2002. Use of the transposon Ac as a gene-searching engine in the maize genome. *The Plant Cell* **14,** 713–726.

Czechowski T, Bari RP, Stitt M, Scheible WR, Udvardi MK, 2004. Real-time RT-PCR profiling of over 1400 Arabidopsis transcription factors: unprecedented sensitivity reveals novel root- and shoot-specific genes. *The Plant Journal* **38,** 366–379.

Delessert C, Kazan K, Wilson IW, Straeten DV, Manners J, Dennis ES, Dolferus R, 2005. The transcription factor ATAF2 represses the expression of pathogenesis-related genes in Arabidopsis. *The Plant Journal* **43,** 745–757.

De Paepe A, Vuylsteke M, Van Hummelen P, Zabeau M, Van Der Straeten D, 2004. Transcriptional profiling by cDNA-AFLP and microarray analysis reveals novel insights into the early response to ethylene in Arabidopsis. *Plant Journal* **39,** 537–559.

Desveaux D, Subramaniam R, Després C, Mess J-N, Lévesque C, Fobert PR, Dangl JL, Brisson N, 2004. A 'Whirly' transcription factor is required for salicylic acid-dependent disease resistance in Arabidopsis. *Developmental Cell* **6,** 229–240.

Devoto A, Ellis C, Magusin A, Chang HS, Chilcott C, Zhu T, Turner JG, 2005. Expression profiling reveals COI1 to be a key regulator of genes involved in wound- and methyl jasmonate-induced secondary metabolism, defence, and hormone interactions. *Plant Molecular Biology* **58,** 497–513.

Diener AC, Ausubel FM, 2005. RESISTANCE TO FUSARIUM OXYSPORUM 1, a dominant Arabidopsis disease-resistance gene, is not race specific. *Genetics* **171,** 305–321.

Dietrich RA, Richberg MH, Schmidt R, Dean C, Dangl JL, 1997. A novel zinc finger protein is encoded by the Arabidopsis LSD1 gene and functions as a negative regulator of plant cell death. *Cell* **88,** 685–694.

Douchkov D, Nowara D, Zierold U, Schweizer P, 2005. A high-throughput gene-silencing system for the functional assessment of defence-related genes in barley epidermal cells. *Molecular Plant–Microbe Interactions* **18,** 755–761.

Dowd C, Wilson IW, McFadden H, 2004. Gene expression profile changes in cotton root and hypocotyl tissues in response to infection with *Fusarium oxysporum* f. sp. *vasinfectum*. *Molecular Plant–Microbe Interactions* **17,** 654–667.

Ebbole DJ, Jin Y, Thon M, Pan H, Bhattarai E, Thomas T, Dean R, 2004. Gene discovery and gene expression in the rice blast fungus, *Magnaporthe grisea*: analysis of expressed sequence tags. *Molecular Plant–Microbe Interactions* **17,** 1337–1347.

Eckey C, Korell M, Leib K, Biedenkopf D, Jansen C, Langen G, Kogel KH, 2004. Identification of powdery mildew-induced barley genes by cDNA-AFLP: functional assessment of an early expressed MAP kinase. *Plant Molecular Biology* **55,** 1–15.

Ekengren SK, Liu Y, Schiff M, Dinesh-Kumar SP, Martin GB, 2003. Two MAPK cascades, NPR1, and TGA transcription factors play a role in Pto-mediated disease resistance in tomato. *The Plant Journal* **36,** 905–917.

Ellis C, Karafyllidis I, Wasternack C, Turner JG, 2002. The Arabidopsis mutant *cev1* links cell wall signaling to jasmonate and ethylene responses. *The Plant Cell* **14,** 1557–1566.

Epple P, Mack AA, Morris VR, Dangl JL, 2003. Antagonistic control of oxidative stress-induced cell death in Arabidopsis by two related, plant-specific zinc finger proteins. *Proceedings of the National Academy of Sciences USA* **100,** 6831–6836.

Eulgem, T, 2005. Regulation of the Arabidopsis defence transcriptome. *Trends in Plant Science* **10,** 71–78.

Fabro G, Kovacs I, Pavet V, Szabados L, Alvarez ME, 2004. Proline accumulation and AtP5CS2 gene activation are induced by plant–pathogen incompatible interactions in Arabidopsis. *Molecular Plant–Microbe Interactions* **17,** 343–350.

Fan W, Dong X, 2002. In vivo interaction between NPR1 and transcription factor TGA2 leads to salicylic acid-mediated gene activation in Arabidopsis. *The Plant Cell* **14,** 1377–1389.

Fang Y, Macool DJ, Xue Z, Heppard EP, Hainey CF, Tingey SV, Miao GH, 2002. Development of a high-throughput yeast two-hybrid screening system to study protein–protein interactions in plants. *Molecular Genetics and Genomics* **267,** 142–153.

Frye CA, Tang D, Innes RW, 2001. Negative regulation of defence responses in plants by a conserved MAPKK kinase. *Proceedings of the National Academy of Sciences USA* **98,** 373–378.

Fujiwara S, Tanaka N, Kaneda T, Takayama S, Isogai A, Che FS, 2004. Rice cDNA microarray-based gene expression profiling of the response to flagellin perception in cultured rice cells. *Molecular Plant–Microbe Interactions* **17,** 986–998.

Gibly A, Bonshtien A, Balaji V, Debbie P, Martin GB, Sessa G, 2004. Identification and expression profiling of tomato genes differentially regulated during a resistance response to *Xanthomonas campestris* pv. *vesicatoria*. *Molecular Plant–Microbe Interactions* **17,** 1212–1222.

Glazebrook J, Chen W, Estes B, Chang H-S, Nawrath C, Métraux J-P, Zhu T, Katagiri F, 2003. Topology of the network integrating salicylate and jasmonate signal transduction derived from global expression phenotyping. *The Plant Journal* **34,** 217–228.

Glombitza S, Dubuis PH, Thulke O, Welzl G, Bovet L, Gotz M, Affenzeller M, Geist B, Hehn A, Asnaghi C, Ernst D, Seidlitz HK, Gundlach H, Mayer KF, Martinoia E, Werck-Reichhart D, Mauch F, Schaffner AR, 2004. Crosstalk and differential response to abiotic and biotic stressors reflected at the transcriptional level of effector genes from secondary metabolism. *Plant Molecular Biology* **54,** 817–835.

Golem S, Culver JN, 2003. Tobacco mosaic virus induced alterations in the gene expression profile of *Arabidopsis thaliana. Molecular Plant–Microbe Interactions* **16,** 681–688.

Grant JJ, Chini A, Basu D, Loake GJ, 2003. Targeted activation tagging of the Arabidopsis NBS-LRR gene, ADR1, conveys resistance to virulent pathogens. *Molecular Plant–Microbe Interactions* **16,** 669–680.

Gruhler A, Schulze WX, Matthiesen R, Mann M, Jensen ON, 2005. Stable isotope labeling of *Arabidopsis thaliana* cells and quantitative proteomics by mass spectrometry. *Molecular and Cellular Proteomics* **4,** 1697–1709.

Gurr SJ, Rushton PJ, 2005. Engineering plants with increased disease resistance: how are we going to express it? *Trends in Biotechnology* **23,** 283–290.

He X, Anderson JC, del Pozo O, Gu Y-Q, Tang X, Martin GB, 2004. Silencing of subfamily I of protein phosphatase 2A catalytic subunits results in activation of plant defence responses and localized cell death. *The Plant Journal* **38,** 563–577.

Hein I, Barciszewska-Pacak M, Hrubikova K, Williamson S, Dinesen M, Soenderby IE, Sundar S, Jarmolowski A, Shirasu K, Lacomme C, 2005. Virus-induced gene silencing-based functional characterization of genes associated with powdery mildew resistance in barley. *Plant Physiology* **138,** 2155–2164.

Hirai MY, Klein M, Fujikawa Y, Yano M, Goodenowe DB, Yamazaki Y, Kanaya S, Nakamura Y, Kitayama M, Suzuki H, Sakurai N, Shibata D, Tokuhisa J, Reichelt M, Gershenzon J, Papenbrock J, Saito K, 2005. Elucidation of gene-to-gene and metabolite-to-gene networks in Arabidopsis by integration of metabolomics and transcriptomics. *Journal of Biological Chemistry* **280,** 25590–25595.

Hirano H, Islam N, Kawasaki H, 2004. Technical aspects of functional proteomics in plants. *Phytochemistry* **65,** 1487–1498.

Holzberg S, Brosio P, Gross C, Pogue GP, 2002. Barley stripe mosaic virus-induced gene silencing in a monocot plant. *The Plant Journal* **30,** 315–327.

Huitema E, Vleeshouwers VGAA, Francis DM, Kamoun S, 2003. Active defence responses associated with non-host resistance of *Arabidopsis thaliana* to the oomycete pathogen *Phytophthora infestans. Molecular Plant Pathology* **4,** 487–500.

Immink RG, Angenent GC, 2002. Transcription factors do it together: the hows and whys of studying protein–protein interactions. *Trends in Plant Science* **7,** 531–534.

Ito T, Motohashi R, Kuromori T, Mizukado S, Sakurai T, Kanahara H, Seki M, Shinozaki K, 2002. A new resource of locally transposed dissociation elements for screening gene-knockout lines in silico on the Arabidopsis genome. *Plant Physiology* **129,** 1695–1699.

Jantasuriyarat C, Gowda M, Haller K, Hatfield J, Lu G, Stahlberg E, Zhou B, Li H, Kim H, Yu Y, Dean RA, Wing RA, Soderlund C, Wang GL, 2005. Large-scale identification of expressed sequence tags involved in rice and rice blast fungus interaction. *Plant Physiology* **138,** 105–115.

Jarosch B, Jansen M, Schaffrath U, 2003. Acquired resistance functions in *mlo* barley, which is hypersusceptible to *Magnaporthe grisea. Molecular Plant–Microbe Interactions* **16,** 107–114.

Jirage D, Tootle TL, Reuber TL, Frost LN, Feys BJ, Parker JE, Ausubel FM, Glazebrook J, 1999. *Arabidopsis thaliana* PAD4 encodes a lipase-like gene that is important for salicylic acid signaling. *Proceedings of the National Academy of Sciences USA* **96,** 13583–13588.

Kazan K, Schenk PM, Wilson I, Manners JM, 2001. DNA microarrays: new tools in the analysis of plant defence responses. *Molecular Plant Pathology* **2,** 177–185.

Kemp BP, Beeching JR, Cooper RM, 2005. cDNA-AFLP reveals genes differentially expressed during the hypersensitive response of cassava. *Molecular Plant Pathology* **6,** 113–123.

Kersten B, Burkle L, Kuhn EJ, Giavalisco P, Konthur Z, Lueking A, Walter G, Eickhoff H, Schneider U, 2002. Large-scale plant proteomics. *Plant Molecular Biology* **48,** 133–141.

Kim S, Ahn IP, Lee YH, 2001. Analysis of genes expressed during rice-*Magnaporthe grisea* interactions. *Molecular Plant–Microbe Interactions* **14,** 1340–1346.

Kim ST, Cho KS, Yu S, Kim SG, Hong JC, Han CD, Bae DW, Nam MH, Kang KY, 2003. Proteomic analysis of differentially expressed proteins induced by rice blast fungus and elicitor in suspension-cultured rice cells. *Proteomics* **3,** 2368–2378.
Kolesnik T, Szeverenyi I, Bachmann D, Kumar CS, Jiang S, Ramamoorthy R, Cai M, Ma ZG, Sundaresan V, Ramachandran S, 2004. Establishing an efficient Ac/Ds tagging system in rice: large-scale analysis of Ds flanking sequences. *The Plant Journal* **37,** 301–314.
Krause M, Durner J, 2004. Harpin inactivates mitochondria in Arabidopsis suspension cells. *Molecular Plant–Microbe Interactions* **17,** 131–139.
Kruger WM, Pritsch C, Chao S, Muehlbauer GJ, 2002. Functional and comparative bioinformatic analysis of expressed genes from wheat spikes infected with *Fusarium graminearum. Molecular Plant–Microbe Interactions* **15,** 445–455.
Kusaba M, 2004. RNA interference in crop plants. *Current Opinion in Biotechnology* **15,** 139–143.
Laugesen S, Bergoin A, Rossignol M, 2004. Deciphering the plant phosphoproteome: tools and strategies for a challenging task. *Plant Physiology and Biochemistry* **42,** 929–936.
Lei Z, Elmer AM, Watson BS, Dixon RA, Mendes PJ, Sumner LW, 2005. A two-dimensional electrophoresis proteomic reference map and systematic identification of 1367 proteins from a cell suspension culture of the model legume *Medicago truncatula. Molecular and Cellular Proteomics* **4,** 1812–1825.
Li J, Brader G, Palva ET, 2004. The WRKY70 transcription factor: a node of convergence for jasmonate-mediated and salicylate-mediated signals in plant defence. *The Plant Cell* **16,** 319–331.
Llorente F, Alonso-Blanco C, Sánchez-Rodriguez C, Jorda L, Molina A, 2005. ERECTA receptor-like kinase and heterotrimeric G protein from Arabidopsis are required for resistance to the necrotrophic fungus *Plectosphaerella cucumerina. The Plant Journal* **43,** 165–180.
Lopez C, Soto M, Restrepo S, Piegu B, Cooke R, Delseny M, Tohme J, Verdier V, 2005. Gene expression profile in response to *Xanthomonas axonopodis* pv. *manihotis* infection in cassava using a cDNA microarray. *Plant Molecular Biology* **57,** 393–410.
Lorenzo O, Chico JM, Sánchez-Serrano JJ, Solano R, 2004. JASMONATE-INSENSITIVE1 encodes a MYC transcription factor essential to discriminate between different jasmonate-regulated defence responses in Arabidopsis. *The Plant Cell* **16,** 1938–1950.
Lorenzo O, Piqueras R, Sánchez-Serrano JJ, Solano R, 2003. ETHYLENE RESPONSE FACTOR1 integrates signals from ethylene and jasmonate pathways in plant defense. *The Plant Cell* **15,** 165–178.
Lu G, Jantasuriyarat C, Zhou B, Wang GL, 2004. Isolation and characterization of novel defence response genes involved in compatible and incompatible interactions between rice and *Magnaporthe grisea. Theoretical and Applied Genetics* **108,** 525–534.
Luo M, Dang P, Bausher MG, Holbrook CC, Lee RD, Lynch RE, Guo BZ, 2005. Identification of transcripts involved in resistance responses to leaf spot disease caused by *Cercosporidium personatum* in peanut (*Arachis hypogaea*). *Phytopathology* **95,** 381–387.
Maldonado AM, Doerner P, Dixon RA, Lamb CJ, Cameron RK, 2002. A putative lipid transfer protein involved in systemic resistance signalling in Arabidopsis. *Nature* **419,** 399–403.
Maleck K, Levine A, Eulgem T, Morgan A, Schmid J, Lawton KA, Dangl JL, Dietrich RA, 2000. The transcriptome of *Arabidopsis thaliana* during systemic acquired resistance. *Nature Genetics* **26,** 403–410.
Marathe R, Guan Z, Anandalakshmi R, Zhao H, Dinesh-Kumar SP, 2004. Study of *Arabidopsis thaliana* resistome in response to cucumber mosaic virus infection using whole genome microarray. *Plant Molecular Biology* **55,** 501–520.
May BP, Liu H, Vollbrecht E, Senior L, Rabinowicz PD, Roh D, Pan X, Stein L, Freeling M, Alexander D, Martienssen R, 2003. Maize-targeted mutagenesis: A knockout resource for maize. *Proceedings of the National Academy of Sciences USA* **100,** 11541–11546.
McCarty DR, Settles AM, Suzuki M, Tan BC, Latshaw S, Porch T, Robin K, Baier J, Avigne W, Lai JS, Messing J, Koch KE, Hannah LC, 2005. Steady-state transposon mutagenesis in inbred maize. *The Plant Journal* **44,** 52–61.
McGrath KC, Dombrecht B, Manners JM, Schenk PM, Edgar CI, Maclean DJ, Scheible WR, Udvardi MK, Kazan K, 2005. Repressor- and activator-type ethylene response factors functioning in jasmonate signaling and disease resistance identified via a genome-wide screen of Arabidopsis transcription factor gene expression. *Plant Physiology* **139,** 949–959.
Mengiste T, Chen X, Salmeron J, Dietrich R, 2003. The *BOTRYTIS SUSCEPTIBLE1* gene encodes an R2R3MYB transcription factor protein that is required for biotic and abiotic stress responses in Arabidopsis. *The Plant Cell* **15,** 2551–2565.

Menke FL, van Pelt JA, Pieterse CMJ, Klessig DF, 2004. Silencing of the mitogen-activated protein kinase MPK6 compromises disease resistance in Arabidopsis. *The Plant Cell* **16,** 897–907.

Moy P, Qutob D, Chapman BP, Atkinson I, Gijzen M, 2004. Patterns of gene expression upon infection of soybean plants by *Phytophthora sojae. Molecular Plant–Microbe Interactions* **17,** 1051–1062.

Narusaka Y, Narusaka M, Seki M, Ishida J, Nakashima M, Kamiya A, Enju A, Sakurai T, Satoh M, Kobayashi M, Tosa Y, Park P, Shinozaki K, 2003. The cDNA microarray analysis using an Arabidopsis *pad3* mutant reveals the expression profiles and classification of genes induced by *Alternaria brassicicola* attack. *Plant and Cell Physiology* **44,** 377–387.

Navarro L, Zipfel C, Rowland O, Keller I, Robatzek S, Boller T, Jones JD, 2004. The transcriptional innate immune response to flg22. Interplay and overlap with Avr gene-dependent defence responses and bacterial pathogenesis. *Plant Physiology* **135,** 1113–1128.

Ndimba BK, Chivasa S, Hamilton JM, Simon WJ, Slabas AR, 2003. Proteomic analysis of changes in the extracellular matrix of Arabidopsis cell suspension cultures induced by fungal elicitors. *Proteomics* **3,** 1047–1059.

Newton RP, Brenton AG, Smith CJ, Dudley E, 2004. Plant proteome analysis by mass spectrometry: principles, problems, pitfalls and recent developments. *Phytochemistry* **65,** 1449–1485.

Nishimura MT, Stein M, Hou BH, Vogel JP, Edwards H, Somerville SC, 2003. Loss of a callose synthase results in salicylic acid-dependent disease resistance. *Science* **301,** 969–972.

Oh IS, Park AR, Bae MS, Kwon SJ, Kim YS, Lee JE, Kang NY, Lee S, Cheong H, Park OK, 2005. Secretome analysis reveals an Arabidopsis lipase involved in defence against *Alternaria brassicicola. The Plant Cell* **17,** 2832–2847.

Pajerowska KM, Parker JE, Gebhardt C, 2005. Potato homologs of *Arabidopsis thaliana* genes functional in defence signalling-identification, genetic mapping, and molecular cloning. *Molecular Plant–Microbe Interactions* **18,** 1107–1119.

Parker JE, Holub EB, Frost LF, Falk A, Gunn ND, Daniels MJ, 1996. Characterization of *eds1*, a mutation in Arabidopsis suppressing resistance to *Peronospora parasitica* specified by several different RPP genes. *The Plant Cell* **8,** 2033–2046.

Parkinson H, Sarkans U, Shojatalab M, Abeygunawardena N, Contrino S, Coulson R, Farne A, Garcia Lara GG, Holloway E, Kapushesky M, Lilja P, Mukherjee G, Oezcimen A, Rayner T, Rocca-Serra P, Sharma A, Sansone S, Brazma A, 2005. ArrayExpress – a public repository for microarray gene expression data at the EBI. *Nucleic Acids Research* **33,** D553–D555.

Pedley KF, Martin GB, 2005. Role of mitogen-activated protein kinases in plant immunity. *Current Opinion in Plant Biology* **8,** 541–547.

Petersen M, Brodersen P, Naested H, Andreasson E, Lindhart U, Johansen B, Nielsen HB, Lacy M, Austin MJ, Parker JE, Sharma SB, Klessig DF, Martienssen R, Mattsson O, Jensen AB, Mundy J, 2000. Arabidopsis map kinase 4 negatively regulates systemic acquired resistance. *Cell* **103,** 1111–1120.

Pieterse CMJ, Van Loon LC, 2004. NPR1: the spider in the web of induced resistance signaling pathways. *Current Opinion in Plant Biology* **7,** 456–464.

Piffanelli P, Zhou F, Casais C, Orme J, Jarosch B, Schaffrath U, Collins NC, Panstruga R, Schulze-Lefert P, 2002. The barley MLO modulator of defense and cell death is responsive to biotic and abiotic stress stimuli. *Plant Physiology* **129,** 1076–1085.

Polverari A, Molesini B, Pezzotti M, Buonaurio R, Marte M, Delledonne M, 2003. Nitric oxide-mediated transcriptional changes in *Arabidopsis thaliana. Molecular Plant–Microbe Interactions* **16,** 1094–1105.

Prusinkiewicz P, 2004. Modeling plant growth and development. *Current Opinion in Plant Biology* **7,** 79–83.

Qutob D, Hraber PT, Sobral BWS, Gijzen M, 2000. Comparative analysis of expressed sequences in *Phytophthora sojae. Plant Physiology* **123,** 243–253.

Ramonell K, Berrocal-Lobo M, Koh S, Wan J, Edwards H, Stacey G, Somerville S, 2005. Loss-of-function mutations in chitin responsive genes show increased susceptibility to the powdery mildew pathogen *Erysiphe cichoracearum. Plant Physiology* **138,** 1027–1036.

Ramonell KM, Zhang B, Ewing RM, Chen Y, Xu D, Stacey G, Somerville S, 2002. Microarray analysis of chitin elicitation in *Arabidopsis thaliana. Molecular Plant Pathology* **3,** 301–311.

Rensink WA, Buell CR, 2005. Microarray expression profiling resources for plant genomics. *Trends in Plant Science* **10,** 603–609.

Rentel MC, Lecourieux D, Ouaked F, Usher SL, Petersen L, Okamoto H, Knight H, Peck SC, Grierson CS, Hirt H, Knight MR, 2004. OXI1 kinase is necessary for oxidative burst-mediated signalling in Arabidopsis. *Nature* **427,** 858–861.

Restrepo S, Myers KL, del Pozo O, Martin GB, Hart AL, Buell CR, Fry WE, Smart CD, 2005. Gene profiling of a compatible interaction between *Phytophthora infestans* and *Solanum tuberosum* suggests a role for carbonic anhydrase. *Molecular Plant–Microbe Interactions* **18,** 913–922.

Rose JK, Bashir S, Giovannoni JJ, Jahn MM, Saravanan RS, 2004. Tackling the plant proteome: practical approaches, hurdles and experimental tools. *The Plant Journal* **39,** 715–733.

Sallaud C, Gay C, Larmande P, Bès M, Piffanelli P, Piégu B, Droc G, Regad F, Bourgeois E, Meynard D, Périn C, Sabau X, Ghesquière A, Glaszmann JC, Delseny M, Guiderdoni E, 2004. High throughput T-DNA insertion mutagenesis in rice: a first step towards *in silico* reverse genetics. *The Plant Journal* **39,** 450–464.

Sallaud C, Meynard D, van Boxtel J, Gay C, Bès M, Brizard JP, Larmande P, Ortega D, Raynal M, Portefaix M, Ouwerkerk PB, Rueb S, Delseny M, Guiderdoni E, 2003. Highly efficient production and characterization of T-DNA plants for rice (*Oryza sativa* L.) functional genomics. *Theoretical and Applied Genetics* **106,** 1396–1408.

Salzman RA, Brady JA, Finlayson SA, Buchanan CD, Summer EJ, Sun F, Klein PE, Klein RR, Pratt LH, Cordonnier-Pratt MM, Mullet JE, 2005. Transcriptional profiling of sorghum induced by methyl jasmonate, salicylic acid, and aminocyclopropane carboxylic acid reveals cooperative regulation and novel gene responses. *Plant Physiology* **138,** 352–368.

Sasaki Y, Asamizu E, Shibata D, Nakamura Y, Kaneko T, Awai K, Amagai M, Kuwata C, Tsugane T, Masuda T, Shimada H, Takamiya K, Ohta H, Tabata S, 2001. Monitoring of methyl jasmonate-responsive genes in Arabidopsis by cDNA macro-array: self-activation of jasmonic acid biosynthesis and cross talk with other phytohormone signaling pathways. *DNA Research* **8,** 153–161.

Scheideler M, Schlaich NL, Fellenberg K, Beissbarth T, Hauser NC, Vingron M, Slusarenko AJ, Hoheisel JD, 2002. Monitoring the switch from housekeeping to pathogen defence metabolism in *Arabidopsis thaliana* using cDNA arrays. *Journal of Biological Chemistry* **277,** 10555–10561.

Schenk PM, Kazan K, Manners JM, Anderson JP, Simpson RS, Wilson IW, Somerville SC, Maclean DJ, 2003. Systemic gene expression in Arabidopsis during an incompatible interaction with *Alternaria brassicicola. Plant Physiology* **132,** 999–1010.

Schenk PM, Kazan K, Wilson I, Anderson JP, Richmond T, Somerville SC, Manners JM. 2000. Coordinated plant defence responses in Arabidopsis revealed by microarray analysis. *Proceedings of the National Academy of Sciences USA* **97,** 11655–11660.

Schenk PM, Vickers CE, Manners JM, 2004. Rapid cloning of novel genes and promoters for functional analysis in transgenic cells. *Transgenics* **4,** 151–156.

Scofield SR, Huang L, Brandt AS, Gill BS, 2005. Development of a virus-induced gene-silencing system for hexaploid wheat and its use in functional analysis of the Lr21-mediated leaf rust resistance pathway. *Plant Physiology* **138,** 2165–2173.

Senthil G, Liu H, Puram VG, Clark A, Stromberg A, Goodin MM, 2005. Specific and common changes in *Nicotiana benthamiana* gene expression in response to infection by enveloped viruses. *Journal of General Virology* **86,** 2615–2625.

Sessions A, Burke E, Presting G, Aux G, McElver J, Patton D, Dietrich B, Ho P, Bacwaden J, Ko C, Clarke JD, Cotton D, Bullis D, Snell J, Miguel T, Hutchison D, Kimmerly B, Mitzel T, Katagiri F, Glazebrook J, Law M, Goff SA, 2002. A high-throughput Arabidopsis reverse genetics system. *The Plant Cell* **14,** 2985–2994.

Sha Y, Li S, Pei Z, Luo L, Tian Y, He C, 2004. Generation and flanking sequence analysis of a rice T-DNA tagged population. *Theoretical and Applied Genetics* **108,** 306–314.

Shen L, Gong J, Caldo RA, Nettleton D, Cook D, Wise RP, Dickerson JA, 2005. BarleyBase – an expression profiling database for plant genomics. *Nucleic Acids Research* **33,** D614–D618.

Shepherd RW, Bass WT, Houtz RL, Wagner GJ, 2005. Phylloplanins of tobacco are defensive proteins deployed on aerial surfaces by short glandular trichomes. *The Plant Cell* **17,** 1851–1861.

Singh KB, Foley RC, Onate-Sanchez L, 2002. Transcription factors in plant defence and stress responses. *Current Opinion in Plant Biology* **5,** 430–436.

Skiba B, Ford R, Pang ECK, 2005. Construction of a cDNA library of *Lathyrus sativus* inoculated with *Mycosphaerella pinodes* and the expression of potential defence-related expressed sequence tags (ESTs). *Physiological and Molecular Plant Pathology* **66,** 55–67.

Slade AJ, Knauf VC, 2005. TILLING moves beyond functional genomics into crop improvement. *Transgenic Research* **14,** 109–115.

Spoel SH, Koornneef A, Claessens SM, Korzelius JP, Van Pelt JA, Mueller MJ, Buchala AJ, Métraux, J-P, Brown R, Kazan K, Van Loon LC, Dong X, Pieterse CMJ, 2003. NPR1 modulates cross-talk between salicylate- and jasmonate-dependent defense pathways through a novel function in the cytosol. *The Plant Cell* **15,** 760–770.

Staswick PE, Yuen GY, Lehman CC, 1998. Jasmonate signaling mutants of Arabidopsis are susceptible to the soil fungus *Pythium irregulare. The Plant Journal* **15,** 747–754.

Strizhov N, Li Y, Rosso MG, Viehoever P, Dekker KA, Weisshaar B, 2003. High-throughput generation of sequence indexes from T-DNA mutagenized *Arabidopsis thaliana* lines. *BioTechniques* **35,** 1164–1168.

Stukkens Y, Bultreys A, Grec S, Trombik T, Vanham D, Boutry M, 2005. NpPDR1, a pleiotropic drug resistance-type ATP-binding cassette transporter from *Nicotiana plumbaginifolia*, plays a major role in plant pathogen defence. *Plant Physiology* **139,** 341–352.

Subramanian S, Graham MY, Yu O, Graham TL, 2005. RNA interference of soybean isoflavone synthase genes leads to silencing in tissues distal to the transformation site and to enhanced susceptibility to *Phytophthora sojae. Plant Physiology* **137,** 1345–1353.

Swidzinski JA, Sweetlove LJ, Leaver CJ, 2002. A custom microarray analysis of gene expression during programmed cell death in *Arabidopsis thaliana. The Plant Journal* **30,** 431–446.

Tao Y, Xie Z, Chen W, Glazebrook J, Chang HS, Han B, Zhu T, Zou G, Katagiri F, 2003. Quantitative nature of Arabidopsis responses during compatible and incompatible interactions with the bacterial pathogen *Pseudomonas syringae. The Plant Cell* **15,** 317–330.

Thatcher LF, Anderson JP, Singh KB, 2005. Plant defence responses: what have we learnt from Arabidopsis? *Functional Plant Biology* **32,** 1–19.

Thomma BPHJ, Eggermont K, Tierens KFM-J, Broekaert WF, 1999. Requirement of functional ethylene-insensitive2 gene for efficient resistance of Arabidopsis to infection by *Botrytis cinerea. Plant Physiology* **121,** 1093–1101.

Thurston G, Regan S, Rampitsch C, Xing T, 2005. Proteomic and phosphoproteomic approaches to understand plant–pathogen interactions. *Physiological and Molecular Plant Pathology* **66,** 3–11.

Upadhyaya NM, Zhou XR, Zhu QH, Ramm K, Wu LM, Eamens A, Sivakumar R, Kato T, Yun DW, Santhoshkumar C, Narayanan KK, Peacock JW, Dennis ES, 2002. An iAc/Ds gene and enhancer trapping system for insertional mutagenesis in rice. *Functional Plant Biology* **29,** 547–559.

Vailleau F, Daniel X, Tronchet M, Montillet JL, Triantaphylides C, Roby D, 2002. A R2R3-MYB gene, AtMYB30, acts as a positive regulator of the hypersensitive cell death program in plants in response to pathogen attack. *Proceedings of the National Academy of Sciences USA* **99,** 10179–10184.

Van Wees SC, Chang HS, Zhu T, Glazebrook J, 2003. Characterization of the early response of Arabidopsis to *Alternaria brassicicola* infection using expression profiling. *Plant Physiology* **132,** 606–617.

Ventelon-Debout M, Delalande F, Brizard JP, Diemer H, Van Dorsselaer A, Brugidou C, 2004. Proteome analysis of cultivar-specific deregulations of *Oryza sativa indica* and *O. sativa japonica* cellular suspensions undergoing rice yellow mottle virus infection. *Proteomics* **4,** 216–225.

Ventelon-Debout M, Nguyen TT, Wissocq A, Berger C, Laudie M, Piégu B, Cooke R, Ghesquière A, Delseny M, Brugidou C, 2003. Analysis of the transcriptional response to Rice Yellow Mottle Virus infection in *Oryza sativa indica* and *japonica* cultivars. *Molecular Genetics and Genomics* **270,** 253–262.

Verhagen BWM, Glazebrook J, Zhu T, Chang HS, van Loon LC, Pieterse CMJ, 2004. The transcriptome of rhizobacteria-induced systemic resistance in Arabidopsis. *Molecular Plant–Microbe Interactions* **17,** 895–908.

Veronese P, Chen X, Bluhm B, Salmeron J, Dietrich R, Mengiste T, 2004. The BOS loci of Arabidopsis are required for resistance to *Botrytis cinerea* infection. *The Plant Journal* **40,** 558–574.

Vijayan P, Shockey J, Levesque CA, Cook RJ, Browse J, 1998. A role for jasmonate in pathogen defence of Arabidopsis. *Proceedings of the National Academy of Sciences USA* **95,** 7209–7214.

Vogel JP, Raab TK, Schiff C, Somerville SC, 2002. *PMR6*, a pectate lyase-like gene required for powdery mildew susceptibility in Arabidopsis. *The Plant Cell* **14,** 2095–2106.

Walker TS, Bais HP, Halligan KM, Stermitz FR, Vivanco JM, 2003. Metabolic profiling of root exudates of *Arabidopsis thaliana. Journal of Agricultural and Food Chemistry* **51,** 2548–2554.

Wang D, Weaver ND, Kesarwani M, Dong X, 2005a. Induction of protein secretory pathway is required for systemic acquired resistance. *Science* **308,** 1036–1040.

Wang YQ, Ohara Y, Nakayashiki H, Tosa Y, Mayama S, 2005b. Microarray analysis of the gene expression profile induced by the endophytic plant growth-promoting rhizobacteria, *Pseudomonas fluorescens* FPT9601-T5 in Arabidopsis. *Molecular Plant–Microbe Interactions* **18,** 385–396.

Wang Y, Yang L, Xu H, Li Q, Ma Z, Chu C, 2005c. Differential proteomic analysis of proteins in wheat spikes induced by *Fusarium graminearum. Proteomics* **5,** 4496–4503.

Weigel RR, Pfitzner UM, Gatz C, 2005. Interaction of NIMIN1 with NPR1 modulates *PR* gene expression in Arabidopsis. *The Plant Cell* **17,** 1279–1291.

Whitham SA, Quan S, Chang HS, Cooper B, Estes B, Zhu T, Wang X, Hou YM, 2003. Diverse RNA viruses elicit the expression of common sets of genes in susceptible *Arabidopsis thaliana* plants. *The Plant Journal* **33,** 271–283.

Wildermuth MC, Dewdney J, Wu G, Ausubel FM, 2001. Isochorismate synthase is required to synthesize salicylic acid for plant defence. *Nature* **414,** 562–565.

Xia Y, Suzuki H, Borevitz J, Blount J, Guo Z, Patel K, Dixon RA, Lamb C, 2004. An extracellular aspartic protease functions in Arabidopsis disease resistance signaling. *EMBO Journal* **23,** 980–988.

Xie DX, Feys BF, James S, Nieto-Rostro M, Turner JG, 1998. COI1: an Arabidopsis gene required for jasmonate-regulated defence and fertility. *Science* **280,** 1091–1094.

Yamazaki Y, Jaiswal P, 2005. Biological ontologies in rice databases. An introduction to the activities in Gramene and Oryzabase. *Plant and Cell Physiology* **46,** 63–68.

Yazaki J, Kojima K, Suzuki K, Kishimoto N, Kikuchi S, 2004. The Rice PIPELINE: a unification tool for plant functional genomics. *Nucleic Acids Research* **32,** D383–D387.

Zeidler D, Zahringer U, Gerber I, Dubery I, Hartung T, Bors W, Hutzler P, Durner J, 2004. Innate immunity in *Arabidopsis thaliana*: lipopolysaccharides activate nitric oxide synthase (NOS) and induce defence genes. *Proceedings of the National Academy of Sciences USA* **101,** 15811–15816.

Zhang B, Ramonell K, Somerville S, Stacey G, 2002. Characterization of early, chitin-induced gene expression in Arabidopsis. *Molecular Plant–Microbe Interactions* **15,** 963–970.

Zhang Y, Tessaro MJ, Lassner M, Li X, 2003. Knockout analysis of Arabidopsis transcription factors TGA2, TGA5, and TGA6 reveals their redundant and essential roles in systemic acquired resistance. *The Plant Cell* **15,** 2647–2653.

Zhao J, Davis LC, Verpoorte R, 2005. Elicitor signal transduction leading to production of plant secondary metabolites. *Biotechnology Advances* **23,** 283–333.

Zhao YF, Thilmony R, Bender CL, Schaller A, He SY, Howe GA, 2003. Virulence systems of *Pseudomonas syringae* pv. *tomato* promote bacterial speck disease in tomato by targeting the jasmonate signaling pathway. *The Plant Journal* **36,** 485–499.

Zierold, U., Scholz, U. and Schweizer, P. (2005) Transcriptome analysis of mlo-mediated resistance in the epidermis of barley. *Molecular Plant Pathology* **6,** 139–151.

Zimmerli L, Stein M, Lipka V, Schulze-Lefert P, Somerville S, 2004. Host and non-host pathogens elicit different jasmonate/ethylene responses in Arabidopsis. *The Plant Journal* **40,** 633–646.

Zimmermann P, Hirsch-Hoffmann M, Hennig L, Gruissem W, 2004. GENEVESTIGATOR: Arabidopsis microarray database and analysis toolbox. *Plant Physiology* **136,** 2621–2632.

Zipfel C, Robatzek S, Navarro L, Oakeley EJ, Jones JD, Felix G, Boller T, 2004. Bacterial disease resistance in Arabidopsis through flagellin perception. *Nature* **428,** 764–767.

Zou J, Rodriguez-Zas S, Aldea M, Li M, Zhu J, Gonzalez DO, Vodkin LO, DeLucia E, Clough SJ, 2005. Expression profiling soybean response to *Pseudomonas syringae* reveals new defense-related genes and rapid HR-specific down regulation of photosynthesis. *Molecular Plant–Microbe Interactions* **8,** 1161–1174.

Chapter 4
Signalling cascades involved in induced resistance

Corné M.J. Pieterse and L. C. Van Loon

Phytopathology, Institute of Environmental Biology, Utrecht University,
The Netherlands

4.1 Introduction

Plant innate immunity is based on a surprisingly complex response that is highly flexible in its capacity to recognize and counteract different invaders. To effectively combat invasion by microbial pathogens and herbivorous insects, plants make use of pre-existing physical and chemical barriers, as well as inducible defence mechanisms that become activated upon attack (see Chapter 6). Apart from reacting locally, plants can mount a systemic response, establishing an enhanced defensive capacity in parts distant from the site of primary attack. This systemically induced response protects the plant against subsequent invaders. Several biologically induced, systemic defence responses have been characterized in detail, such as systemic acquired resistance (SAR), which is triggered by pathogens causing limited infection, such as hypersensitive necrosis (Durrant & Dong, 2004), rhizobacteria induced systemic resistance (ISR), which is activated upon colonization of roots by selected strains of non-pathogenic rhizobacteria (Van Loon *et al.*, 1998; Pieterse *et al.*, 2003; see also Chapter 8), and wound induced defence, which is typically elicited upon tissue damage, such as caused by insect feeding (Kessler & Baldwin, 2002; Howe, 2005; see also Chapter 5).

Although different types of induced resistance are at least partially controlled by distinct signalling pathways, they all share the characteristic that they have broad spectrum effectiveness. In many cases, this enhanced defensive capacity cannot be attributed to direct activation of defence related genes. Instead, the broad spectrum protection is commonly based on a faster and stronger activation of basal defence mechanisms when an induced plant is exposed to either microbial pathogens or herbivorous insects. It is therefore hypothesized that the broad spectrum characteristic of induced resistance is largely based on this conditioning of the tissue to react more effectively to a stress condition. By analogy with a phenotypically similar phenomenon in animals and humans, this enhanced capacity to express basal defence mechanisms is called 'priming' (Conrath *et al.*, 2002).

The plant hormones jasmonic acid (JA), salicylic acid (SA) and ethylene (ET) are major regulators of induced resistance (Pieterse & Van Loon, 1999; Glazebrook, 2001; Thomma *et al.*, 2001). Plants respond with the production of a specific blend of these alarm signals upon pathogen or insect attack. The production of these signals varies greatly in quantity, composition and timing, and results in the activation of differential sets of defence related genes that eventually determine the nature of the defence response against the attacker encountered (Reymond & Farmer, 1998; Rojo *et al.*, 2003; De Vos *et al.*, 2005). Global expression profiling of various Arabidopsis-attacker interactions revealed substantial crosstalk between SA-, JA- and ET-dependent defence pathways (Glazebrook *et al.*, 2003; De Vos *et al.*, 2005). Cross-communication between these pathways provides a powerful regulatory potential that allows the plant to fine-tune its defence responses. Other plant hormones, such as abscisic acid (ABA), brassinosteroids and auxins have been reported to also play a role in induced defence against pathogens, but their significance is understood less well (Jameson, 2000; Audenaert *et al.*, 2002a; Krishna, 2003; Nakashita *et al.*, 2003; Thaler & Bostock, 2004; Ton & Mauch-Mani, 2004; Mauch-Mani & Mauch, 2005; Ton *et al.*, 2005).

In this chapter, we aim to review the current status of induced disease resistance signalling research. We will focus on the roles of salicylic acid (SA), jasmonic acid (JA) and ethylene (ET) in the signalling cascades involved in the different types of induced resistance. In addition, we will cover two emerging new topics in induced resistance research: pathway crosstalk and priming.

4.2 SA, JA and ET: important signals in primary defence

Apart from their roles in plant development, SA, JA and ET have repeatedly been implicated in the regulation of primary defence responses. In many cases, infection by microbial pathogens and attack by herbivorous insects is associated with enhanced production of these hormones and a concomitant activation of distinct sets of defence related genes (Maleck *et al.*, 2000; Schenk *et al.*, 2000; Reymond *et al.*, 2004; De Vos *et al.*, 2005). Moreover, exogenous application of these compounds often results in an enhanced level of resistance (Van Wees *et al.*, 1999). Depending on the host – pathogen interaction, JA, SA, and ET appear to be differentially involved in basal resistance. It has been proposed that the defence signalling pathways that are induced are influenced by the mode of attack of the pathogen, i.e. whether it requires living plant cells (biotrophs) or kills host cells and feeds on the dead tissue (necrotrophs) (Parbery, 1996; Glazebrook, 2005). SA-dependent defence responses are usually associated with a form of programmed cell death known as the hypersensitive response. This response can restrict the growth of biotrophic pathogens by killing the infected cells. In fact, this type of defence is effective against a wide range of biotrophs, but usually fails to protect against, or can even be beneficial for, necrotrophic pathogens (Govrin & Levine, 2000; Thomma *et al.*, 2001). JA-dependent defence responses, which are not associated with cell death, are generally considered to provide an alternative defence against necrotrophs (McDowell & Dangl, 2000). Compelling evidence for the role of SA, JA and ET came from recent genetic analyses of plant mutants and transgenics that are affected in the biosynthesis or perception of these compounds.

4.2.1 SA

A central role for SA became apparent with the use of NahG transformants. NahG plants constitutively express the bacterial *NahG* gene, encoding salicylate hydroxylase, which converts SA into inactive catechol. Tobacco and *Arabidopsis thaliana* NahG plants show enhanced disease susceptibility to a broad range of oomycete, fungal, bacterial and viral pathogens (Delaney *et al.*, 1994; Kachroo *et al.*, 2000). Genetic screens in Arabidopsis to unravel plant defence pathways have identified recessive mutants affected in SA signalling that also show enhanced susceptibility to pathogen infection. For instance, the *sid1*, *sid2* and *pad4* mutants are defective in SA accumulation in response to pathogen infection. As a result, these mutants display enhanced susceptibility to the bacterial pathogen *Pseudomonas syringae* pv. *tomato* DC3000 (*Pst* DC3000) and the oomycete pathogen *Hyaloperonospora parasitica* (Zhou *et al.*, 1998; Nawrath & Métraux, 1999; Wildermuth *et al.*, 2001), confirming the importance of SA in basal resistance against these different types of pathogens.

4.2.2 JA

JA was similarly demonstrated to play a role in basal resistance. For example, both the *jar1* mutant, with reduced sensitivity to methyl jasmonate (MeJA), and the *fad3fad7fad8* triple mutant, which is defective in JA biosynthesis, exhibit susceptibility to normally non-pathogenic soil borne oomycetes of the genus *Pythium* (Staswick *et al.*, 1998; Vijayan *et al.*, 1998). In another study, mutant *fad3fad7fad8* showed extremely high mortality from attack by larvae of the common saprophagous fungal gnat, *Bradysia impatiens* (McConn *et al.*, 1997), demonstrating an important role of JA in primary defence against herbivorous insects. Recently, increased susceptibility of *jar1* to *Fusarium oxysporum* (Berrocal-Lobo & Molina, 2004) and impairment of induced resistance against cucumber mosaic virus in *fad3fad7fad8* mutants have been reported (Ryu *et al.*, 2004). The JA insensitive mutant *coi1* shows enhanced susceptibility to the bacterial leaf pathogen *Erwinia carotovora* (Norman-Setterblad *et al.*, 2000) and the necrotrophic fungi *Alternaria brassicicola* and *Botrytis cinerea* (Thomma *et al.*, 1998). Conversely, overexpression of a JA carboxyl methyl transferase increased endogenous levels of MeJA and resulted in a higher resistance to *B. cinerea* (Seo *et al.*, 2001). Moreover, constitutive activation of the JA signalling pathway in the Arabidopsis mutant *cev1* resulted in enhanced resistance to *P. syringae* and the mildew fungi *Erysiphe cichoracearum, Erysiphe orontii,* and *Oidium lycopersicum* (Ellis *et al.*, 2002). All these examples clearly point to a role of JA in resistance against pathogens with diverse lifestyles, challenging the general notion that JA-dependent defence responses are predominantly effective against necrotrophic pathogens.

4.2.3 ET

The role of ET in plant resistance seems more ambiguous (Van Loon *et al.*, 2006). In some cases, ET is involved in disease resistance, whereas in other cases it is associated with symptom development. For instance, several ET insensitive mutants of Arabidopsis have been reported to exhibit enhanced disease susceptibility to *B. cinerea* (Thomma *et al.*, 1999), *Pst* DC3000 (Pieterse *et al.*, 1998) and *E. carotovora* (Norman-Setterblad *et al.*, 2000), indicating that ET-dependent defences contribute to basal resistance against these pathogens. A similar phenomenon was observed in soybean mutants with reduced sensitivity to ET, which developed more severe symptoms in response to infection by the fungal pathogens *Septoria glycines* and *Rhizoctonia solani* (Hoffman *et al.*, 1999). In addition, Knoester *et al.* (1998) reported that ET insensitive tobacco transformed with the mutant ET receptor gene *etr1-1* from Arabidopsis displayed susceptibility to the normally non-pathogenic oomycete *Pythium sylvaticum*. Thus, ET plays a role in non-host resistance as well. In other cases, reduced ET sensitivity was associated with tolerance. For instance, ET insensitive tomato genotypes allowed wild-type levels of growth of virulent *Pst* DC3000 and *Xanthomonas campestris* pv. *vesicatoria*, but developed less severe symptoms of disease (Lund *et al.*, 1998; Ciardi *et al.*, 2000). A similar phenomenon was observed in the Arabidopsis ET insensitive *ein2* mutant, which displayed increased tolerance to virulent strains of both *Pst* DC3000 and *X. campestris* pv. *campestris* (Bent *et al.*, 1992). In addition, soybean mutants with reduced sensitivity to ET developed similar or less severe disease symptoms in response to the bacterial pathogen *P. syringae* pv. *glycinea* and the oomycete *Phytophthora sojae* (Hoffman *et al.*, 1999). In these interactions, ET is primarily involved in symptom development, rather than disease resistance.

4.3 SA, JA and ET: important signals in induced disease resistance

Upon primary infection or insect attack, plants develop enhanced resistance against subsequent invaders. A classic example of such a systemically induced resistance is activated after primary infection with a necrotizing pathogen, rendering distant, uninfected plant parts more resistant towards a broad spectrum of virulent pathogens, including viruses, bacteria and fungi (Kuć, 1982; see also Chapter 1). This form of induced resistance is often referred to as systemic acquired resistance (SAR, Ross, 1961) and has been demonstrated in many plant–pathogen interactions (Ryals *et al.*, 1996; Sticher *et al.*, 1997). Pathogen induced SAR is typically characterized by a restriction of pathogen growth and a suppression of disease symptom development compared to non-induced plants infected by the same pathogen (Hammerschmidt, 1999). Another, phenotypically similar form of induced resistance is rhizobacteria induced systemic resistance (ISR), which is activated upon colonization of plant roots by selected strains of non-pathogenic rhizobacteria (Van Loon *et al.*, 1998; Chapter 8). Although the terms SAR and ISR are taken to be synonymous (Hammerschmidt *et al.*, 2001), for convenience we distinguish between pathogen- and rhizobacteria-induced resistance by using the term SAR for the pathogen-induced type and ISR for the rhizobacteria-induced type of resistance. Figure 4.1 illustrates the broad spectrum effectiveness of both types of biologically induced resistance in Arabidopsis. Pathogen induced SAR requires SA, whereas rhizobacteria mediated ISR is almost always dependent on JA and ET signalling (Van Loon & Bakker, 2005). In the past decade, many components of the corresponding signalling cascades have been elucidated.

4.3.1 Systemic acquired resistance

The onset of SAR is associated with increased levels of SA both locally at the site of infection and systemically in distant tissues (Mauch-Mani & Métraux, 1998). Moreover, SAR is associated with the coordinate activation of a specific set of genes encoding pathogenesis related (PR) proteins, some of which possess antimicrobial activity (Van Loon, 1997). Exogenous application of SA, or its functional analogues 2,6-dichloroisonicotinic acid (INA) or benzothiadiazole (BTH) induces SAR and activates the same set of *PR* genes (Ryals *et al.*, 1996; see also Chapter 2). Transgenic NahG plants that cannot accumulate SA, and the recessive mutants *sid1*, *sid2* and *pad4*, which are compromised in pathogen induced SA accumulation, are incapable of developing SAR and do not show

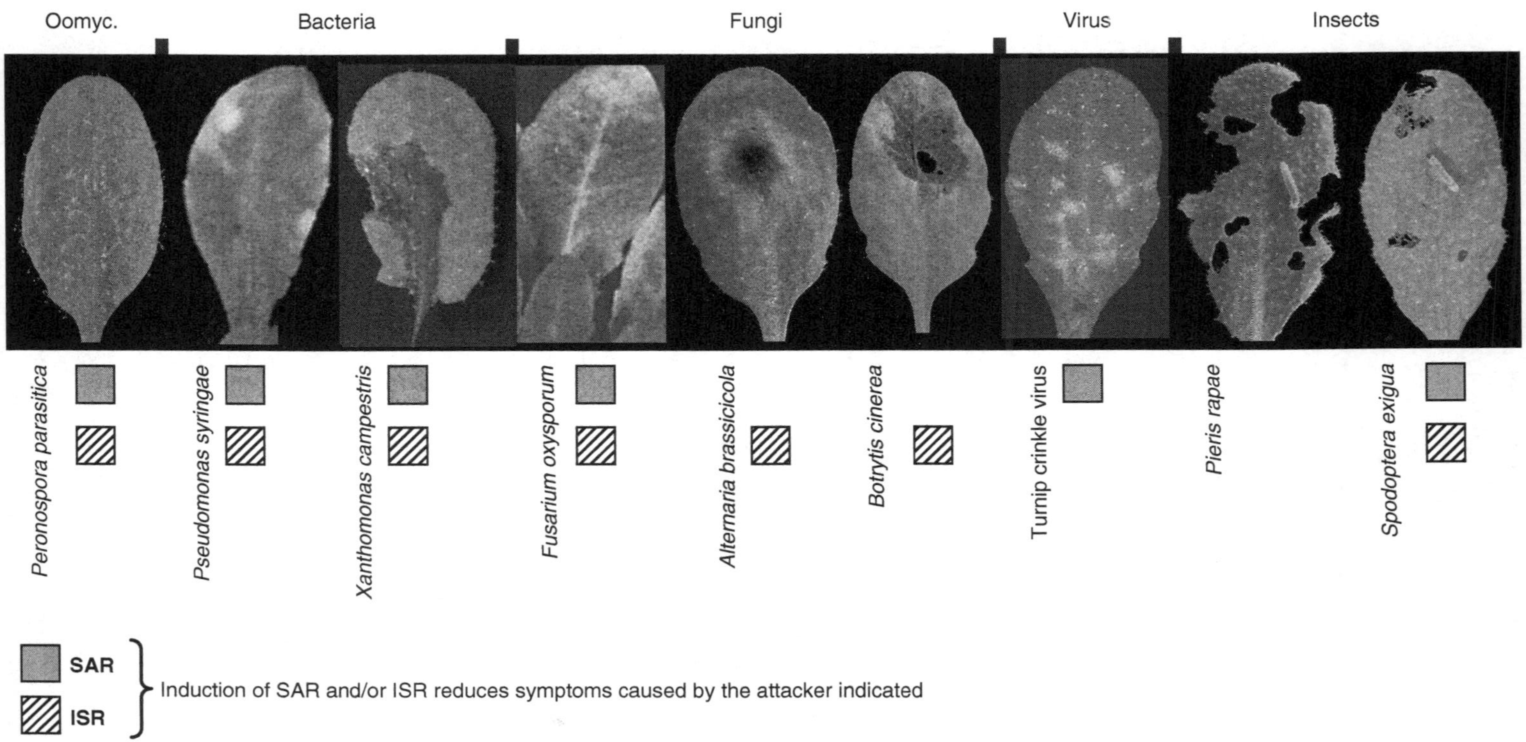

Figure 4.1 Spectrum of effectiveness of pathogen-induced SAR and rhizobacteria-mediated ISR in Arabidopsis. Photographs show typical symptoms caused by the respective pathogens and insects. SAR was induced by infiltrating three leaves per plant with avirulent *Pseudomonas syringae* pv. *tomato* DC3000(*avrRpt2*) bacteria, two days before challenge. ISR was induced by growing plants in soil containing ISR-inducing *Pseudomonas fluorescens* WCS417r bacteria. The effectiveness of SAR and ISR against these pathogens is indicated with shaded (SAR) or hatched (ISR) squares and was assessed on the basis of symptom severity (Pieterse *et al.*, 1996; Van Wees *et al.*, 1997; Ton *et al.*, 2002c; V. van Oosten, M. Dicke, J.A. van Pelt and C.M.J. Pieterse, unpublished results).

PR gene activation upon pathogen infection (Gaffney *et al.*, 1993; Lawton *et al.*, 1995; Zhou *et al.*, 1998; Nawrath & Métraux, 1999). All together, this indicates that SA is a necessary intermediate in the SAR signalling pathway.

Many conditions have been described to induce SAR as well as defence related proteins (Van Loon, 2000). Particularly, the expression of a *PR-1* gene or protein is usually taken as a molecular marker to indicate that SAR was induced. All *PR-1* genes in plants appear to be inducible by SA, and endogenous production or exogenous application of SA has been shown to be both necessary and sufficient to elicit the induced state (Vernooij *et al.*, 1994). Pathogen induced synthesis of SA in tobacco is considered to occur from benzoate, whereas the evidence in Arabidopsis points to isochorismate as the immediate precursor (Wildermuth *et al.*, 2001). Although SA can be transported in the plant, reciprocal graftings of transgenic NahG plants, in which SA is degraded, and non-transformed plants as rootstocks or scions, demonstrated that SA is not the translocated signal in SAR (Vernooij *et al.*, 1994). Similar graftings between transgenic ET insensitive tobacco plants expressing a mutant ET receptor gene from Arabidopsis as rootstock and non-transformed control plants as scion showed little or no SAR induction in the scion, indicating that ET perception is necessary for the generation, release or transport of the mobile signal to distant tissues. Upon arrival of the mobile signal, the latter tissues must start producing SA, which induces the defence related proteins locally (Verberne *et al.*, 2003). The nature of the mobile signal has remained elusive so far. An Arabidopsis mutant, *dir1*, impaired specifically in the systemic character of SAR, implicates involvement of a lipid transfer protein (Maldonado *et al.*, 2002), suggesting that the mobile signal may contain a lipid moiety.

4.3.1.1 NPR1: a crucial regulatory protein of SAR

Transduction of the SA signal to activate *PR* gene expression and SAR requires the function of NPR1, also known as NIM1 (Cao *et al.*, 1994; Delaney *et al.*, 1995; Shah *et al.*, 1997). NPR1 is a regulatory protein identified in Arabidopsis through several genetic screens for SAR compromised mutants (Dong, 2004; Pieterse & Van Loon, 2004). During induction of SAR, NPR1 is translocated into the nucleus (Kinkema *et al.*, 2000). NPR1 acts as a modulator of *PR* gene expression but does not bind DNA directly (Després *et al.*, 2000). Yeast two-hybrid analyses indicated that NPR1 acts through members of the TGA subclass of the basic Leu zipper (bZIP) family of transcription factors (TGAs) that are implicated in the activation of target *PR* genes (Zhang *et al.*, 1999; Després *et al.*, 2000; Zhou *et al.*, 2000). Electromobility shift assays showed that NPR1 substantially increases binding of TGA2 to SA responsive promoter elements in the Arabidopsis *PR-1* gene (Després *et al.*, 2000), suggesting that NPR1-mediated DNA binding of TGAs is important for *PR* gene activation. Recently, microarray analyses showed that in addition to controlling the expression of *PR* genes, NPR1 also controls the expression of protein secretory pathway genes. Up-regulation of these genes is essential for SAR, because mutations in some of them diminished the secretion of PR proteins, with a concomitant reduction in the level of resistance (Wang *et al.*, 2005).

4.3.1.2 NPR1–TGA interactions in vivo

Evidence that binding between NPR1 and TGAs occurs *in planta* was provided in several studies. Subramaniam and co-workers used a protein fragment complementation assay to demonstrate interactions between NPR1 and TGA2 *in vivo*, and showed that the SA induced interaction is predominantly localized in the nucleus (Subramaniam *et al.*, 2001). Fan & Dong (2002) followed a genetic approach, using Arabidopsis transgenics that over-expressed the C-terminal domain of TGA2. This mutant TGA2 protein was capable of interacting with NPR1, but lacked the DNA binding activity important for TGA function. Accumulation of the dominant-negative mutant TGA2 protein in a wild type background led to dose-dependent abolition of TGA function in an NPR1-dependent manner. The resulting phenotype resembled that of mutant *npr1* plants in that the ability to express *PR-1* in response to the SA-analogue 2,6-dichloroisonicotinic acid (INA) was impaired, and the susceptibility to infection by *Pseudomonas syringae* pv. *maculicola* was enhanced. Chromatin immunoprecipitation experiments revealed that *in vivo* both TGA2 and TGA3 are recruited in a SA- and NPR1-dependent manner to SA responsive elements in the *PR-1* promoter (Johnson *et al.*, 2003), supporting the notion that both these transcription factors can act as positive regulators of defence-related gene expression.

4.3.1.3 TGA function and redox regulation

Knockout analysis of single, double, and triple mutants of *TGA2*, *TGA5* and *TGA6* in various combinations established that these three TGAs play an essential and partially redundant role in the activation of *PR* gene expression and SAR in Arabidopsis (Zhang *et al.*, 2003). The seven known Arabidopsis TGAs show differential binding activity towards NPR1 in yeast two-hybrid assays, with TGA2, TGA3 and TGA6 showing the strongest binding (Zhang *et al.*, 1999; Després *et al.*, 2000; Zhou *et al.*, 2000). TGA1 and TGA4 did not bind to NPR1 in yeast assays. However, by using an *in planta* transient expression assay mechanistically similar to the yeast two-hybrid system, Després and co-workers demonstrated that TGA1 does interact with NPR1 in Arabidopsis leaves upon SA treatment (Després *et al.*, 2003). In the same study, yeast two-hybrid assays with chimeric TGA1 proteins in which various domains were exchanged with TGA2 revealed that a 30 amino acid segment is important for NPR1 interaction. Amino acid sequence comparison with other TGAs revealed that both TGA1 and TGA4 contain two Cys residues in this region that are missing in the TGAs that interact with NPR1 in yeast. Mutation of these Cys residues to Asn and Ser transformed TGA1 and TGA4 into proteins capable of interacting with NPR1 in yeast. Because the Cys residues can form disulfide bridges that might prevent interaction of TGA1 and TGA4 with NPR1, Després *et al.* (2003) tested whether the *in vivo* redox state of TGA1 affects NPR1 binding. Upon treatment of Arabidopsis leaves with SA, the Cys residues of TGA1 were reduced, thereby facilitating interaction with NPR1 and subsequent enhancement of binding of TGA1 to SA responsive promoter elements.

4.3.1.4 Redox changes: connection between the SA signal and NPR1 functioning

NPR1 plays an important role in the SA mediated activation of defence related genes by enhancing DNA binding of TGAs to SA responsive elements in their promoters. But how does NPR1 transduce the SA signal? Previously, experiments with NPR1/NIM1 over-expressers demonstrated that high levels of NPR1 proteins per se do not induce *PR*-gene expression or resistance, indicating that NPR1 needs to be activated by a factor acting downstream of SA (Cao *et al.*, 1998; Friedrich *et al.*, 2001). Observations that NPR1 proteins from different plant species contain conserved Cys residues capable of forming inter- or intra-molecular disulfide bonds, and that a mutation in one of these Cys residues resulted in a mutant *npr1* phenotype, led Mou *et al.* (2003) to the hypothesis that NPR1 protein conformation might be sensitive to SA-induced changes in cellular redox status. Induction of SAR was indeed shown to be associated with a change in redox state, possibly caused by accumulation of antioxidants. Under these conditions, NPR1 was reduced from an inactive oligomeric complex to an active monomeric form. The latter appeared to be required for *PR-1* gene activation, as inhibition of NPR1 reduction prevented *PR-1* gene expression. Mutation of the two Cys residues critical for NPR1 oligomer formation led to constitutive monomerization and nuclear localization of NPR1, as well as constitutive *PR-1* gene expression. Thus, cellular redox changes induced as a result of SA action connect the SA signal with NPR1 activity during SAR. Figure 4.2 summarizes the important steps in SAR signalling.

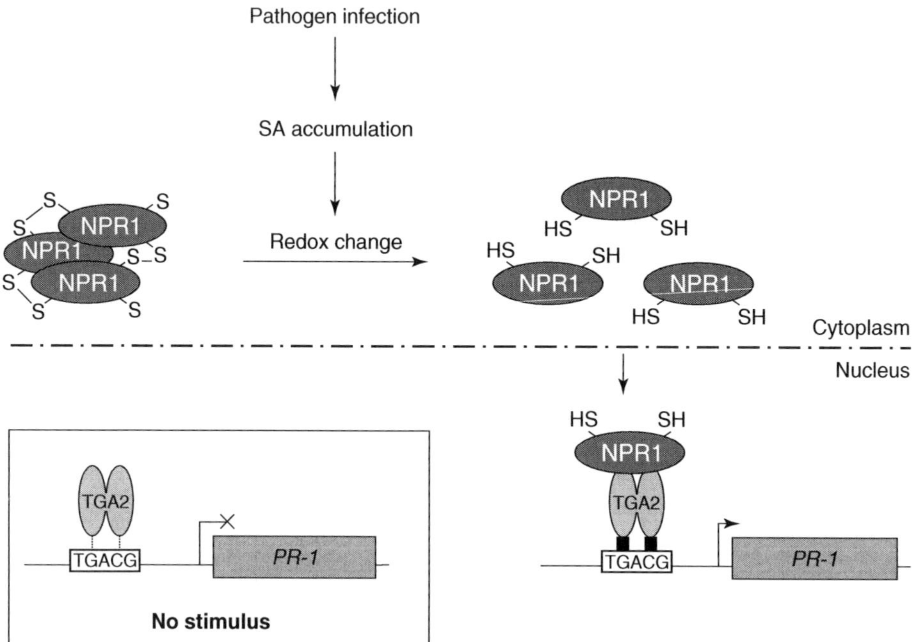

Figure 4.2 Model for SAR signalling illustrating the role of SA-mediated redox changes, NPR1, and TGA transcription factors in SAR-related gene expression. In non-induced cells, oxidized NPR1 is present as inactive oligomers that remain in the cytosol. Binding of TGAs to the cognate SA-responsive promoter elements (TGACG) does not activate *PR-1* gene expression (insert). Upon infection by a necrotizing pathogen, SA accumulates and plant cells attain a more reducing environment, possibly due to the accumulation of antioxidants. Under these conditions, NPR1 oligomers are reduced to an active monomeric state through reduction of intermolecular disulfide bonds. Monomeric NPR1 is translocated into the nucleus where it interacts with TGAs, such as TGA2. The binding of NPR1 to TGAs increases the DNA-binding activity of these transcription factors to the cognate *cis* element (black boxes), resulting in the activation of *PR-1* gene expression (adapted from Pieterse & Van Loon, 2004).

4.3.2 Rhizobacteria-induced systemic resistance

Plants produce exudates and lysates at their root surface, where rhizobacteria are attracted in large numbers (Lynch & Whipps, 1991; Lugtenberg *et al.*, 2001; Walker *et al.*, 2003). Selected strains of non-pathogenic rhizobacteria appear to be plant growth promoting, because they possess the capability to stimulate plant growth (Kloepper *et al.*, 1980; Pieterse & Van Loon, 1999; Bloemberg & Lugtenberg, 2001). Although direct effects on plant growth have been reported (Lynch, 1976; Van Peer & Schippers, 1989), growth promotion results mainly from the suppression of soil borne pathogens and other deleterious micro-organisms (Schippers *et al.*, 1987). Fluorescent *Pseudomonas* spp. are among the most effective plant growth promoting rhizobacteria and have been shown to be responsible for the reduction of soil borne diseases in naturally disease suppressive soils (Raaijmakers & Weller, 1998; Weller *et al.*, 2002; Duff *et al.*, 2003). This type of natural biological control can result from competition for nutrients, siderophore mediated competition for iron, antibiosis or the production of lytic enzymes (Bakker *et al.*, 1991; Van Loon & Bakker, 2003). Apart from such direct antagonistic effects on soil borne pathogens, some rhizobacterial strains are also capable of reducing disease incidence in above ground plant parts through a plant mediated defence mechanism called ISR (Van Loon *et al.*, 1998; Chapter 8). Like SAR, rhizobacteria mediated ISR has been demonstrated in many plant species, e.g. bean, carnation, cucumber, radish, tobacco, tomato and the model plant *Arabidopsis thaliana*, and is effective against a broad spectrum of plant pathogens, including fungi, bacteria and viruses (Van Loon *et al.*, 1998).

Several bacterially derived compounds have been implicated in the elicitation of ISR (Van Loon *et al.*, 1998; Bakker *et al.*, 2003; Van Loon & Bakker, 2005). Elicitors comprise cell wall components such as lipopolysaccharides and flagella, as well as metabolites, such as siderophores and antibiotics (Van Peer & Schippers, 1992; Leeman *et al.*, 1995b; Van Wees *et al.*, 1997; Bakker *et al.*, 2003; Iavicoli *et al.*, 2003). Whereas a receptor for bacterial flagellin has been identified (Gomez-Gomez & Boller, 2000), putative receptors for bacterial cell wall preparations have not been isolated. However, the striking homologies with sensitive perception mechanisms for pathogen associated molecular patterns (PAMPS) that function in the innate immune response of plants and animals (Nürnberger *et al.*, 2004) suggest that rhizobacteria are recognized by general immune surveillance mechanisms.

4.3.2.1 ISR in Arabidopsis: discovery of an SA independent signalling cascade

To study the signal transduction pathway of rhizobacteria mediated ISR, an Arabidopsis based model system was developed. In this model system, the non-pathogenic rhizobacterial strain *Pseudomonas fluorescens* WCS417r is used as the inducing agent (Pieterse *et al.*, 1996). WCS417r has been shown to trigger ISR in several plant species, e.g. carnation, radish, tomato and bean (Pieterse *et al.*, 2002). Colonization of Arabidopsis roots by ISR-inducing WCS417r bacteria protects the plants against different types of pathogens, including the bacterial leaf pathogens *Pst* DC3000, *X. campestris* pv. *armoraciae*, and *E. carotovora* pv. *carotovora*, the fungal root pathogen *Fusarium oxysporum* f. sp. *raphani*, the fungal leaf pathogens *A. brassicicola* and *B. cinerea,* and the oomycete leaf pathogen *H. parasitica* (Pieterse *et al.*, 1996; Van Wees *et al.*, 1997; Ton *et al.*, 2002a; H.J.A.Van Pelt & C.M.J. Pieterse, unpublished results).

Research on the molecular mechanism of rhizobacteria-mediated ISR was initially focused on the role of PR-proteins, as the accumulation of these proteins was considered to be strictly correlated with induced disease resistance. However, radish plants of which the roots were treated with ISR-inducing WCS417r did not accumulate PR proteins, although these plants clearly showed enhanced resistance against fusarium wilt disease (Hoffland *et al.*, 1995). Similarly, Arabidopsis plants expressing WCS417r-mediated ISR showed enhanced resistance against *F. oxysporum* f. sp. *raphani* and *Pst* DC3000, but this did not coincide with the activation of the SAR marker genes *PR-1*, *PR-2* and *PR-5* (Pieterse *et al.*, 1996; Van Wees *et al.*, 1997). Determination of SA levels in ISR-expressing Arabidopsis plants revealed that ISR is not associated with increased accumulation of SA (Pieterse *et al.*, 2000). Moreover, WCS417r-mediated ISR was expressed normally in SA-non-accumulating Arabidopsis NahG plants (Pieterse *et al.*, 1996; Van Wees *et al.*, 1997). This led to the conclusion that WCS417r-mediated ISR is an SA-independent resistance mechanism and that WCS417r-mediated ISR and pathogen induced SAR are regulated by distinct signalling pathways. SA independent ISR has been shown not only in Arabidopsis (Van Wees *et al.*, 1997; Iavicoli *et al.*, 2003; Ryu *et al.*, 2003) but also in tobacco (Press *et al.*, 1997; Zhang *et al.*, 2002), and tomato (Yan *et al.*, 2002). This wide range of induction of ISR indicates that the ability of these *Pseudomonas* strains to activate an SA-independent pathway controlling systemic resistance is common to a broad range of plants.

Not all ISR-inducing rhizobacteria trigger an enhanced defensive capacity via an SA-independent pathway. Under iron limiting conditions, certain rhizobacterial strains produce SA as a siderophore (Meyer *et al.*, 1992; Visca *et al.*, 1993). An enhanced resistance elicited by *P. fluorescens* CHA0 in tobacco might be fully explained by the bacterial production of SA, which could elicit a SAR response. Treatment of tobacco roots with CHA0 triggers accumulation of SA-inducible PR-proteins in the leaves (Maurhofer *et al.*, 1994). Moreover, transformation of the SA-biosynthetic gene cluster of CHA0 into *P. fluorescens* P3 improved the systemic resistance-inducing capacity of this strain (Maurhofer *et al.*, 1998). Another strain that elicits an SA-dependent enhanced defensive capacity is *Pseudomonas aeruginosa* 7NSK2. An SA-deficient mutant of this bacterium failed to induce resistance in bean and tobacco (De Meyer & Höfte, 1997). Moreover, 7NSK2 was unable to induce resistance in NahG tobacco plants against tobacco mosaic virus (De Meyer *et al.*, 1999). An SA-overproducing mutant of 7NSK2 was shown to trigger the SA-dependent SAR pathway by producing SA at the root surface (De Meyer & Höfte, 1997). However, Audenaert *et al.* (2002b) showed that a combination of the secondary siderophore pyochelin and the antibiotic pyocyanin is required to induce enhanced resistance by wild-type 7NSK2. SA is an intermediate in the formation of pyochelin, and the combination of pyochelin and pyocyanin is toxic to root cells, thereby setting off the SAR response.

4.3.2.2 Genetic dissection of the SA-independent ISR signalling cascade

ISR-inducing rhizobacteria show little specificity in their colonization of roots of different plant species (Van Loon *et al.*, 1998). In contrast, the ability to induce ISR appears to be dependent on the bacterium/host combination. For instance, *P. fluorescens* WCS374r is capable of inducing ISR in radish, but not in Arabidopsis (Leeman *et al.*, 1995a;

Van Wees *et al.*, 1997). Conversely, Arabidopsis is responsive to *Pseudomonas putida* WCS358r, whereas radish is not (Van Peer *et al.*, 1991; Van Peer & Schippers, 1992; Leeman *et al.*, 1995a; Van Wees *et al.*, 1997). WCS417r is capable of inducing ISR in both Arabidopsis and radish (Van Wees *et al.*, 1997), as well as in other species, i.e. carnation (Van Peer *et al.*, 1991), radish (Leeman *et al.*, 1995a), tomato (Duijff *et al.*, 1998) and bean (Bigirimana & Höfte, 2002), but not in *Eucalyptus* (Ran *et al.*, 2005). Besides differences in inducibility between species, there can also be differences within species. Arabidopsis accessions Columbia (Col-0) and Landsberg *erecta* (L*er*-0) are responsive to ISR induction by WCS417r, but accessions Wassilewskija (Ws-0) and RLD1 are not (Van Wees *et al.*, 1997; Ton *et al.*, 1999, 2001). Both these accessions are compromised in a common trait governing a step between the recognition of the bacterium and the expression of ISR. These data clearly indicate that ISR is genetically determined.

Since SA was not involved in WCS417r-elicited ISR, the Arabidopsis JA-response mutant *jar1* and the ET-response mutant *etr1* were tested for their ability to express ISR. Both mutants were unable to mount resistance against *Pst* DC3000 after colonization of the roots by WCS417r (Pieterse *et al.*, 1998), indicating that ISR requires responsiveness to both JA and ET. Another indication for the involvement of the JA-signalling pathway came from the analysis of Arabidopsis mutant *eds8*, which was previously shown to exhibit enhanced susceptibility to *P. syringae* (Glazebrook *et al.*, 1996). This mutant was impaired in both WCS417r-mediated ISR (Ton *et al.*, 2002d) and JA-signalling (Ton *et al.*, 2002c; Glazebrook *et al.*, 2003). To further elucidate the role of ET in the ISR signalling pathway, a large set of well characterized ET-signalling mutants was analysed. None of these mutants showed an ISR response against *Pst* DC3000 after colonization of the roots by WCS417r (Knoester *et al.*, 1999). These results confirmed that an intact ET-signalling pathway is required for the establishment of ISR. Particularly interesting was the analysis of the *eir1* mutant, which is ET-insensitive in the roots, but not in the shoot. This *eir1* mutant was incapable of showing ISR after root colonization by WCS417r. In contrast, after leaf infiltration with WCS417r, it did show ISR, indicating that responsiveness to ET is required at the site of rhizobacterial induction (Knoester *et al.*, 1999).

Further evidence for the involvement of the ET-response pathway came from the identification of the Arabidopsis *ISR1* locus (Ton *et al.*, 1999). Genetic analysis of the progeny of a cross between the WCS417r-responsive ecotype Col-0 and the ISR-impaired ecotype RLD1 revealed a single locus, designated *ISR1*, to be important in the expression of ISR against several different pathogens (Ton *et al.*, 2002b). Accessions with the recessive *isr1* allele have reduced sensitivity to ET and enhanced susceptibility to *Pst* DC3000 (Ton *et al.*, 2001). These results strongly indicate that the Arabidopsis *ISR1* locus encodes a novel component in the ET-signal transduction pathway that is important for both basal resistance and ISR in Arabidopsis.

4.3.2.3 Dual role for NPR1 in SAR and ISR

To investigate a possible involvement of the SAR regulatory protein NPR1 in ISR signalling, the Arabidopsis npr1 mutant was tested in the ISR bioassay. Surprisingly, the *npr1* mutant was incapable of showing WCS417r-mediated ISR (Pieterse *et al.*, 1998; Van Wees *et al.*, 2000). This result clearly showed that WCS417r-mediated ISR, like SA-dependent SAR, is an NPR1-dependent defence response. Further analysis of the ISR signal-transduction pathway revealed that NPR1 acts downstream of the JA- and ET-dependent steps (Pieterse *et al.*, 1998). Because SAR is associated with NPR1-dependent *PR*-gene expression, and ISR is not, the action of NPR1 in ISR must be different from that in SAR. These different activities are not mutually exclusive because simultaneous activation of ISR and SAR can lead to an enhanced defensive activity compared to that observed with either type of induced resistance alone (Van Wees *et al.*, 2000). These results suggest that the NPR1 protein is important in regulating and intertwining different hormone-dependent defence pathways.

4.3.2.4 ISR is associated with priming for enhanced defence

In Arabidopsis, both JA and ET activate specific sets of defence-related genes (Schenk *et al.*, 2000), but when applied exogenously, each can induce resistance (Pieterse *et al.*, 1998; Van Wees *et al.*, 1999). To investigate how far ISR is associated with these changes in JA/ET-responsive gene expression, Van Wees *et al.* (1999) monitored the expression of a set of well characterized JA- and/or ET-responsive, defence-related genes (i.e. *LOX1*, *LOX2*, *VSP2*, *PDF1.2*, *HEL*, *CHI-B* and *PAL1*) in Arabidopsis plants expressing WCS417r-mediated ISR. None of these genes was up-regulated in induced plants, neither locally in the roots nor systemically in the leaves. This suggested that the resistance attained was not associated with major increases in the levels of either JA or ET. Indeed, analysis of JA and ET levels in leaves of ISR-expressing plants revealed no changes in the production of these signal molecules (Pieterse *et al.*, 2000; Hase *et al.*, 2003). Therefore, it had to be assumed that the JA and ET dependency of ISR is based on an enhanced sensitivity to these hormones, rather than on an increase in their production.

To identify ISR-related genes, the transcriptional response of over 8000 Arabidopsis genes was monitored during WCS417r-mediated ISR (Verhagen *et al.*, 2004). However, systemically in the leaves, none of the ~8000 genes tested showed a consistent change in expression in response to effective colonization of the roots by WCS417r, indicating that the onset of ISR in the leaves is not associated with detectable changes in gene expression. However, after challenge inoculation of WCS417r-induced plants with the bacterial leaf pathogen *Pst* DC3000, 81 genes showed an augmented expression pattern in ISR-expressing leaves compared to inoculated control leaves, suggesting that ISR-expressing plants are primed to respond faster and/or more strongly upon pathogen attack. The majority of the primed genes was predicted to be regulated by JA and/or ET signalling, confirming earlier findings that colonization of the roots by WCS417r primed Arabidopsis plants for augmented expression of the JA- and/or ET-responsive genes *VSP2*, *PDF1.2* and *HEL* (Van Wees *et al.*, 1999; Hase *et al.*, 2003). Priming is a phenomenon that has been shown to be associated with different types of induced resistance (Conrath *et al.*, 2002). It provides the plant with an enhanced capacity for rapid and effective activation of cellular defence responses once a pathogen is contacted, and it allows the plant to react more effectively to any invader encountered by boosting the defences that are activated in the host. This mechanism could also explain the broad-spectrum action of induced resistance.

The first evidence that priming for potentiated expression of plant defence responses plays an important role in rhizobacteria-mediated ISR came from experiments with carnation. Upon colonization of the roots by WCS417, carnation plants developed an enhanced defensive capacity against *Fusarium oxysporum* f. sp. *dianthi*. Before challenge inoculation, no increase in phytoalexin levels could be detected in induced plants, but upon subsequent inoculation with *F. oxysporum*, phytoalexin levels in ISR-expressing plants rose significantly faster than upon challenge of non-induced plants (Van Peer *et al.*, 1991). In bean, *Bacillus pumilus* SE34 induced ISR against the root-rot fungus *F. oxysporum* f. sp. *pisi* (Benhamou *et al.*, 1996). By itself, colonization of the roots by the rhizobacterium did not induce morphological alterations of root tissue. However, upon challenge with *F. oxysporum*, root cell walls of ISR-expressing plants were rapidly strengthened at sites of attempted fungal penetration by appositions containing large amounts of callose and phenolic materials, thereby effectively preventing fungal ingress (Benhamou *et al.*, 1996). Other ISR-inducing rhizobacteria have also been demonstrated to enhance the plant's defensive capacity by priming for potentiated defence-related gene expression (e.g. De Meyer *et al.*, 1999; Ahn *et al.*, 2002; Kim *et al.*, 2004; Tjamos *et al.*, 2005), indicating that priming is a common feature in rhizobacteria-mediated ISR. Priming for defence may combine advantages of enhanced disease protection with low metabolic costs. Recently,

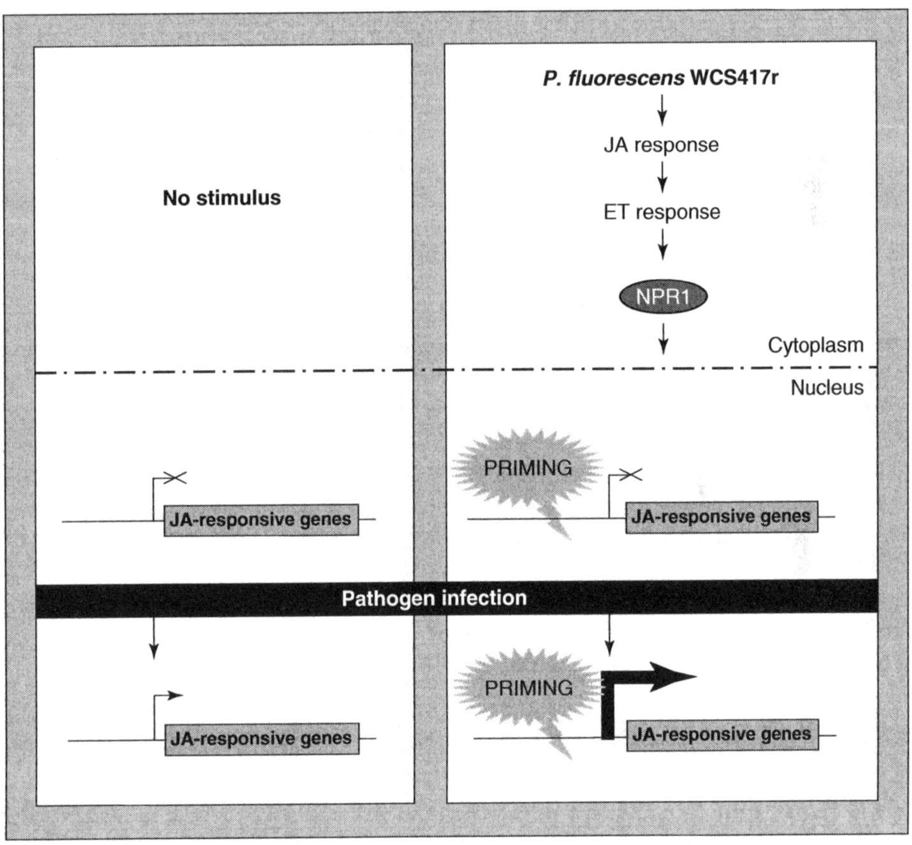

Figure 4.3 Model for the signal-transduction pathway controlling rhizobacteria-mediated ISR in Arabidopsis. Colonization of the roots by *P. fluorescens* WCS417r leads to enhanced defensive capacity against a broad spectrum of plant pathogens. For the expression of ISR, responsiveness to the plant hormones JA and ET are required, as well as the regulatory protein NPR1. The induced state is not associated with major changes in defence-related gene expression (as opposed to SAR). However, ISR-expressing plants are primed to express a specific set of JA-responsive genes faster and to a higher level upon pathogen infection.

Van Hulten *et al.* (2006) examined the costs and benefits of priming in comparison to activated defence in Arabidopsis. The study revealed that the benefits of priming-mediated resistance outweigh the costs under conditions of pathogen pressure, suggesting an evolutionary advantage of this mechanism of induced resistance over constitutive activation of defence responses. Figure 4.3 summarizes the key steps in the ISR signalling pathway.

4.4 Crosstalk between signalling pathways

In the induction of systemic resistance in Arabidopsis against *Pst* DC3000, SA-inducible SAR and JA/ET-dependent ISR can act additively. However, both pathways can also interact antagonistically, indicating that signalling pathways cross-communicate (Reymond & Farmer, 1998; Pieterse & Van Loon, 1999; Felton & Korth, 2000; Feys & Parker, 2000; Dicke & Van Poecke, 2002; Kunkel & Brooks, 2002; Rojo *et al.*, 2003; Bostock, 2005). For instance, activation of SA-dependent SAR has been shown to suppress JA signalling in plants, thereby prioritizing SA-dependent resistance to microbial pathogens over JA-dependent defence that is, in general, more effective against insect herbivory (Stout *et al.*, 1999; Thaler *et al.*, 1999; Felton & Korth, 2000; Thaler *et al.*, 2002; Bostock, 2005). Pharmacological and genetic experiments have indicated that SA-mediated suppression of JA-inducible gene expression plays an important role in this process (Peña-Cortés *et al.*, 1993; Van Wees *et al.*, 1999; Glazebrook *et al.*, 2003). Crosstalk can sometimes work in both directions, as evidenced by occasional suppression of SA responses by JA (Niki *et al.*, 1998; Kunkel & Brooks, 2002; Glazebrook *et al.*, 2003).

4.4.1 Complexity of the plant's induced resistance signalling network

To understand how plants integrate pathogen- and insect-induced signals into specific defence responses, De Vos *et al.* (2005) monitored the dynamics of SA, JA and ET signalling in *Arabidopsis* after attack by a set of microbial pathogens and herbivorous insects with different modes of attack. *Arabidopsis* plants were exposed to microbial pathogens (*Pst* DC3000 and *A. brassicicola*), tissue chewing caterpillars (*Pieris rapae*), cell content feeding thrips (*Frankliniella occidentalis*), or phloem feeding aphids (*Myzus persicae*). Monitoring the 'signal signature' in each plant–attacker combination showed that the kinetics of SA, JA and ET production vary greatly in both quantity and timing. Analysis of global gene expression profiles demonstrated that the signal signature characteristic of each *Arabidopsis*–attacker combination is orchestrated into a surprisingly complex set of transcriptional alterations in which, in all cases, stress related genes are over-represented. Comparison of transcript profiles revealed that consistent changes induced by pathogens and insects with very different modes of attack can show considerable overlap. Of all consistent changes induced by *A. brassicicola*, *P. rapae* and *F. occidentalis*, more than 50% were also induced consistently by *Pst* DC3000. However, although these four attackers all stimulated JA biosynthesis, the majority of the changes in JA-responsive gene expression were attacker specific. Hence, SA, JA and ET play a primary role in the orchestration of the plant's defence response, but other regulatory mechanisms, such as pathway crosstalk or additional attacker-induced signals, eventually shape the highly complex attacker-specific defence response.

4.4.2 Trade-offs between different types of induced resistance

Several studies have shown that activation of a particular defence pathway by one particular pathogen or insect negatively affects resistance to other groups of pathogens or insects. For instance, Moran (1998) demonstrated that in cucumber, pathogen-induced SAR against the fungus *Colletotrichum orbiculare* was associated with reduced resistance against feeding by the spotted cucumber beetle *Diabrotica undecimpunctata howardi* and enhanced reproduction of the melon aphid *Aphis gossypii*. A similar phenomenon was observed by Preston *et al.* (1999), who demonstrated that TMV-inoculated tobacco plants expressing SAR were more suitable for grazing by the tobacco hornworm *Manduca sexta* than non-induced control plants. Conversely, Felton *et al.* (1999) demonstrated that transgenic tobacco plants with reduced SA levels as a result of silencing of the *PAL* gene exhibited reduced SAR against TMV but enhanced herbivore-induced resistance to *Heliothis virescens* larvae. In contrast, *PAL*-overexpressing tobacco plants showed a strong reduction in herbivore-induced insect resistance, while TMV-induced SAR was enhanced in these plants.

Application of the SAR inducer acibenzolar-*S*-methyl (BTH) has been shown to negatively affect insect resistance as well. For instance, BTH induced resistance against the bacterial pathogen *P. syringae* pv. *tomato*, but improved suitability of tomato leaves for feeding by leaf chewing larvae of the corn earworm *Helicoverpa zea* (Stout *et al.*, 1999). A similar phenomenon was observed by Thaler *et al.* (1999), who showed that application of BTH to field-grown tomato plants compromised resistance to the beet armyworm (*Spodoptera exigua*). In most cases, reduced insect resistance observed in SAR-expressing plants is attributed to the inhibition of JA production by BTH or increased SA levels.

4.4.3 Concomitant expression of induced defence pathways

Whereas negative interactions between pathogen and insect resistance have been clearly demonstrated, other studies failed to demonstrate such a negative relationship. For instance, Ajlan & Potter (1992) found that inoculation of the lower leaves of tobacco with TMV had no effect on population growth of tobacco aphids (*Myzus nicotianae*). Similarly, Inbar *et al.* (1998) found no negative effect of BTH application on population growth of whiteflies (*Bemisia argentifolii*) and leaf miners (*Liriomyza* spp.). However, Stout *et al.* (1999) showed that inoculation of tomato leaves with the bacterial pathogen *P. syringae* pv. *tomato* induced resistance against both *P. syringae* pv. *tomato* and the corn earworm in distal plant parts. Conversely, feeding by the insect *H. zea* likewise induced resistance against both *P. syringae* pv. *tomato* and itself.

A demonstration of induced resistance effective simultaneously against pathogens and insects in the field was provided by Zehnder *et al.* (2001). In cucumber, induction of rhizobacteria-mediated ISR against the insect-transmitted bacterial wilt disease, caused by *Erwinia tracheiphila*, was associated with reduced feeding of the cucumber beetle vector. It appeared that induction of ISR was associated with reduced concentrations of cucurbitacin, a secondary plant metabolite and powerful feeding stimulant for cucumber beetles. Induction of ISR against *E. tracheiphila* was also effective in the absence of beetle vectors, suggesting that ISR protects cucumber against bacterial wilt not only by reducing beetle feeding and transmission of the pathogen, but also through the induction of defence responses that are active against the pathogen itself. These observations indicate that negative interactions between induced pathogen and insect resistance are by no means general.

4.4.4 Key players in pathway crosstalk

The antagonistic effect of SA on JA signalling was recently shown to be controlled by a novel function of the defence regulatory protein NPR1 (Spoel *et al.*, 2003). The nuclear localization of NPR1 that is essential for SA-mediated *PR*-gene expression appeared not to be required for the suppression of JA signalling. Thus, crosstalk between SA and JA is modulated through a novel function of NPR1 in the cytosol (Spoel *et al.*, 2003). The mode of action of NPR1 in the cytosol is unknown, but it is tempting to speculate that it interferes with the previously identified SCF^{COI1} ubiquitin-ligase complex (Devoto *et al.*, 2002; Xu *et al.*, 2002) that regulates JA responsive gene expression through targeted ubiquitination and subsequent proteasome-mediated degradation of a negative regulator of JA signalling.

Additional key elements involved in pathway crosstalk have been identified. For instance, the Arabidopsis transcription factor WRKY70 was shown to act as both an activator of SA-responsive genes and a repressor of JA-inducible genes, thereby integrating signals from these antagonistic pathways (Li *et al.*, 2004). In addition, the transcription factors ERF1 and MYC2 were found to integrate signals from the JA and ET pathway in activating defence-related genes that are responsive to both JA and ET (Lorenzo *et al.*, 2003, 2004). Crosstalk between defence signalling pathways is thought to provide the plant with a powerful regulatory potential, which helps the plant to 'decide' on the most appropriate defensive strategy, depending on the type of attacker it is encountering. Yet, it may also allow attackers to manipulate plants to their own benefit by shutting down induced defence through influences on the signalling network (Kahl *et al.*, 2000).

4.5 Outlook

Plant diseases are responsible for large crop losses in agriculture. Conventional disease control is based on resistance breeding and application of chemical agents. Classic resistance breeding depends on the availability of resistance genes, which often show limited durability. The use of chemical agents and their persistence in soil are potentially harmful to the environment, notably when chemicals are applied repeatedly in large amounts such as in the control of soil-borne fungal pathogens. Moreover, both these disease control strategies are directed against a single or a small group of plant pathogens. Induced disease resistance is an attractive alternative form of plant protection, as it is based on the activation of extant resistance mechanisms in the plant and is effective against a broad spectrum of plant pathogens (Kuć, 1982; Van Loon *et al.*, 1998).

Previously, Van Wees *et al.* (2000) demonstrated that simultaneous activation of ISR and SAR results in an enhanced level of induced protection against *Pst* DC3000. It appeared that the JA/ET-dependent ISR pathway and the SA-dependent SAR pathway act independently and additively to increase protection against this particular pathogen. Moreover, ISR and SAR confer differential protection against biotrophic and necrotrophic pathogens (Ton *et al.*, 2002c). Thus, combining both types of induced resistance can protect the plant against a complementary spectrum of pathogens and can result in an additive level of induced protection against pathogens that are resisted through both the JA/ET- and the SA-dependent pathways, such as *Pst* DC3000. Hence, combining SAR and ISR provides an attractive tool for improvement of disease control.

Knowledge of defence signalling pathways has been proven to be instrumental for the development of new strategies for broad-spectrum disease resistance. Examples are genetic engineering of the SAR pathway, and the development of defence signal-mimicking chemicals, such as BTH. However, crosstalk between SA- and JA-dependent defence pathways may be a burden when enhanced pathogen resistance is associated with reduced resistance against insects. Fortunately, negative crosstalk between SA- and JA-dependent defences appears to be confined to specific inducer–plant–attacker combinations. Only in cases in which the inducer strongly activates the SAR pathway does there seem to be an antagonistic effect on resistance against attackers that are resisted through JA-dependent defences. In other cases, there seems to be little or no antagonism, and SA- and JA-dependent defences can be expressed concomitantly to boost the plant's potential to resist invaders. Thus, the general notion that SA-dependent pathogen resistance and JA-dependent insect resistance are mutually exclusive needs to be adjusted.

Future research on the molecular mechanisms of induced resistance and crosstalk between plant defence pathways will provide more insight into how plants are able to integrate signals into appropriate defences. Ultimately, this will not only provide fundamental insights into how plants cope with different enemies, but also be instrumental in developing strategies for biologically based, environmentally friendly and durable crop protection.

4.6 Acknowledgements

C.M.J.P. received funding from the Centre for BioSystems Genomics, which is part of the Netherlands Genomics Initiative, and the Netherlands Organization for Scientific Research (NWO grants 865.04.002 and 863.04.019).

4.7 References

Ahn I-P, Park K, Kim C-H, 2002. Rhizobacteria-induced resistance perturbs viral disease progress and triggers defense-related gene expression. *Molecules and Cells* **13,** 302–308.

Ajlan AM, Potter DA, 1992. Lack of effect of tobacco mosaic virus-induced systemic acquired resistance on arthropod herbivores in tobacco. *Phytopathology* **82,** 647–651.

Audenaert K, De Meyer GB, Höfte MM, 2002a. Abscisic acid determines basal susceptibility of tomato to *Botrytis cinerea* and suppresses salicylic acid-dependent signaling mechanisms. *Plant Physiology* **128,** 491–501.

Audenaert K, Pattery T, Cornelis P, Höfte M, 2002b. Induction of systemic resistance to *Botrytis cinerea* in tomato by *Pseudomonas aeruginosa* 7NSK2: Role of salicylic acid, pyochelin, and pyocyanin. *Molecular Plant–Microbe Interactions* **15,** 1147–1156.

Bakker PAHM, Van Peer R, Schippers B, 1991. Suppression of soil-borne plant pathogens by fluorescent pseudomonads: mechanisms and prospects. In: Beemster ABR, Bollen GJ, Gerlagh M, Ruissen MA, Schippers B, Tempel A, eds. *Biotic Interactions and Soil-Borne Diseases*. Amsterdam: Elsevier Scientific, 217–230.

Bakker PAHM, Ran LX, Pieterse CMJ, Van Loon LC, 2003. Understanding the involvement of rhizobacteria-mediated induction of systemic resistance in biocontrol of plant diseases. *Canadian Journal of Plant Pathology* **25,** 5–9.

Benhamou N, Kloepper JW, Quadt-Hallman A, Tuzun S, 1996. Induction of defense-related ultrastructural modifications in pea root tissues inoculated with endophytic bacteria. *Plant Physiology* **112,** 919–929.

Bent AF, Innes RW, Ecker JR, Staskawicz BJ, 1992. Disease development in ethylene-insensitive *Arabidopsis thaliana* infected with virulent and avirulent *Pseudomonas* and *Xanthomonas* pathogens. *Molecular Plant–Microbe Interactions* **5,** 372–378.

Berrocal-Lobo M, Molina A, 2004. Ethylene response factor 1 mediates Arabidopsis resistance to the soilborne fungus *Fusarium oxysporum*. *Molecular Plant–Microbe Interactions* **17,** 763–770.

Bigirimana J, Höfte M, 2002. Induction of systemic resistance to *Colletotrichum lindemuthianum* in bean by a benzothiadiazole derivative and rhizobacteria. *Phytoparasitica* **30,** 159–168.

Bloemberg GV, Lugtenberg BJJ, 2001. Molecular basis of plant growth promotion and biocontrol by rhizobacteria. *Current Opinion in Plant Biology* **4,** 343–350.

Bostock RM, 2005. Signal crosstalk and induced resistance: straddling the line between cost and benefit. *Annual Review of Phytopathology* **43,** 545–580.
Cao H, Bowling SA, Gordon AS, Dong X, 1994. Characterization of an Arabidopsis mutant that is non-responsive to inducers of systemic acquired resistance. *The Plant Cell* **6,** 1583–1592.
Cao H, Li X, Dong X, 1998. Generation of broad-spectrum disease resistance by overexpression of an essential regulatory gene in systemic acquired resistance. *Proceedings of the National Academy of Sciences USA* **95,** 6531–6536.
Ciardi JA, Tieman DM, Lund ST, Jones JB, Stall RE, Klee HJ, 2000. Response to *Xanthomonas campestris* pv. *vesicatoria* in tomato involves regulation of ethylene receptor gene expression. *Plant Physiology* **123,** 81–92.
Conrath U, Pieterse CMJ, Mauch-Mani B, 2002. Priming in plant–pathogen interactions. *Trends in Plant Science* **7,** 210–216.
De Meyer G, Höfte M, 1997. Salicylic acid produced by the rhizobacterium *Pseudomonas aeruginosa* 7NSK2 induces resistance to leaf infection by *Botrytis cinerea* on bean. *Phytopathology* **87,** 588–593.
De Meyer G, Audenaert K, Höfte M, 1999. *Pseudomonas aeruginosa* 7NSK2-induced systemic resistance in tobacco depends on in planta salicylic acid accumulation but is not associated with *PR1a* expression. *European Journal of Plant Pathology* **105,** 513–517.
De Vos M, Van Oosten VR, Van Poecke RMP, Van Pelt JA, Pozo MJ, Mueller MJ, Buchala AJ, Métraux JP, Van Loon LC, Dicke M, Pieterse CMJ, 2005. Signal signature and transcriptome changes of *Arabidopsis* during pathogen and insect attack. *Molecular Plant–Microbe Interactions* **18,** 923–937.
Delaney TP, Uknes S, Vernooij B, Friedrich L, Weymann K, Negrotto D, Gaffney T, Gut-Rella M, Kessmann H, Ward E, Ryals J, 1994. A central role of salicylic acid in plant disease resistance. *Science* **266,** 1247–1250.
Delaney TP, Friedrich L, Ryals JA, 1995. Arabidopsis signal transduction mutant defective in chemically and biologically induced disease resistance. *Proceedings of the National Academy of Sciences USA* **92,** 6602–6606.
Després C, DeLong C, Glaze S, Liu E, Fobert PR, 2000. The Arabidopsis NPR1/NIM1 protein enhances the DNA binding activity of a subgroup of the TGA family of bZIP transcription factors. *The Plant Cell* **12,** 279–290.
Després C, Chubak C, Rochon A, Clark R, Bethune T, Desveaux D, Fobert PR, 2003. The Arabidopsis NPR1 disease resistance protein is a novel cofactor that confers redox regulation of DNA binding activity to the basic domain/leucine zipper transcription factor TGA1. *The Plant Cell* **15,** 2181–2191.
Devoto A, Nieto-Rostro M, Xie DX, Ellis C, Harmston R, Patrick E, Davis J, Sherratt L, Coleman M, Turner JG, 2002. COI1 links jasmonate signalling and fertility to the SCF ubiquitin-ligase complex in Arabidopsis. *The Plant Journal* **32,** 457–466.
Dicke M, Van Poecke RMP, 2002. Signaling in plant–insect interactions: signal transduction in direct and indirect plant defence. In: Scheel D, Wasternack C, eds. *Plant Signal Transduction: Frontiers in Molecular Biology*. Oxford: Oxford University Press, 289–316.
Dong X, 2004. NPR1, all things considered. *Current Opinion in Plant Biology* **7,** 547–552.
Duff B, Schouten A, Raaijmakers JM, 2003. Pathogen self-defense: Mechanisms to counteract microbial antagonism. *Annual Review of Phytopathology* **41,** 501–538.
Duijff BJ, Pouhair D, Olivain C, Alabouvette C, Lemanceau P, 1998. Implication of systemic induced resistance in the suppression of fusarium wilt of tomato by *Pseudomonas fluorescens* WCS417r and by nonpathogenic *Fusarium oxysporum* Fo47. *European Journal of Plant Pathology* **104,** 903–910.
Durrant WE, Dong X, 2004. Systemic acquired resistance. *Annual Review of Phytopathology* **42,** 185–209.
Ellis C, Karafyllidis L, Turner JG, 2002. Constitutive activation of jasmonate signaling in Arabidopsis mutant correlates with enhanced resistance to *Erysiphe cichoracearum, Pseudomonas syringae*, and *Myzus persicae. Molecular Plant–Microbe Interactions* **15,** 1025–1030.
Fan WH, Dong XN, 2002. In vivo interaction between NPR1 and transcription factor TGA2 leads to salicylic acid-mediated gene activation in Arabidopsis. *The Plant Cell* **14,** 1377–1389.
Felton GW, Korth KL, Bi JL, Wesley SV, Huhman DV, Mathews MC, Murphy JB, Lamb C, Dixon RA, 1999. Inverse relationship between systemic resistance of plants to microorganisms and to insect herbivory. *Current Biology* **9,** 317–320.
Felton GW, Korth KL, 2000. Trade-offs between pathogen and herbivore resistance. *Current Opinion in Plant Biology* **3,** 309–314.
Feys BJ, Parker JE, 2000. Interplay of signaling pathways in plant disease resistance. *Trends in Genetics* **16,** 449–455.

Friedrich L, Lawton K, Dietrich R, Willits M, Cade R, Ryals J, 2001. NIM1 overexpression in Arabidopsis potentiates plant disease resistance and results in enhanced effectiveness of fungicides. *Molecular Plant–Microbe Interactions* **14,** 1114–1124.

Gaffney T, Friedrich L, Vernooij B, Negrotto D, Nye G, Uknes S, Ward E, Kessmann H, Ryals J, 1993. Requirement of salicylic acid for the induction of systemic acquired resistance. *Science* **261,** 754–756.

Glazebrook J, 2001. Genes controlling expression of defense responses in Arabidopsis – 2001 status. *Current Opinion in Plant Biology* **4,** 301–308.

Glazebrook J, 2005. Contrasting mechanisms of defense against biotrophic and necrotrophic pathogens. *Annual Review in Phytopathology* **43,** 205–227.

Glazebrook J, Rogers EE, Ausubel FM, 1996. Isolation of Arabidopsis mutants with enhanced disease susceptibility by direct screening. *Genetics* **143,** 973–982.

Glazebrook J, Chen W, Estes B, Chang H-S, Nawrath C, Métraux J-P, Zhu T, Katagiri F, 2003. Topology of the network integrating salicylate and jasmonate signal transduction derived from global expression phenotyping. *The Plant Journal* **34,** 217–228.

Gomez-Gomez L, Boller T, 2000. FLS2: An LRR receptor-like kinase involved in the perception of the bacterial elicitor flagellin in Arabidopsis. *Molecular Cell* **5,** 1003–1012.

Govrin EM, Levine A, 2000. The hypersensitive reaction facilitates plant infection by the necrotrophic fungus *Botrytis cinerea. Current Biology* **10,** 751–757.

Hammerschmidt R, 1999. Induced disease resistance: how do induced plants stop pathogens? *Physiological and Molecular Plant Pathology* **55,** 77–84.

Hammerschmidt R, Métraux J-P, Van Loon LC, 2001. Inducing resistance: a summary of papers presented at the First International Symposium on Induced Resistance to Plant Diseases, Corfu, May 2000. *European Journal of Plant Pathology* **107,** 1–6.

Hase S, Van Pelt JA, Van Loon LC, Pieterse CMJ, 2003. Colonization of Arabidopsis roots by *Pseudomonas fluorescens* primes the plant to produce higher levels of ethylene upon pathogen infection. *Physiological and Molecular Plant Pathology* **62,** 219–226.

Hoffland E, Pieterse CMJ, Bik L, Van Pelt JA, 1995. Induced systemic resistance in radish is not associated with accumulation of pathogenesis-related proteins. *Physiological and Molecular Plant Pathology* **46,** 309–320.

Hoffman T, Schmidt JS, Zheng X, Bent AF, 1999. Isolation of ethylene-insensitive soybean mutants that are altered in pathogen susceptibility and gene-for-gene disease resistance. *Plant Physiology* **119,** 935–949.

Howe GA, 2005. Jasmonates as signals in the wound response. *Journal of Plant Growth Regulation* **23,** 223–237.

Iavicoli A, Boutet E, Buchala A, Métraux J-P, 2003. Induced systemic resistance in *Arabidopsis thaliana* in response to root inoculation with *Pseudomonas fluorescens* CHA0. *Molecular Plant–Microbe Interactions* **16,** 851–858.

Inbar M, Doostdar H, Sonoda RM, Leibee GL, Mayer RT, 1998. Elicitors of plant defensive systems reduce insect densities and disease incidence. *Journal of Chemical Ecology* **24,** 135–149.

Jameson PE, 2000. Cytokinins and auxins in plant–pathogen interactions – An overview. *Journal of Plant Growth Regulation* **32,** 369–380.

Johnson C, Boden E, Arias J, 2003. Salicylic acid and NPR1 induce the recruitment of trans-activating TGA factors to a defense gene promoter in Arabidopsis. *The Plant Cell* **15,** 1846–1858.

Kachroo P, Yoshioka K, Shah J, Dooner KD, Klessig DF, 2000. Resistance to turnip crinkle virus in Arabidopsis is regulated by two host genes and is salicylic acid dependent but NPR1, ethylene, and jasmonate independent. *The Plant Cell* **12,** 677–690.

Kahl J, Siemens DH, Aerts RJ, Gabler R, Kuhnemann F, Preston CA, Baldwin IT, 2000. Herbivore-induced ethylene suppresses a direct defense but not a putative indirect defense against an adapted herbivore. *Planta* **210,** 336–342.

Kessler A, Baldwin IT, 2002. Plant responses to insect herbivory: The emerging molecular analysis. *Annual Review of Plant Biology* **53,** 299–328.

Kim MS, Kim YC, Cho BH, 2004. Gene expression analysis in cucumber leaves primed by root colonization with *Pseudomonas chlororaphis* O6 upon challenge-inoculation with *Corynespora cassiicola. Plant Biology* **6,** 105–108.

Kinkema M, Fan W, Dong X, 2000. Nuclear localization of NPR1 is required for activation of *PR* gene expression. *The Plant Cell* **12,** 2339–2350.

Kloepper JW, Leong J, Teintze M, Schroth MN, 1980. Enhanced plant growth by siderophores produced by plant growth-promoting rhizobacteria. *Nature* **286,** 885–886.

Knoester M, Van Loon LC, Van den Heuvel J, Hennig J, Bol JF, Linthorst HJM, 1998. Ethylene-insensitive tobacco lacks nonhost resistance against soil-borne fungi. *Proceedings of the National Academy of Sciences USA* **95,** 1933–1937.
Knoester M, Pieterse CMJ, Bol JF, Van Loon LC, 1999. Systemic resistance in Arabidopsis induced by rhizobacteria requires ethylene-dependent signaling at the site of application. *Molecular Plant–Microbe Interactions* **12,** 720–727.
Krishna P, 2003. Brassinosteroid-mediated stress responses. *Journal of Plant Growth Regulation* **22,** 289–297.
Kuć J, 1982. Induced immunity to plant disease. *Bioscience* **32,** 854–860.
Kunkel BN, Brooks DM, 2002. Cross talk between signaling pathways in pathogen defense. *Current Opinion in Plant Biology* **5,** 325–331.
Lawton K, Weymann K, Friedrich L, Vernooij B, Uknes S, Ryals J, 1995. Systemic acquired resistance in Arabidopsis requires salicylic acid but not ethylene. *Molecular Plant–Microbe Interactions* **8,** 863–870.
Leeman M, Van Pelt JA, Den Ouden FM, Heinsbroek M, Bakker PAHM, Schippers B, 1995a. Induction of systemic resistance by *Pseudomonas fluorescens* in radish cultivars differing in susceptibility to fusarium wilt, using a novel bioassay. *European Journal of Plant Pathology* **101,** 655–664.
Leeman M, Van Pelt JA, Den Ouden FM, Heinsbroek M, Bakker PAHM, Schippers B, 1995b. Induction of systemic resistance against fusarium wilt of radish by lipopolysaccharides of *Pseudomonas fluorescens*. *Phytopathology* **85,** 1021–1027.
Li J, Brader G, Palva ET, 2004. The WRKY70 transcription factor: a node of convergence for jasmonate-mediated and salicylate-mediated signals in plant defense. *The Plant Cell* **16,** 319–331.
Lorenzo O, Piqueras R, Sánchez-Serrano JJ, Solano R, 2003. ETHYLENE RESPONSE FACTOR1 integrates signals from ethylene and jasmonate pathways in plant defense. *The Plant Cell* **15,** 165–178.
Lorenzo O, Chico JM, Sánchez-Serrano JJ, Solano R, 2004. *JASMONATE-INSENSITIVE1* encodes a MYC transcription factor essential to discriminate between different jasmonate-regulated defense responses in Arabidopsis. *The Plant Cell* **16,** 1938–1950.
Lugtenberg BJJ, Dekkers L, Bloemberg GV, 2001. Molecular determinants of rhizosphere colonization by *Pseudomonas*. *Annual Review of Phytopathology* **39,** 461–490.
Lund ST, Stall RE, Klee HJ, 1998. Ethylene regulates the susceptible response to pathogen infection in tomato. *The Plant Cell* **10,** 371–382.
Lynch JM, 1976. Products of soil microorganisms in relation to plant growth. *Critical Reviews in Microbiology* **5,** 67–107.
Lynch JM, Whipps JM, 1991. Substrate flow in the rhizosphere. In: Keister DL, Cregan PB, eds. *The Rhizosphere and Plant Growth*. Dordrecht: Kluwer, 15–24.
Maldonado AM, Doerner P, Dixon RA, Lamb CJ, Cameron RK, 2002. A putative lipid transfer protein involved in systemic resistance signalling in *Arabidopsis*. *Nature* **419,** 399–403.
Maleck K, Levine A, Eulgem T, Morgan A, Schmid J, Lawton KA, Dangl JL, Dietrich RA, 2000. The transcriptome of *Arabidopsis thaliana* during systemic acquired resistance. *Nature Genetics* **26,** 403–410.
Mauch-Mani B, Métraux J-P, 1998. Salicylic acid and systemic acquired resistance to pathogen attack. *Annals of Botany* **82,** 535–540.
Mauch-Mani B, Mauch F, 2005. The role of abscisic acid in plant–pathogen interactions. *Current Opinion in Plant Biology* **8,** 409–414.
Maurhofer M, Hase C, Meuwly P, Métraux J-P, Défago G, 1994. Induction of systemic resistance to tobacco necrosis virus by the root-colonizing *Pseudomonas fluorescens* strain CHA0: Influence of the *gacA* gene and pyoverdine production. *Phytopathology* **84,** 139–146.
Maurhofer M, Reimmann C, Schmidli-Sacherer P, Heeb SD, Défago G, 1998. Salicylic acid biosynthesis genes expressed in *Pseudomonas fluorescens* strain P3 improve the induction of systemic resistance in tobacco against tobacco necrosis virus. *Phytopathology* **88,** 678–684.
McConn J, Creelman RA, Bell E, Mullet JE, Browse J, 1997. Jasmonate is essential for insect defense in Arabidopsis. *Proceedings of the National Academy of Sciences USA* **94,** 5473–5477.
McDowell JM, Dangl JL, 2000. Signal transduction in the plant immune response. *Trends in Biochemical Sciences* **25,** 79–82.
Meyer J-M, Azelvandre P, Georges C, 1992. Iron metabolism in *Pseudomonas*: salicylic acid, a siderophore of *Pseudomonas fluorescens* CHA0. *Biofactors* **4,** 23–27.
Moran P, 1998. Plant-mediated interactions between insects and a fungal plant pathogen and the role of plant chemical responses to infection. *Oecologia* **115,** 523–530.

Mou Z, Fan WH, Dong XN, 2003. Inducers of plant systemic acquired resistance regulate NPR1 function through redox changes. *Cell* **113,** 935–944.

Nakashita H, Yasuda M, Nitta T, Asami T, Fujioka S, Arai Y, Sekimata K, Takatsuto S, Yamaguchi I, Yoshida S, 2003. Brassinosteroid functions in a broad range of disease resistance in tobacco and rice. *The Plant Journal* **33,** 887–898.

Nawrath C, Métraux J-P, 1999. Salicylic acid induction-deficient mutants of Arabidopsis express PR-2 and PR-5 and accumulate high levels of camalexin after pathogen inoculation. *The Plant Cell* **11,** 1393–1404.

Niki T, Mitsuhara I, Seo S, Ohtsubo N, Ohashi Y, 1998. Antagonistic effect of salicylic acid and jasmonic acid on the expression of pathogenesis-related (PR) protein genes in wounded mature tobacco leaves. *Plant and Cell Physiology* **39,** 500–507.

Norman-Setterblad C, Vidal S, Palva TE, 2000. Interacting signal pathways control defense gene expression in Arabidopsis in response to cell wall-degrading enzymes from *Erwinia carotovora. Molecular Plant–Microbe Interactions* **13,** 430–438.

Nürnberger T, Brunner F, Kemmerling B, Piater L, 2004. Innate immunity in plants and animals: striking similarities and obvious differences. *Immunological Reviews* **198,** 249–266.

Parbery DG, 1996. Trophism and the ecology of fungi associated with plants. *Biological Reviews Cambridge Philosophical Society* **71,** 473–527.

Peña-Cortés H, Albrecht T, Prat S, Weiler EW, Willmitzer L, 1993. Aspirin prevents wound-induced gene expression in tomato leaves by blocking jasmonic acid biosynthesis. *Planta* **191,** 123–128.

Pieterse CMJ, Van Wees SCM, Hoffland E, Van Pelt JA, Van Loon LC, 1996. Systemic resistance in Arabidopsis induced by biocontrol bacteria is independent of salicylic acid accumulation and pathogenesis-related gene expression. *The Plant Cell* **8,** 1225–1237.

Pieterse CMJ, Van Wees SCM, Van Pelt JA, Knoester M, Laan R, Gerrits H, Weisbeek PJ, Van Loon LC, 1998. A novel signaling pathway controlling induced systemic resistance in Arabidopsis. *The Plant Cell* **10,** 1571–1580.

Pieterse CMJ, Van Loon LC, 1999. Salicylic acid-independent plant defence pathways. *Trends in Plant Science* **4,** 52–58.

Pieterse CMJ, Van Pelt JA, Ton J, Parchmann S, Mueller MJ, Buchala AJ, Métraux J-P, Van Loon LC, 2000. Rhizobacteria-mediated induced systemic resistance (ISR) in *Arabidopsis* requires sensitivity to jasmonate and ethylene but is not accompanied by an increase in their production. *Physiological and Molecular Plant Pathology* **57,** 123–134.

Pieterse CMJ, Van Wees SCM, Ton J, Van Pelt JA, Van Loon LC, 2002. Signalling in rhizobacteria-induced systemic resistance in *Arabidopsis thaliana. Plant Biology* **4,** 535–544.

Pieterse CMJ, Van Pelt JA, Verhagen BWM, Ton J, Van Wees SCM, Léon-Kloosterziel KM, Van Loon LC, 2003. Induced systemic resistance by plant growth-promoting rhizobacteria. *Symbiosis* **35,** 39–54.

Pieterse CMJ, Van Loon LC, 2004. NPR1: the spider in the web of induced resistance signaling pathways. *Current Opinion in Plant Biology* **7,** 456–464.

Press CM, Wilson M, Tuzun S, Kloepper JW, 1997. Salicylic acid produced by *Serratia marcescens* 91–166 is not the primary determinant of induced systemic resistance in cucumber or tobacco. *Molecular Plant–Microbe Interactions* **10,** 761–768.

Preston CA, Lewandowski C, Enyedi AJ, Baldwin IT, 1999. Tobacco mosaic virus inoculation inhibits wound-induced jasmonic acid-mediated responses within but not between plants. *Planta* **209,** 87–95.

Raaijmakers JM, Weller DM, 1998. Natural plant protection by 2,4-diacetylphloroglucinol-producing *Pseudomonas* spp. in take-all decline soils. *Molecular Plant–Microbe Interactions* **11,** 144–152.

Ran LX, Li ZN, Wu GJ, Van Loon LC, Bakker PAHM, 2005. Induction of systemic resistance against bacterial wilt in *Eucalyptus urophylla* by fluorescent *Pseudomonas* spp. *European Journal of Plant Pathology* **113,** 59–70.

Reymond P, Farmer EE, 1998. Jasmonate and salicylate as global signals for defense gene expression. *Current Opinion in Plant Biology* **1,** 404–411.

Reymond P, Bodenhausen N, Van Poecke RMP, Krishnamurthy V, Dicke M, Farmer EE, 2004. A conserved transcriptional pattern in response to a specialist and a generalist herbivore. *The Plant Cell* **16,** 3132–3147.

Rojo E, Solano R, Sánchez-Serrano JJ, 2003. Interactions between signaling compounds involved in plant defense. *Journal of Plant Growth Regulation* **22,** 82–98.

Ross AF, 1961. Systemic acquired resistance induced by localized virus infections in plants. *Virology* **14,** 340–358.

Ryals JA, Neuenschwander UH, Willits MG, Molina A, Steiner H-Y, Hunt MD, 1996. Systemic acquired resistance. *The Plant Cell* **8,** 1808–1819.

Ryu C-M, Hu C-H, Reddy MS, Kloepper JW, 2003. Different signaling pathways of induced resistance by rhizobacteria in *Arabidopsis thaliana* against two pathovars of *Pseudomonas syringae. New Phytologist* **160,** 413–420.

Ryu C-M, Murphy JF, Mysore KS, Kloepper JW, 2004. Plant growth-promoting rhizobacteria systemically protect *Arabidopsis thaliana* against *Cucumber mosaic virus* by a salicylic acid and NPR1-independent and jasmonic acid-dependent signaling pathway. *The Plant Journal* **39,** 381–392.

Schenk PM, Kazan K, Wilson I, Anderson JP, Richmond T, Somerville SC, Manners JM, 2000. Coordinated plant defense responses in Arabidopsis revealed by microarray analysis. *Proceedings of the National Academy of Sciences USA* **97,** 11655–11660.

Schippers B, Bakker AW, Bakker PAHM, 1987. Interactions of deleterious and beneficial rhizosphere microorganisms and the effect of cropping practices. *Annual Review of Phytopathology* **115,** 339–358.

Seo HS, Song JT, Cheong J-J, Lee Y-H, Lee Y-W, Hwang I, Lee JS, Choi YD, 2001. Jasmonic acid carboxyl methyltransferase: A key enzyme for jasmonate-regulated plant responses. *Proceedings of the National Academy of Sciences USA* **98,** 4788–4793.

Shah J, Tsui F, Klessig DF, 1997. Characterization of a salicylic acid-insensitive mutant (*sai1*) of *Arabidopsis thaliana*, identified in a selective screen utilizing the SA-inducible expression of the *tms2* gene. *Molecular Plant–Microbe Interactions* **10,** 69–78.

Spoel SH, Koornneef A, Claessens SMC, Korzelius JP, Van Pelt JA, Mueller MJ, Buchala AJ, Métraux J-P, Brown R, Kazan K, Van Loon LC, Dong X, Pieterse CMJ, 2003. NPR1 modulates cross-talk between salicylate- and jasmonate-dependent defense pathways through a novel function in the cytosol. *The Plant Cell* **15,** 760–770.

Staswick PE, Yuen GY, Lehman CC, 1998. Jasmonate signaling mutants of Arabidopsis are susceptible to the soil fungus *Pythium irregulare. The Plant Journal* **15,** 747–754.

Sticher L, Mauch-Mani B, Métraux J-P, 1997. Systemic acquired resistance. *Annual Review of Phytopathology* **35,** 235–270.

Stout MJ, Fidantsef AL, Duffey SS, Bostock RM, 1999. Signal interactions in pathogen and insect attack: systemic plant-mediated interactions between pathogens and herbivores of the tomato, *Lycopersicon esculentum. Physiological and Molecular Plant Pathology* **54,** 115–130.

Subramaniam R, Desveaux D, Spickler C, Michnick SW, Brisson N, 2001. Direct visualization of protein interactions in plant cells. *Nature Biotechnology* **19,** 769–772.

Thaler JS, Fidantsef AL, Duffey SS, Bostock RM, 1999. Trade-offs in plant defense against pathogens and herbivores: a field demonstration of chemical elicitors of induced resistance. *Journal of Chemical Ecology* **25,** 1597–1609.

Thaler JS, Fidantsef AL, Bostock RM, 2002. Antagonism between jasmonate- and salicylate-mediated induced plant resistance: effects of concentration and timing of elicitation on defense-related proteins, herbivore, and pathogen performance in tomato. *Journal of Chemical Ecology* **28,** 1131–1159.

Thaler J, Bostock RM, 2004. Interactions between abscisic-acid-mediated responses and plant resistance to pathogens and insects. *Ecology* **85,** 48–58.

Thomma BPHJ, Eggermont K, Penninckx IAMA, Mauch-Mani B, Vogelsang R, Cammue BPA, Broekaert WF, 1998. Separate jasmonate-dependent and salicylate-dependent defense-response pathways in Arabidopsis are essential for resistance to distinct microbial pathogens. *Proceedings of the National Academy of Sciences USA* **95,** 15107–15111.

Thomma BPHJ, Eggermont K, Tierens KFM, Broekaert WF, 1999. Requirement of functional *ethylene-insensitive 2* gene for efficient resistance of Arabidopsis to infection by *Botrytis cinerea. Plant Physiology* **121,** 1093–1102.

Thomma BPHJ, Penninckx IAMA, Broekaert WF, Cammue BPA, 2001. The complexity of disease signaling in *Arabidopsis. Current Opinion in Immunology* **13,** 63–68.

Tjamos SE, Flemetakis E, Paplomatas EJ, Katinakis P, 2005. Induction of resistance to *Verticillium dahliae* in *Arabidopsis thaliana* by the biocontrol agent K-165 and pathogenesis-related proteins gene expression. *Molecular Plant–Microbe Interactions* **18,** 555–561.

Ton J, Pieterse CMJ, Van Loon LC, 1999. Identification of a locus in Arabidopsis controlling both the expression of rhizobacteria-mediated induced systemic resistance (ISR) and basal resistance against *Pseudomonas syringae* pv. *tomato. Molecular Plant–Microbe Interactions* **12,** 911–918.

Ton J, Davison S, Van Wees SCM, Van Loon LC, Pieterse CMJ, 2001. The Arabidopsis *ISR1* locus controlling rhizobacteria-mediated induced systemic resistance is involved in ethylene signaling. *Plant Physiology* **125,** 652–661.

Ton J, Van Pelt JA, Van Loon LC, Pieterse CMJ, 2002a. Differential effectiveness of salicylate-dependent and jasmonate/ethylene-dependent induced resistance in Arabidopsis. *Molecular Plant–Microbe Interactions* **15,** 27–34.

Ton J, Van Pelt JA, Van Loon LC, Pieterse CMJ, 2002b. The Arabidopsis *ISR1* locus is required for rhizobacteria-mediated induced systemic resistance against different pathogens. *Plant Biology* **4,** 224–227.

Ton J, Van Pelt JA, Van Loon LC, Pieterse CMJ, 2002c. Differential effectiveness of salicylate-dependent and jasmonate/ethylene-dependent induced resistance in *Arabidopsis*. *Molecular Plant–Microbe Interactions* **15,** 27–34.

Ton J, De Vos M, Robben C, Buchala AJ, Métraux J-P, Van Loon LC, Pieterse CMJ, 2002d. Characterisation of Arabidopsis enhanced disease susceptibility mutants that are affected in systemically induced resistance. *The Plant Journal* **29,** 11–21.

Ton J, Mauch-Mani B, 2004. β-amino-butyric acid-induced resistance against necrotrophic pathogens is based on ABA-dependent priming for callose. *The Plant Journal* **38,** 119–130.

Ton J, Jakab G, Toquin V, Flors V, Iavicoli A, Maeder MN, Métraux J-P, Mauch-Mani B, 2005. Dissecting the β-aminobutyric acid-induced priming phenomenon in Arabidopsis. *The Plant Cell* **17,** 987–999.

Van Hulten M, Pelser M, Van Loon LC, Pieterse CMJ, Ton J, 2006. Costs and benefits of priming for defense in Arabidopsis. *Proceedings of the National Academy of Sciences USA* **103**, 5602–5607.

Van Loon LC, 1997. Induced resistance and the role of pathogenesis-related proteins. *European Journal of Plant Pathology* **103,** 753–765.

Van Loon LC, 2000. Systemic induced resistance. In: Slusarenko AJ, Fraser RSS, Van Loon LC, eds. *Mechanisms of Resistance to Plant Diseases*. Dordrecht: Kluwer Academic Publishers, 521–574.

Van Loon LC, Bakker PAHM, Pieterse CMJ, 1998. Systemic resistance induced by rhizosphere bacteria. *Annual Review of Phytopathology* **36,** 453–483.

Van Loon LC, Bakker PAHM, 2003. Signalling in rhizobacteria-plant interactions. In: De Kroon J, Visser EJW, eds. *Root Ecology*. Berlin: Springer, 287–330.

Van Loon LC, Bakker PAHM, 2005. Induced systemic resistance as a mechanism of disease suppression by rhizobacteria. In: Siddiqui ZA, ed *PGPR: Biocontrol and Biofertilization*. Dordrecht: Springer, 39–66.

Van Loon LC, Geraats BPJ, Linthorst HJM, 2006. Ethylene as a modulator of disease resistance in plants. *Trends in Plant Science* **11,** 184–191.

Van Peer R, Schippers B, 1989. Plant growth responses to bacterization and rhizosphere microbial development in hydroponic cultures. *Canadian Journal of Microbiology* **35,** 456–463.

Van Peer R, Niemann GJ, Schippers B, 1991. Induced resistance and phytoalexin accumulation in biological control of fusarium wilt of carnation by *Pseudomonas* sp. strain WCS417r. *Phytopathology* **81,** 728–734.

Van Peer R, Schippers B, 1992. Lipopolysaccharides of plant growth-promoting *Pseudomonas* sp. strain WCS417r induce resistance in carnation to fusarium wilt. *Netherlands Journal of Plant Pathology* **98,** 129–139.

Van Wees SCM, Pieterse CMJ, Trijssenaar A, Van't Westende YAM, Hartog F, Van Loon LC, 1997. Differential induction of systemic resistance in Arabidopsis by biocontrol bacteria. *Molecular Plant–Microbe Interactions* **10,** 716–724.

Van Wees SCM, Luijendijk M, Smoorenburg I, Van Loon LC, Pieterse CMJ, 1999. Rhizobacteria-mediated induced systemic resistance (ISR) in Arabidopsis is not associated with a direct effect on expression of known defense-related genes but stimulates the expression of the jasmonate-inducible gene *Atvsp* upon challenge. *Plant Molecular Biology* **41,** 537–549.

Van Wees SCM, De Swart EAM, Van Pelt JA, Van Loon LC, Pieterse CMJ, 2000. Enhancement of induced disease resistance by simultaneous activation of salicylate- and jasmonate-dependent defense pathways in *Arabidopsis thaliana*. *Proceedings of the National Academy of Sciences USA* **97,** 8711–8716.

Verberne MC, Hoekstra J, Bol JF, Linthorst HJM, 2003. Signaling of systemic acquired resistance in tobacco depends on ethylene perception. *The Plant Journal* **35,** 27–32.

Verhagen BWM, Glazebrook J, Zhu T, Chang H-S, Van Loon LC, Pieterse CMJ, 2004. The transcriptome of rhizobacteria-induced systemic resistance in *Arabidopsis*. *Molecular Plant–Microbe Interactions* **17,** 895–908.

Vernooij B, Friedrich L, Morse A, Reist R, Kolditz-Jawhar R, Ward E, Uknes S, Kessmann H, Ryals J, 1994. Salicylic acid is not the translocated signal responsible for inducing systemic acquired resistance but is required in signal transduction. *The Plant Cell* **6,** 959–965.

Vijayan P, Shockey J, Lévesque CA, Cook RJ, Browse J, 1998. A role for jasmonate in pathogen defense of *Arabidopsis*. *Proceedings of the National Academy of Sciences USA* **95,** 7209–7214.
Visca P, Ciervo A, Sanfilippo V, Orsi N, 1993. Iron-regulated salicylate synthesis by *Pseudomonas* spp. *Journal of General Microbiology* **139,** 1995–2001.
Walker TS, Bais HP, Halligan KM, Stermitz FR, Vivanco JR, 2003. Metabolic profiling of root exudates of *Arabidopsis thaliana*. *Journal of Agricultural and Food Chemistry* **51,** 2548–2554.
Wang D, Weaver ND, Kesarwani M, Dong X, 2005. Induction of protein secretory pathway is required for systemic acquired resistance. *Science* **308,** 1036–1040.
Weller DM, Raaijmakers JM, McSpadden Gardener BB, Thomashow LS, 2002. Microbial populations responsible for specific soil suppressiveness to pathogens. *Annual Review of Phytopathology* **40,** 309–348.
Wildermuth MC, Dewdney J, Wu G, Ausubel FM, 2001. Isochorismate synthase is required to synthesize salicylic acid for plant defence. *Nature* **414,** 562–565.
Xu L, Liu F, Lechner E, Genschik P, Crosby WL, Ma H, Peng W, Huang D, Xie D, 2002. The SCF^{COI1} ubiquitin-ligase complexes are required for jasmonate response in Arabidopsis. *The Plant Cell* **14,** 1919–1935.
Yan Z, Reddy MS, Ryu C-M, McInroy JA, Wilson M, Kloepper JW, 2002. Induced systemic protection against tomato late blight elicited by plant growth-promoting rhizobacteria. *Phytopathology* **92,** 1329–1333.
Zehnder GW, Murphy JF, Sikora EJ, Kloepper JW, 2001. Application of rhizobacteria for induced resistance. *European Journal of Plant Pathology* **107,** 39–50.
Zhang S, Moyne A-L, Reddy MS, Kloepper JW, 2002. The role of salicylic acid in induced systemic resistance elicited by plant growth-promoting rhizobacteria against blue mold of tobacco. *Biological Control* **25,** 288–296.
Zhang Y, Fan W, Kinkema M, Li X, Dong X, 1999. Interaction of NPR1 with basic leucine zipper protein transcription factors that bind sequences required for salicylic acid induction of the *PR-1* gene. *Proceedings of the National Academy of Sciences USA* **96,** 6523–6528.
Zhang YL, Tessaro MJ, Lassner M, Li X, 2003. Knockout analysis of Arabidopsis transcription factors TGA2, TGA5, and TGA6 reveals their redundant and essential roles in systemic acquired resistance. *The Plant Cell* **15,** 2647–2653.
Zhou JM, Trifa Y, Silva H, Pontier D, Lam E, Shah J, Klessig DF, 2000. NPR1 differentially interacts with members of the TGA/OBF family of transcription factors that bind an element of the *PR-1* gene required for induction by salicylic acid. *Molecular Plant–Microbe Interactions* **13,** 191–202.
Zhou N, Tootle TL, Tsui F, Klessig DF, Glazebrook J, 1998. PAD4 functions upstream from salicylic acid to control defense responses in Arabidopsis. *The Plant Cell* **10,** 1021–1030.

Chapter 5
Types and mechanisms of rapidly induced plant resistance to herbivorous arthropods

Michael J. Stout

Department of Entomology, Louisiana State University Agricultural Center, Baton Rouge, USA

5.1 Introduction: induced resistance in context

The interactions between plants and the herbivorous arthropods that feed on plants are multilayered, complex and dynamic. These interactions are mediated by the biochemical, physicochemical, physiological and morphological traits of plants. Some of these traits enable the plants that express them to evade, reduce or minimize the impacts of injury by herbivores. Plant resistance is ordinarily an outcome of the expression of multiple resistance-related traits by a plant. As a composite trait, plant resistance is best measured by quantifying yield or fitness gain in a resistant plant relative to a susceptible counterpart in an environment containing the herbivore. In practice, however, plant yield or fitness is often not measured, and herbivore performance or levels of putative resistance-related traits are used as imperfect indicators of plant resistance.

The levels of expression of most, if not all, of the resistance-related traits of plants are affected by a host of biotic and abiotic factors such as nutrient, light and water availability, soil type and density of surrounding plants. Consequently, plant resistance varies depending on the environment of the plant. The term 'induced resistance' has come to be used to refer to one type of plant phenotypic plasticity in which initial attack by an arthropod herbivore causes an increase in the resistance of the plant to subsequent herbivores (Karban & Baldwin, 1997). The term is used in opposition to the term constitutive resistance – resistance not affected by prior herbivory – although it must be emphasized that the level of 'constitutive' resistance expressed by a plant may be affected by factors other than prior herbivory. Herbivore-induced changes in plant resistance can occur within hours, days or weeks of initial attack ('rapid induced resistance') or, in long-lived plants, over longer timescales ('delayed induced resistance') (Karban & Baldwin, 1997). The spatial extent of induced resistance within a plant can vary from extremely localized to plant systemic. Herbivory can also induce changes that affect plant resistance by affecting plants and arthropods at a distance from the damaged plant.

This chapter presents an overview of the mechanisms of rapidly induced resistance in plants, by which is meant the relationship between the biochemical, physiological and morphological changes that occur following herbivory and the changes in plant resistance that also occur following herbivory. Broadly considered, the relationship is a causal one; that is, the increases in levels of expression of resistance-related traits that occur following herbivory are responsible for induced resistance. However, the complex nature of the biochemical and morphological changes that occur following herbivory makes it difficult to assign causal roles to specific changes. The challenges associated with understanding cause–effect relationships in induced resistance, and the approaches used to overcome these challenges, are a major focus of this chapter.

This chapter proceeds by first comparing the threats posed by arthropod herbivores and pathogenic micro-organisms. This comparison serves as a reminder that mechanisms of induced resistance to arthropods are part of an integrated system of plant defences against the diverse abiotic and biotic threats faced by plants. A survey of the various types of rapidly induced resistance to arthropods that have been described in the literature follows. This survey is illustrated with selected examples of inducible biochemical and morphological traits that are associated with induced resistance. The chapter concludes with a more detailed consideration of the complex causal basis of induced resistance and of some of the approaches used to elucidate cause–effect relationships in induced resistance.

An understanding of the types and mechanisms of induced resistance to arthropods is important for several reasons. Most directly, understanding the mechanisms of induced resistance may lead to the development of strategies for using induced resistance to protect crop plants. Studies of the mechanisms of induced resistance may also provide general insights into mechanisms of plant resistance, because the mechanisms of induced resistance appear to be fundamentally similar to mechanisms of constitutive resistance (Gatehouse, 2002). This will, in turn, facilitate the development of crop cultivars that possess broad-spectrum and durable resistance to pests. Finally, understanding the types and mechanisms of induced plant resistance present in a plant may help resolve the importance of induced resistance to overall plant resistance. This latter issue has rarely been addressed, and so it is not altogether clear how much induced resistance contributes to the overall resistance of plants. The greater resistance of the lettuce (*Lactuca sativa*) variety Valmaine to *Diabrotica balteata* was attributed both to differences among varieties in the chemical and physical properties of a constitutive trait (latex) and to differences among cultivars in their inducibility (Huang *et al.*, 2003), suggesting that inducible resistance is an important component of cultivar resistance to this pest in lettuce. Suppression of biochemical responses to herbivory in wild tobacco plants, *Nicotiana attenuata*, rendered the plants susceptible to insects not ordinarily found on plants in which induced responses are operative, suggesting that induced resistance is a partial determinant of the spectrum of arthropods that feed on wild tobacco (Kessler *et al.*, 2004). The extent to which these insights from lettuce and wild tobacco extend to arthropod–plant interactions in general is unclear.

5.2 Comparison of the threats posed by pathogens and herbivores

Inducible resistance has been recognized as an important component of plant resistance to pathogenic micro-organisms for over 40 years (see Chapter 1), and thus the study of inducible resistance to pathogens serves as an important counterpoint to the study of inducible resistance to arthropod herbivores (Vallad & Goodman, 2004). It will thus be helpful to consider how the threats posed by pathogenic micro-organisms and herbivorous arthropods differ and to consider some of the implications of these differences for plant resistance strategies before embarking upon a discussion of types of induced resistance to arthropods.

A key difference between pathogens and herbivores is the greater degree of physiological and behavioural 'autonomy' (Kessler & Baldwin, 2002) shown by herbivores: they are larger and more mobile than pathogens, and they possess relatively complex peripheral and central nervous systems. Accordingly, arthropod herbivores can locate potential host plants at a distance and employ non-random behaviours to increase the probability of coming into contact with a potential host (Bernays & Chapman, 1994). Arthropod herbivores are also capable of evaluating the acceptability of a potential host after they begin feeding and can move away from a plant or feeding site if it is unsuitable in some way. Thus, even those arthropods whose arrival at a plant is largely the result of undirected movement – for example, weak-flying aphids that are carried to potential hosts by wind – exercise some control over landing or arrestment of movement, and even those insects whose feeding sites are determined by the ovipositional preferences of their mother are capable of moving to a new feeding site.

The greater autonomy of herbivores relative to pathogens has several consequences (Baldwin & Preston, 1999; Kessler & Baldwin, 2002). First, there is a behavioural component in plant resistance to arthropods, largely absent in plant resistance to pathogens, in which plants interfere with host-finding and feeding behaviours of herbivores. In addition, the scale at which plant responses to pathogens are effective differs from the spatial scale at which plant responses to herbivores are effective. Because mobile insects can move away from a feeding site that has become unacceptable, the extremely localized increases in plant resistance effective against many pathogens are likely to be effective only against the smallest or least mobile of insects. Finally, the expanded spatial scale of plant–herbivore interactions has apparently resulted in greater involvement of natural enemies in plant–herbivore interactions than in plant–pathogen interactions.

A second difference between pathogenic micro-organisms and arthropod herbivores relates to the ways they extract nutrients from their host plants, although in this regard there is also considerable diversity within each group. Arthropod herbivores can be divided into chewing arthropods and piercing/sucking arthropods, the latter group containing cell content feeders and sap feeders. Many piercing/sucking insects form intimate and long lasting associations with their hosts, whereas chewing arthropods are usually more mobile (Walling, 2000). Microbial pathogens can be classified as biotrophs or necrotrophs. Biotrophs maintain themselves on living plant cells and therefore must evade recognition or suppress plant resistance mechanisms, whereas necrotrophs kill plant cells and absorb nutrients from the dead cells (Dangl & Jones, 2001; Stout *et al.*, 2006).

The method by which an attacker obtains nutrients from its host plant clearly influences the nature of a plant's response to the attacker. Past comparisons of responses to pathogens and herbivores have often emphasized the apparent distinctions that exist among plant responses to pathogens and herbivores. In particular, much emphasis has been placed on an important dichotomy in plant signalling pathways following pathogen and herbivore attack (Walling, 2000; Gatehouse, 2002; Kessler & Baldwin, 2002). In many plants, wounding and some types of herbivory activate a signalling pathway involving jasmonic acid (JA), leading to the expression of resistance-related traits such as proteinase inhibitors, whereas some types of pathogen infection activate a signalling pathway involving salicylic acid (SA), leading to the expression of a distinct set of resistance-related traits. These two signal transduction pathways appear to be mutually inhibitory (Kessler & Baldwin, 2002; Stout *et al.*, 2006; see also Chapter 4).

Recent research, however, indicates that this dichotomy between responses to arthropods and pathogens is not as definitive as once thought and suggests that the feeding style of the attacker is a more important determinant of plant response than is the taxonomic identity of the attacker. In tomato (*Lycopersicon esculentum*) and several other plant species, damage by chewing arthropods results in the activation of the JA-mediated responses, whereas damage by piercing/sucking arthropods results in the activation of SA-mediated responses (Fidantsef *et al.*, 1999; Walling, 2000). Arthropods that feed on entire cell contents may cause the activation of both JA- and SA-mediated pathways (Grinberg *et al.*, 2005). In Arabidopsis, infection by some pathogens resulted in an induction of JA-mediated responses, infection by others resulted in induction of SA-mediated responses, and infection by yet others resulted in activation of both SA- and JA-mediated responses (Thomma *et al.*, 2001). Thus, in the few systems that have been investigated in detail, piercing/sucking arthropods appear to induce responses similar to those induced by biotrophic pathogens, while responses to necrotrophs share some features with responses to chewing herbivores.

It must also be noted here that many pathogens and herbivores influence the nature of the plant response by injecting or secreting chemical substances into the host during the feeding or infection process. Many pathogens and some arthropods release enzymes that are necessary for infection or feeding (e.g. polygalacturonases) but that also trigger responses in the host plant. Other substances released by pathogens or herbivores may participate in specific receptor–ligand interactions and may induce or suppress responses in the host plant (Hahn, 1996). Products of avirulence genes in pathogens, for example, induce a chain of events leading ultimately to a hypersensitive response (see below) and plant resistance (Lam *et al.*, 2001). Fatty acid amides found in the oral secretions of several Lepidopteran species are potent elicitors of volatile emissions from plants that attract the natural enemies of herbivores (indirect induced defence; see below) (Tumlinson & Lait, 2005). Glucose oxidase, a salivary enzyme from *Helicoverpa zea*, suppresses the wound induction of nicotine in tobacco (Musser *et al.*, 2005).

The ability of plants to respond differently to different attackers is often assumed to have functional significance; that is, the responses induced by a given attacker are assumed to be those responses that are most efficacious against the attacker. Evidence for this hypothesis is mixed. In Arabidopsis, there is limited correspondence between the ability of a pathogen to induce a response and the effectiveness of the response against the inducing pathogen (Thomma *et al.*, 2001). In tomato, SA-mediated responses, which are induced by aphid feeding, were recently shown to reduce the population growth of aphids; however, JA-mediated responses, which are not induced by aphid feeding, were also shown to reduce aphid population growth (Cooper *et al.*, 2004).

5.3 Types of induced resistance

5.3.1 Hypersensitive responses

The resistance of some plants to some pathogenic micro-organisms involves the induction of a rapid and localized programmed cell death response at the site of attempted infection

(Lam *et al.*, 2001). This programmed cell death response is called a hypersensitive response (HR) and is governed by specific resistance genes in the plant that recognize the presence or products of corresponding avirulence genes in the pathogen in a so-called gene-for-gene interaction (Dangl & Jones, 2001; see also Chapter 6). The HR functions in pathogen resistance by physically isolating the would-be pathogen to necrotic tissue, thereby depriving it of nutrients and water. Antimicrobial compounds that accumulate in and around the site of attempted infection may also be involved in suppressing pathogen spread. The HR is primarily effective against biotrophic pathogens that cannot utilize dead tissues.

A phenomenon similar to the HR has been implicated in the resistance of some plants to arthropods with a piercing/sucking mode of feeding. In a thorough study, Ollerstam *et al.* (2002) found an HR-like response in leaves of willow, *Salix viminalis*, attacked by first-instar gall midges (*Dasineura marginemtorquens*). The response occurred within 12 hours of eclosion of eggs, was more extensive in resistant willow genotypes than in susceptible genotypes, resulted in 100% mortality of larvae within 40 hours of egg hatch, and was associated with the accumulation of phenolic compounds. Similarly, resistant varieties of wheat (*Triticum aestivum*) attacked by early-instar Hessian flies, *Mayetiola destructor*, exhibited extensive regions of cell death at the sites of larval feeding (Grover, 1995). Larvae attempting to feed at sites in which this reaction occurred probably starved to death.

Some have questioned whether HR responses to arthropods and pathogens are strictly analogous phenomena, since HR-like responses to insect feeding often differ in important ways from pathogen-induced HRs (Ollerstam *et al.*, 2002). Hypersensitive responses to arthropod feeding often develop more slowly than pathogen-induced responses, and often develop in plant cultivars both resistant and susceptible to the inducing insects. Moreover, gene-for-gene interactions between plants and arthropods do not always involve an HR-like response, whereas gene-for-gene interactions between plants and pathogens almost always do (Kaloshian, 2004). Despite these differences, pathogen-induced HRs and HR-like responses to arthropods are similar in two important respects, that they primarily affect the organism that induces them rather than subsequent organisms and that they involve the death of plant cells at the site of attack.

Prior studies may have underestimated the importance of the HR or HR-like phenomena in plant resistance to some arthropods. In a study conducted by Fernandes & Negreiros (2001), hypersensitivity was the most important source of mortality of gall-forming insects on seven of eight taxonomically disparate plant species in tropical Brazil. Hypersensitivity may also be involved in the resistance of some plants to chewing insects, although this involvement is rarely investigated. At least two studies have shown that hypersensitive responses to oviposition by chewing insects led to insect death via desiccation or detachment of eggs (Hilker & Meiners, 2002).

5.3.2 Direct induced resistance

Direct induced resistance refers to a phenomenon in which rapid changes in plant biochemistry, physiology or morphology directly reduce the quality of the plant as a host for subsequent herbivores. Representatives of all major classes of secondary chemicals have been shown to be inducible, and the levels of many types of primary chemicals are affected by herbivory as well (Karban & Baldwin, 1997). In any one plant, herbivory causes multiple changes in plant primary and secondary chemistry, plant physiology and plant morphology, with different types of herbivory causing different changes. Because these biochemical and morphological traits mediate virtually all aspects of plant–arthropod interactions, changes in these traits can affect multiple aspects of a plant–insect interaction, from host finding to host utilization. Thus, direct induced resistance can be manifested in many ways that depend not only on the plant–insect interaction in view but also on the experimental methods used to investigate the interaction.

Direct induced resistance has been envisioned by some as the parallel of systemic acquired resistance (SAR), a long-lasting, broad-spectrum resistance to pathogens that develops in response to attack by necrosis-inducing pathogens (Vallad & Goodman, 2004). Indeed, direct induced resistance can, like SAR, be systemic in its extent and broad spectrum and long lasting in its effects. In addition, SAR and direct induced resistance are, in many cases, controlled by parallel signalling pathways, the SA-mediated pathway (SAR) and JA-mediated pathway (direct induced resistance). However, the term direct induced resistance as it is used in the current literature appears to encompass a broader range of phenomena than SAR. Many of these phenomena are localized rather than systemic or transient rather than long-lasting. More importantly, the signalling dichotomy foundational to this putative parallel may not be absolute; as already noted, many piercing–sucking insects induce the SA pathway, and many pathogens induce the JA pathway.

Direct induced resistance can be manifested as interference with behaviours associated with host location or oviposition. Bernasconi *et al.* (1998), for example, presented evidence that the corn leaf aphid, *Rhopalosiphum maidis*, was repelled by odours emitted from maize plants damaged and treated with caterpillar regurgitant, and also showed that aphids preferred untreated plants to plants treated with regurgitant in a field choice test. Release of volatile organic compounds from wild tobacco (*N. attenuata*) following herbivory was associated with an approximately threefold reduction in the oviposition rate of *Manduca quinquemaculata* (Kessler & Baldwin, 2001). Similarly, injury to tobacco plants (*N. tabacum*) caused by *Heliothis virescens* larvae resulted in the emission of several nocturnal volatiles, repellence of conspecific female moths, and reduction in oviposition (De Moraes *et al.*, 2001).

Induction may also interfere with aspects of herbivore feeding behaviour. Herbivores often exhibit a reduced preference for leaves from previously damaged plants. In both wild radish and black mustard, for example, feeding by larvae of the genus *Pieris* resulted in reduced amounts of leaf area consumed by subsequent caterpillars. In both cases, reduced feeding was correlated with increased production of trichomes and glucosinolates (Agrawal, 1999; Traw & Dawson, 2002). These changes in feeding preference may alter patterns of herbivory on previously damaged plants. In birch, *Betula pendula*, artificial damage of leaves reduced subsequent damage by grazing insects and resulted in a greater dispersion of herbivore feeding throughout the canopy (Silkstone, 1987). Similarly, activation of induced responses in wild-type tomato plants resulted in greater dispersion of caterpillar feeding damage when compared with damage on mutant tomato plants unable to mount an induced response (Rodriguez-Saona & Thaler, 2005). These changes in behaviour may have a direct impact on the efficiency with which herbivores use their host plants; in addition, these changes in behaviour may have an indirect impact on herbivore performance by increasing the efficiency of natural enemies (indirect induced defence; see below).

Induced resistance is often manifested post-ingestionally as a decline in arthropod 'performance', by which is meant reductions in growth, fecundity, survival and other such indicators. Here, a distinction has been made between those plant characters that exercise their effects by interacting with targets in the body proper of the arthropod and often dramatically reduce the performance of non-adapted insects that ingest them (toxins) and those plant characters that interfere with digestion, nutrient acquisition or nutrient utilization and often merely slow the growth of arthropods (anti-digestive or anti-nutritive chemicals). Prominent examples of herbivore-inducible toxins include nicotine and cardiac glycosides. Nicotine, which is induced in *Nicotiana* spp. by chewing herbivory, is an agonist of certain cholinergic synapses and is thus toxic to a wide variety of animals. High levels of nicotine such as those found in induced tobacco plants correlate with reduced growth of insects adapted to feeding on nicotine-containing plants, and with reduced survivorship in insects not adapted to nicotine (Voelckel *et al.*, 2001; Wink & Theile, 2002). Cardiac glycosides, which are inducible by chewing and sucking herbivores in species of the genus *Asclepias*, are inhibitors of the Na^+/K^+ ATPase pumps in animals (Zalucki *et al.*, 2001; Martel & Malcolm, 2004). High levels of cardenolides were associated with increased mortality of first-instar monarch butterflies (*Danaus plexippus*), a specialist on *Asclepias* sp. (Zalucki *et al.*, 2001).

The paradigmatic examples of inducible secondary chemicals with anti-nutritive or anti-digestive effects are the protease inhibitors (PIs) found in various plant species (Lawrence & Koundal, 2002). Several classes of these inducible proteins have been identified that competitively inhibit the proteolytic activity of the various classes of proteases found in insect guts. The presence of PIs in plants may reduce arthropod growth directly, by reducing the digestion of dietary protein, or indirectly, by creating deficiencies in amino acids in the arthropods that feed on them (Duffey & Stout, 1996; Jongsma & Bolter, 1997; Lawrence & Koundal, 2002). Induction of PIs is correlated with the reduced performance of herbivores in numerous plant–insect systems, the best studied of which are tomato and tobacco.

Another example of an induced response leading to anti-nutritive or anti-digestive effects in an herbivore was recently provided by Pechan *et al.* (2002). These authors demonstrated that a cysteine protease induced in maize leaf tissue within 1 hour of feeding by Lepidopteran larvae was correlated with an approximately 74% reduction in growth of *Spodoptera frugiperda* larvae. Electron microscopy revealed that consumption of leaf tissue containing high levels of the protease resulted in severe damage to peritrophic membrane of larvae. Because the peritrophic membrane likely performs several functions related to digestion in insects, damage to the peritrophic membrane provides a likely explanation for the reduction in growth of larvae feeding on tissues with elevated levels of the cysteine protease.

5.3.3 Indirect induced resistance

Other changes induced in plants by herbivory do not affect subsequent herbivores directly, but rather affect them indirectly by enhancing the effectiveness of carnivorous natural enemies of the herbivore. The best-studied form of indirect induced resistance involves the release by damaged plants of volatile organic compounds that attract predators and parasitoids of the herbivore (Dicke *et al.*, 2003). In addition, induction of morphological structures of importance to predators and parasitoids has been reported, and changes in the feeding behaviour of herbivores on induced plants may increase the efficacy of natural enemies.

The induction of volatile compounds following herbivory is similar to the induction of non-volatile secondary chemicals by herbivory (Paré *et al.*, 1999). The blend of volatiles released following damage can be complex and usually differs both quantitatively and qualitatively from volatile blends released constitutively. The volatiles produced by damaged plants are derived from several biosynthetic pathways, principally the mevalonate, lipoxygenase and shikimic acid pathways. Release of volatiles from damaged plants can occur not only from the site of feeding but also from undamaged portions of damaged plants. Finally, the blends released from damaged plants differ depending on the type of herbivory.

The release of volatile compounds from damaged plants has been shown to increase the attractiveness of the plants to predators and parasitoids of the damaging herbivores. Evidence for increased attractiveness comes largely from laboratory assays using wind tunnels or olfactometers. In a study with cabbage, *Pieris brassicae*, a wind tunnel was used to show that parasitoids (*Cotesia glomerata*) were two to eight times more likely to fly towards damaged cabbage plants than undamaged cabbage plants (Mattiacci *et al.*, 2001a). Olfactometers have been used to demonstrate the attraction of predatory mites to herbivorous mites feeding on several plant species (Dicke *et al.*, 2003). In addition, a few studies have shown that herbivore-induced release of volatiles and attraction of predators and parasitoids occur under ecologically realistic conditions. For example, Thaler (1999) showed that parasitism of *Spodoptera exigua* larvae by *Hyposoter exiguae* wasps was 37% greater on field-grown tomato plants that had been induced by treating them with JA than on control plants.

In some plant–herbivore–parasitoid systems, the volatiles released by damaged plants appear to contain a large amount of information. Closely related herbivores and even different life stages of the same insect species can induce blends of volatiles that are distinguishable by natural enemies. De Moraes *et al.* (1998) showed, under field conditions, that the parasitic wasp *Cardiochiles nigriceps* visited tobacco and cotton plants damaged by a host caterpillar (*Heliothis virescens*) more than they visited tobacco and cotton plants damaged by a non-host caterpillar (*Helicoverpa zea*); these authors also showed that the blend of volatiles released systemically by cotton, tobacco and maize plants differed following herbivory by *H. virescens* and *H. zea*. Systemic emission of volatiles in *Brassica oleracea* cv. *gemmifera* occurs only after prolonged feeding, and only if systemic portions of the plant receive damage in addition to the initial, inducing damage; furthermore, volatile emission ceases within one day if additional damage is not received. These features of volatile induction in *Bioleracea* ensure that volatile signals are produced only when damage is relatively severe (Mattiacci *et al.*, 2001b).

Soybean plants damaged by the stink bug *Euschistus heros* and the caterpillar *Anticarsia gemmatalis* exemplify several of the features of herbivore-induced volatile production outlined above (Moraes *et al.*, 2005). Plants damaged by both *E. heros* and *A. gemmatalis* showed significant increases in total volatile production. The blends of volatiles released from plants damaged by bugs differed qualitatively from the blends released by undamaged plants. Moreover, volatile production differed among plants damaged by *E. heros* and *A. gemmatalis*, and even differed with the sex and life stage of *E. heros* used to damage plants. The egg parasitoid *Telenomus podisi* showed a significant preference in an olfactometer for odours from soybeans damaged by adult and nymphal stink bugs when tested against undamaged plants. Moreover, there also appeared to be specificity in response to different types of damage, as odours from plants damaged by *A. gemmatalis*, which is not a host for *T. podisi*, did not attract the parasitoid.

5.3.4 Plant stress-induced resistance

Direct and indirect induced resistance are thought to be triggered when plants recognize an initial attack as an indicator of increased risk of future attack (Karban *et al.*, 1999). Consistent with this hypothesis, induced resistance is often induced by levels of herbivory too low to affect plant growth or fitness, thus enabling a plant to increase its phenotypic commitment to resistance *before* herbivory increases to damaging levels. However, induced resistance has also been shown following moderate to high levels of herbivory. In such cases, tissue loss or physiological stress caused by herbivory may lead to changes in plant biochemistry, physiology or morphology in addition to those changes that result from activation of signalling pathways by low levels of herbivory. These stress-induced changes can have consequences for subsequent herbivores that are distinct from the consequences of activating resistance-related response pathways.

Tissue removal by herbivores often leads to reductions in the quantity or quality of plant resources for subsequent herbivores. Phloem-feeding insects that alter source–sink relationships and stem-girdling insects that disrupt photo-assimilate transport in the host plant are examples of herbivores that can change the nutritive quality of the host for subsequent herbivores by changing patterns of resource allocation within the plant. Herbivore-induced decreases in phloem amino nitrogen may explain, for example, negative interactions between spatially and temporally separated planthoppers in rice and cordgrass (Ferrenberg & Denno, 2003; Matsumura & Suzuki, 2003). In birch, Johnson *et al.* (2002) found that physical disruption of the midribs of birch leaves by leafminers (*Eriocrania* spp.) reduced survivorship of the aphid *Euceraphis betulae* on damaged leaves, probably because leafminer damage disrupted phloem hydraulics.

High levels of herbivory may also impose physiological stress on plants, resulting in stress-related changes in gene expression and secondary metabolism and in turn to changes in plant quality for subsequent herbivores. Herbivore-induced water and nutrient stress may, for example, underlie some of the effects of root herbivory on above-ground herbivores, although Bezemer *et al.* (2003) recently provided evidence that root feeders can sometimes activate expression of defence-related responses in some plants.

5.3.5 Tolerance

Plant tolerance refers to the ability of some plants to sustain tissue loss without losses in fitness or yield (Stowe *et al.*, 2000). In contrast with other types of plant resistance, expression of tolerance by a plant does not result in a reduction in the amount of tissue lost by the plant or in a reduction in insect performance. The plant traits responsible for reducing the fitness consequences of injury by herbivores are not well understood, but some of the physiological mechanisms thought to be responsible for plant tolerance are only activated or expressed following herbivory and in this sense are inducible (Tiffin, 2000). Notably, many plants respond to defoliation by increasing rates of photosynthesis or nutrient uptake in remaining tissues, or by altering patterns of resource allocation in the plant. However, it is unclear whether these changes which occur following herbivory render the plant more tolerant of subsequent bouts of herbivory, or whether they merely represent the plant's attempt to minimize the fitness effects of the initial (inducing) damage.

5.3.6 Interplant signalling

Interest in the idea that damaged plants may emit volatiles that induce resistance in neighbouring plants has revived in recent years, as part of a growing awareness of the importance of volatiles in mediating ecological interactions. Dolch and Tscharntke (2000) showed that manual defoliation of alders reduced subsequent herbivory on neighbouring, undamaged alders, with the amount of herbivory increasing as distance from the defoliated trees increased. Airborne volatile compounds appeared to be at least partly responsible for interplant transfer of resistance (Tscharntke *et al.*, 2001). Similarly, Karban *et al.* (2000) showed that wild tobacco plants near clipped sagebrush, which produces large amounts of methyl jasmonate, experienced reduced levels of herbivory relative to plants not next to sagebrush. However, tobacco plants were within 15 cm of sagebrush, a fact that raises questions about the scale at which interplant communication may operate. Very recent research in maize (reviewed in Turlings and Ton, 2006) lends support to the idea that herbivore-induced plant volatiles may affect surrounding plants by 'priming' them, such that induction in plants previously exposed to volatiles is stronger and quicker than in plants not previously exposed to volatiles.

5.3.7 Concurrent expression of multiple types of induced resistance

The fact that the types of induced resistance discussed above are distinguishable does not mean that they are exclusive; in fact, co-regulation, simultaneous deployment and concerted action of these various types of induced resistance are probably the norm. Three recent examples will suffice. Tobacco plants damaged by tobacco budworm emit a blend of volatiles during the day that attracts a parasitic wasp, and they emit a different blend of volatiles at night that repels oviposition by budworm females (De Moraes *et al.*, 1998, 2001). Thus, the release of volatiles induced by budworm feeding is apparently coordinated to maximize the benefits received from expression of direct and indirect resistance mechanisms. Wild tobacco plants (*N. attenuata*) damaged by three herbivores in the field emitted a similar blend of induced volatiles (Kessler & Baldwin, 2001). Simulating this volatile emission in undamaged plants (by using pure compounds) resulted in an estimated 92–95% reduction in herbivory by *Manduca quinquemaculata*. Importantly, the authors demonstrated that this reduction was attributable to both increased predation of *M. quinquemaculata* eggs (corresponding to indirect induced resistance) and to decreased oviposition by *M. quinquemaculata* (corresponding to direct induced resistance). Thus, overall induced resistance was due to the concerted action of both direct and indirect types. In tomato, mutant plants deficient in their ability to produce jasmonic acid were also compromised in their ability to express both direct and indirect induced resistance, demonstrating the co-regulation of these two types of induced resistance in tomato (Thaler *et al.*, 2002).

5.4 Establishing the causal basis of induced resistance

The changes in resistance that occur in plants following herbivory can be said, in a general sense, to be caused by the biochemical, physiological and morphological changes that also occur following herbivory. Difficulties arise, however, when attempting to assign causal roles to specific changes in plant biochemistry or morphology and when attempting to understand the importance of a particular biochemical or morphological change relative to other changes. These difficulties exist because induced resistance, like constitutive resistance, typically has an extremely complex causal basis (Duffey & Stout, 1996). Understanding exactly how changes in plant biochemistry and morphology translate into induced resistance remains a major challenge for further studies (Baldwin, 2001).

5.4.1 The complex causal basis of induced resistance

One factor that hinders elucidation of the causal basis of induced resistance is the large number of biochemical changes that occur in damaged plants. A hint of the biochemical extent of induction is provided by studies of gene expression in plants attacked by herbivorous arthropods. Microarray studies using *Arabidopsis* and *N. attenuata* indicate that herbivory by both chewing and sucking insects results in changes in the expression of hundreds of genes, some of which are up-regulated and some of which are down-regulated (Baldwin, 2001; Hermsmeier *et al.*, 2001; Moran *et al.*, 2002; Roda & Baldwin, 2003). Of course, not all of these changes in gene expression translate directly into changes in resistance-related traits; nonetheless, these studies do demonstrate that herbivory causes a comprehensive transcriptional and biochemical reorganization in plants. More direct evidence for the comprehensiveness of induced responses comes from the limited number of biochemical screens of induced plants that have been conducted. In wild tobacco, treatment with jasmonic acid results in the induction of at least eight secondary compounds in three chemical classes (phenolics, alkaloids and terpenoids) as well as increases in activities of proteinaceous PIs and levels of foliar nitrogen and protein (Baldwin, 2001; Keinänen *et al.*, 2001). In tomato leaves, feeding by chewing herbivores such as *Helicoverpazea* induces simultaneous alterations in the levels of at least nine secondary chemicals and proteins with established roles in plant resistance, including increases in the activities of PIs and three oxidative enzymes as well as profound shifts in phenolic metabolism (Stout *et al.*, 1998; M.J. Stout, unpublished data). Many other compounds with less defined roles in plant resistance are also induced by herbivory in tomato (Walling, 2000).

Another impediment to elucidating the causal roles of specific induced traits arises from the interactions that occur among resistance-related traits in plants. Induced morphological and biochemical traits may interact in additive or synergistic fashion, or they may counteract or inhibit one another. In wild parsnip, *Pastinaca sativa*, several furanocoumarins and the methylenedioxyphenyl compound myristicin, all of which are induced by mechanical damage, have been shown to have synergistic effects on specialist and generalist Lepidopteran herbivores (Berenbaum & Neal, 1985; Berenbaum & Zangerl, 1993; Zangerl *et al.*, 1997). Polyphenol oxidase, which is induced in tomato leaves following chewing herbivory, irreversibly interacts with and thereby reduces the activity of PIs, which are also induced in tomato following chewing herbivory (Duffey & Stout, 1996). More generally, changes in levels of plant protein, which commonly occur following herbivory, can obscure the causal role played by other inducible metabolites, because protein levels influence the toxic or growth-reducing qualities of many secondary metabolites.

Temporal and spatial heterogeneity in the expression of induced responses also complicates the elucidation of cause–effect relationships in induced resistance. Different inducible traits may exhibit different temporal patterns of induction or relaxation following damage (Laue *et al.*, 2000), and spatial patterns of induction may likewise differ for different traits. In tomato, direct induced resistance appears to be activated more rapidly than indirect induced resistance following spider mite herbivory (Kant *et al.*, 2004), and different components of the induced response to chewing herbivory (e.g. PIs, polyphenol oxidase, peroxidase and lipoxygenase) exhibit unique patterns of spatial expression (Stout *et al.*, 1998). As a consequence of this spatial and temporal heterogeneity, the contributions of specific induced traits to induced resistance undoubtedly vary with spatial and temporal distance from the site of damage.

Finally, the contribution of a specific induced trait to induced resistance also depends upon the constitutive background in which the induced trait is expressed. A demonstration of this principle (albeit using constitutively expressed metabolites) was recently provided by De Leo *et al.* (2001). These authors showed that constitutive expression of the same MTI-2 PI at approximately the same level in different plants (constitutive backgrounds) had different effects on insect growth and survivorship. Also, in birch leaves, (uncharacterized) responses to previous damage altered the behaviour of *Epirrita autumnata* in ways that made the larvae more vulnerable to predation, but the magnitude of this effect depended on plant architectural complexity, a constitutive plant trait (Kaitaniemi *et al.*, 2004).

Thus, the causal relationship between changes in resistance-related plant traits and induced resistance is not straightforward, because plant resistance in general, and induced resistance in particular, is an emergent property of a plant that results from the combined action of multiple biochemical, morphological and physicochemical traits that interact with one another and that are expressed heterogeneously in space and time. Not every biochemical or morphological change that occurs as a result of herbivory will contribute equally, if at all, to induced resistance; moreover, the contribution that a specific induced trait makes to induced resistance will depend on the context in which that trait is expressed.

5.4.2 Approaches to understanding the causal basis of induced resistance

Given this complexity, what experimental approaches can be used to implicate an induced trait in induced resistance? Historically, much of the evidence for the roles of particular induced traits in induced resistance has been provided by correlations. Correlations can be valuable, particularly when studies are carefully and thoroughly done and when backed by pharmacological evidence for the toxic or growth-reducing properties of the trait in question. However, correlations are not sufficient to definitively establish the causal role of a specific biochemical or morphological change, as they cannot exclude the possibility of spurious correlations arising from induction of multiple, interacting plant traits.

A study by Pohlon & Baldwin (2001) nicely illustrates both the value and limitations of the correlational approach. These authors investigated the relationship between expression of resistance and expression of putative resistance-related traits in wild tobacco by harvesting and flash-freezing leaf material from induced plants every day for five days following induction, and incorporating leaf material into artificial diets for bioassays with

Manduca sexta. This technique allowed them to 'capture' the temporal dynamics of induction, translating a dynamic interaction into a series of static ones. They found the greatest reduction in growth of larvae in insects reared on foliage harvested one day after induction, but reductions in larval growth were also found in larvae reared on diets from foliage harvested two to five days after induction. Interestingly, no significant induction of nicotine and PIs had occurred by 24 hours, suggesting that unmeasured chemical changes were responsible for induced resistance on the first day after induction. From two to five days after induction, a significant correlation was found between PI and nicotine induction and reduction in larval growth. Thus, this study lends support to the hypothesis that the combination of PIs and nicotine is responsible for induced resistance at some time points following induction but also demonstrates that other, unmeasured, factors may cause reductions in growth.

Correlative evidence linking specific induced biochemical and morphological traits to increased resistance has been supplemented in a few plant–arthropod systems by various other types of evidence. Chemical elicitors of induced resistance have been used to uncouple activation of induced biochemical responses from loss of tissue and from other factors (e.g. salival elicitors) that are sometimes associated with actual herbivory. Inhibitors have also been used to suppress induced responses and induced resistance in plants. Elicitors and inhibitors have probably been used to greatest effect in *Nicotiana* spp. and in tomato. In a study with the latter species (Stout *et al.*, 1998), the biochemical responses in leaves to feeding by chewing herbivores (e.g. induction of PIs and polyphenol oxidase) were simulated by exposing plants to methyl jasmonate. Plants so exposed were poorer sources of food for *S. exigua* larvae as indicated by lower larval growth rates. Inhibiting the biochemical responses of leaves to methyl jasmonate by pre-treating plants with SA, an inhibitor of the JA pathway, also inhibited the induction of resistance. Similarly, application of acetylsalicylic acid to leaves wounded by caterpillars inhibited induction of polyphenol oxidase and PIs and also inhibited the induction of resistance. Treatment of wounded, acetylsalicylic acid-treated plants with JA restored induction of biochemical responses and resistance. The consistent association of induced PIs and polyphenol oxidase with resistance under various experimental conditions argues strongly for their collective role in induced resistance. On the other hand, induction of the activities of two other enzymes, peroxidase and lipoxygenase, by various treatments was not consistently associated with induced resistance in tomato (Stout *et al.*, 1998).

Mutant plant lines have been used to investigate the causal basis of induced resistance in a few systems. Li *et al.* (2002) used a mutant tomato line compromised in its ability to biosynthesize jasmonic acid to show that the feeding and fecundity of spider mites, *Tetranychus urticae*, were increased on plants unable to accumulate proteinase inhibitors and other biochemical traits regulated by the octadecanoid pathway. Treatment of plants with jasmonic acid restored the resistance. Similarly, an Arabidopsis mutant deficient in linolenic acid, the fatty acid precursor to jasmonic acid, was extremely vulnerable to a fungal gnat to which wild-type plants were almost completely resistant (McConn *et al.*, 1997). Again, exogenous jasmonate restored resistance to the insect.

Increasing use is also being made of modern molecular genetic tools to investigate cause–effect relationships in induced resistance (Roda & Baldwin, 2003). One conceptually simple approach involves the transfer of a single inducible gene from the plant in which it naturally occurs and expressing the gene constitutively in another plant. PIs, because they are direct products of single genes, are especially suited for this approach, and it has now been used several times (Jongsma & Bolter, 1997). In some cases, expression of a PI in a new host rendered the host more resistant to herbivores, whereas in other cases it did not. For example, leaves from poplar plants transformed with a gene encoding oryzacystatin, a cysteine PI, were less suitable as a food source for *Chrysomela tremulae* (Jongsma & Bolter, 1997), but potato plants transformed with a gene encoding oryzacystatin were no less suitable for Colorado potato beetles. In the latter case, however, beetles did show a decline in the efficiency with which they utilized foliage for food (Cloutier *et al.*, 2000).

Anti-sense suppression of genes involved in defence signalling or biosynthesis of secondary metabolites has also been used to investigate the causal basis of induced resistance in tobacco and tomato. *Nicotiana sylvestris* plants engineered to express an enzyme involved in the biosynthesis of nicotine, putrescine *N*-methyltransferase, in an anti-sense orientation were compromised in their ability to accumulate nicotine following treatment with methyl jasmonate and were also less resistant to *M. sexta* (Voelckel *et al.*, 2001). Larval mass after eight days of feeding was approximately four times greater on anti-sense than on wild-type plants. Expression of a prosystemin anti-sense gene in tomato plants interfered with the induction of PIs following herbivory and reduced the resistance of the transgenic plants to *M. sexta* (Orozco-Cardenas *et al.*, 1993). Transgenic plants supported growth rates of larvae that were approximately three times higher than growth rates on control plants. Finally, in one of a very few studies conducted in the field, anti-sense suppression in *N. attenuata* of a lipoxygenase gene involved in the biosynthesis of jasmonic acid interfered with the induction of nicotine, PIs and volatile sesquiterpenoids (Kessler *et al.*, 2004). The decreased production of the inducible metabolites was associated with increased susceptibility to insect herbivores: *M. sexta* caterpillars were 4.4-fold heavier after nine days of feeding on anti-sense plants relative to control plants, overall levels of natural herbivory were approximately seven times higher on anti-sense plants than on control plants, and anti-sense plants were susceptible to a leafhopper that was never observed feeding on wild-type plants.

None of the alternative approaches used thus far yield unambiguous results. There are two primary sources of this ambiguity, both inevitable consequences of the complex causal basis of plant resistance. First, attempts to manipulate the expression of a specific trait often also affect the context in which the trait is expressed, and changing the identity and concentrations of co-occurring primary and secondary metabolites is bound to affect the activity of the trait in question. The constitutive expression of an inducible trait in a plant in which it is not normally expressed represents an extreme example of this problem. Second, attempts to alter the expression of a trait by manipulating biochemical or signal transduction pathways often affect the expression of traits in addition to the one under study, and thus changes in resistance cannot be unambiguously assigned to the trait in question. The insufficiencies associated with the correlative and manipulative approaches used thus far to investigate the causal basis of induced resistance strongly argue for the idea that evidence for the causal role of a specific trait is best obtained using a combination of approaches (Duffey & Stout, 1996).

5.5 Arthropods as dynamic participants in plant–arthropod interactions

Herbivores as well as plants are dynamic organisms. Accordingly, herbivores often respond to changes in their host plants by altering aspects of their own biology in ways that mitigate the effects of plant responses. Dynamism on the part of arthropod herbivores can alter outcomes of plant–insect interactions and can further obscure cause–effect relationships in induced resistance.

Some of these countermeasures consist of relatively simple changes in insect behaviour. Deterioration of plant quality caused by induction of resistance-related plant traits can cause arthropods to consume more plant tissue (compensatory feeding) or to adjust their feeding sites on the induced plants. Compensatory feeding was observed, for example, in Colorado potato beetles feeding on potato foliage transformed to express a cysteine PI. Beetles on transformed plants consumed 2½ times more leaf tissue on transgenic plants than on control plants, and, probably as a result of this compensatory feeding, beetles feeding on transformed plants suffered no reductions in survival, growth, or reproduction (Cloutier *et al.*, 2000). Incidentally, although these behavioural responses may ameliorate the direct effects of the induced response on the herbivore, they may ultimately benefit the plant if they result in greater exposure of herbivores to predators and parasitoids (i.e. the increased movement of herbivores may contribute to indirect induced resistance).

Arthropods may also respond to the induced responses of their host plants by making physiological adjustments. Probably the best-studied examples of a physiological countermeasure are the responses observed in many Lepidopteran and Coleopteran insects following consumption of PIs. Several species of specialist and generalist herbivores respond to ingestion of PIs by rapidly producing new types of proteinases in their guts that are less sensitive to (i.e. inhibited to a lesser degree by) the PIs that induced them. In the same study in which adult Colorado potato beetles exhibited compensatory feeding behaviour on transgenic, PI-expressing potato plants (Cloutier *et al.*, 2000), adult beetles that had fed on PI-expressing plants also showed an approximately threefold reduction in the sensitivity of gut proteases to inhibition. Similarly, arthropods may also respond to high levels of allelochemicals in their diet by producing greater amounts of detoxicative enzymes such as esterases and cytochrome P450s (Li *et al.*, 2002).

There are undoubtedly many other ways by which herbivores adapt to the induced responses of their host plants. The behavioural responses of arthropods to the constitutive traits of their plants can be very sophisticated, as evidenced by the trenching behaviour of monarch butterfly larvae (*Danaus plexippus*) on latex-producing plants (Zalucki *et al.*, 2001). Analogous sophisticated behavioural responses to induced responses are expected. Moreover, recent evidence demonstrates that insects are capable of responding physiologically not only to the deleterious end products of signal transduction pathways (e.g. PIs, nicotine), but also to the components of the signalling pathways themselves. Li *et al.* (2002) showed that supplementing the diets of fifth-instar *Helicoverpa zea* larvae with physiologically realistic concentrations of JA or SA induced the expression of four cytochrome P450 genes in the insects to levels similar to those induced by ingestion of several plant allelochemicals. The ability of this species to increase the expression of its detoxicative enzymes in response to plant signalling molecules may render them more resistant to plant secondary chemicals even before they are induced in the plant.

5.6 Conclusions

Plant resistance to an herbivore is the outcome of a complex and dynamic interaction. Plants possess a great variety of traits that enable them to reduce, evade or minimize the damage caused by herbivores. Expression of these traits affects multiple and various aspects of the interactions between plants and herbivorous arthropods. Importantly, many of the resistance-related traits of plants are inducible by herbivory; that is, they are expressed at a higher level or to a greater degree as a result of prior herbivory. Several types of induced resistance to insects can be distinguished, including the hypersensitive response, direct induced resistance, indirect induced resistance and plant-stress induced resistance. Induced resistance of all types has a complex causal basis, because the changes in plant biochemistry, physiology and morphology that underlie induced resistance are exceedingly complex. This complexity inheres not only in the large number of biochemical and morphological changes induced by herbivory, but also in the spatial and temporal complexity of these changes, in the interactions that occur among induced plant traits, and in the context dependence of the biological activity of most plant traits. The capacity of insects to adapt to the induced responses of their host plants adds an additional layer of complexity. Investigations of the causal basis of induced resistance should employ a combination of approaches to account for the complex causal basis of induced resistance.

Future investigations of induced resistance to arthropods should seek a more holistic understanding of the phenomenon. For example, the relative importance of inducible and non-inducible plant traits to overall plant resistance to arthropods is a question that requires more attention. Moreover, the experiments needed to fully incorporate the concept of tolerance into the theoretical and experimental literature on induced resistance have not yet been performed. Finally, the ways in which plants coordinate and consolidate their various defence strategies into an integrated whole, capable of dealing simultaneously in an effective manner with multiple biotic and abiotic stresses, are not understood. Answers to these questions will require cooperation between plant pathologists, ecologists, entomologists, plant physiologists and molecular biologists, and will require the use of a variety of experimental approaches in a greater variety of model systems.

5.7 References

Agrawal AA, 1999. Induced responses to herbivory in wild radish: effects on several herbivores and plant fitness. *Ecology* **80,** 1713–1723.

Baldwin IT, 2001. An ecologically motivated analysis of plant–herbivore interactions in native tobacco. *Plant Physiology* **127,** 1449–1458.

Baldwin IT, Preston CA, 1999. The eco-physiological complexity of plant responses to insect herbivores. *Planta* **208,** 137–145.

Berenbaum M, Neal JJ, 1985. Synergism between myristicin and xanthotoxin, a naturally cooccuring plant toxicant. *Journal of Chemical Ecology* **11,** 1349–1358.

Berenbaum MR, Zangerl AR, 1993. Furanocoumarin metabolism in *Papilio polyxenes*: biochemistry, genetic variability, and ecological significance. *Oecologia* **95,** 370–375.

Bernasconi ML, Turlings TCJ, Ambrosetti L, Bassetti P, Dorn S, 1998. Herbivore-induced emissions of maize volatiles repels the corn leaf aphid, *Rhopalosiphum maidis*. *Entomologia Experimentalis et Applicata* **87,** 133–142.

Bernays EA, Chapman RF, 1994. *Host-Plant Selection by Phytophagous Insects*. New York: Chapman & Hall.

Bezemer TM, Wagenaar R, van Dam NM, Wäckers FL, 2003. Interactions between above- and below ground insect herbivores as mediated by the plant defense system. *Oikos* **101,** 555–562.

Cloutier C, Jean C, Fournier M, Yelle S, Michaud D, 2000. Adult Colorado potato beetles compensate for nutritional stress on oryzacystatin I-transgenic potato plants by hypertrophic behavior and over production of insensitive proteases. *Archives of Insect Biochemistry and Physiology* **44,** 69–81.

Cooper WC, Jia L, Goggin FL, 2004. Acquired and R-gene-mediated resistance against the potato aphid in tomato. *Journal of Chemical Ecology* **30,** 2527–2530.

Dangl JL, Jones JDG, 2001. Plant pathogens and integrated defence responses to infection. *Nature* **411,** 826–833.

De Leo F, Bonadé-Bottino M, Ceci LR, Gallerani R, Jouanin L, 2001. Effects of a mustard trypsin inhibitor expressed in different plants on three lepidopteran pests. *Insect Biochemistry and Molecular Biology* **31,** 593–602.

De Moraes CM, Lewis WJ, Paré PW, Alborn HT, Tumlinson JH, 1998. Herbivore-infested plants selectively attract parasitoids. *Nature* **393,** 570–573.

De Moraes CM, Mescher MC, Tumlinson JH, 2001. Caterpillar-induced nocturnal plant volatiles repel conspecific females. *Nature* **410,** 577–580.

Dicke M, van Poecke RMP, de Boer JG, 2003. Inducible indirect defence of plants: from mechanisms to ecological functions. *Basic and Applied Ecology* **4,** 27–42.

Dolch R, Tscharntke T, 2000. Defoliation of alders (*Alnus glutinosa*) affects herbivory by leaf beetles on undamaged neighbors. *Oecologia* **125,** 504–511.

Duffey SS, Stout MJ, 1996. Antinutritive and toxic components of plant defense against insects. *Archives of Insect Biochemistry and Physiology* **32,** 3–37.

Fernandes GW, Negreiros D, 2001. The occurrence and effectiveness of hypersensitive reaction against galling herbivores across host taxa. *Ecological Entomology* **26,** 46–55.

Ferrenberg SM, Denno RF, 2003. Competition as a factor underlying the abundance of an uncommon phytophagous insect, the salt marsh plant hopper *Delphacodes penedetecta, Ecological Entomology* **28,** 58–66.

Fidantsef AL, Stout MJ, Thaler JS, Duffey SS, Bostock RM, 1999. Signal interactions in pathogen and insect attack: expression of lipoxygenase, proteinase inhibitor II, and pathogenesis-related protein P4 in the tomato, *Lycopersicon esculentum. Physiological and Molecular Plant Pathology* **54,** 97–114.

Gatehouse JA, 2002. Plant resistance towards insect herbivores: a dynamic interaction. *New Phytologist* **156,** 145–169.

Grinberg M, Perl-Treves R, Palevsky E, Shomer I, Soroker V, 2005. Interaction between cucumber plants and the broad mite, *Polyphagotarsonemus latus*: from damage to defense gene expression. *Entomologia Experimentalis et Applicata* **115,** 135–144.

Grover Jr PB, 1995. Hypersensitive response of wheat to the Hessian fly. *Entomologia Experimentalis et Applicata* **74,** 283–294.

Hahn MG, 1996. Microbial elicitors and their receptors in plants. *Annual Review of Phytopathology* **34,** 387–412.

Hermsmeier D, Schittko U, Baldwin IT, 2001. Molecular interactions between the specialist herbivore *Manduca sexta* (Lepidoptera, Sphingidae) and its natural host *Nicotiana attenuata*. I. Large scale changes in the accumulation of growth and defense-related plant mRNAs. *Plant Physiology* **125,** 683–700.

Hilker M, Meiners T, 2002. Induction of plant responses to oviposition and feeding by herbivorous arthropods: a comparison. *Entomologia Experimentalis et Applicata* **104,** 181–192.

Huang J, McAuslane HJ, Nuessly GS, 2003. Resistance in lettuce to *Diabrotica balteata* (Coleoptera: Chrysomelidae): the roles of latex and inducible defense. *Environmental Entomology* **32,** 9–16.

Johnson SN, Mayhew PJ, Douglas AE, Hartley SE, 2002. Insects as leaf engineers: can leaf-miners alter leaf structure for birch aphids? *Functional Ecology* **16,** 575–584.

Jongsma MA, Bolter C, 1997. The adaptation of insects to plant protease inhibitors. *Journal of Insect Physiology* **43,** 885–895.

Kaitaniemi P, Vehviläinen H, Rouhomäki K, 2004. Movement and disappearance of mountain birch defoliators are influenced by the interactive effects of plant architecture and induced resistance. *Ecological Entomology* **29,** 437–446.

Kaloshian I, 2004. Gene-for-gene disease resistance: bridging insect pest and pathogen defense. *Journal of Chemical Ecology* **30,** 2419–2438.

Kant MR, Ament K, Sabelis MW, Haring MA, Schuurink RC, 2004. Differential timing of spider mite-induced direct and indirect defenses in tomato plants. *Plant Physiology* **135,** 483–495.

Karban R, Agrawal AA, Thaler JS, Adler LS, 1999. Induced plant responses and information content about risk of herbivory. *Trends in Ecology and Evolution* **14,** 443–447.

Karban R, Baldwin IT, 1997. *Induced Responses to Herbivory*. Chicago: The University of Chicago Press.

Karban R, Baldwin IT, Baxter KJ, Laue G, Felton GW, 2000. Communication between plants: induced resistance in wild tobacco plants following clipping of neighboring sagebrush. *Oecologia* **125,** 66–71.

Keinänen M, Oldham NJ, Baldwin IT, 2001. Rapid HPLC screening of jasmonate-induced increases in tobacco alkaloids, phenolics, and diterpene glycosides in *Nicotiana attenuata. Journal of Agricultural and Food Chemistry* **49,** 3553–3558.

Kessler A, Baldwin IT, 2001. Defensive function of herbivore-induced plant volatile emissions in nature. *Science* **291,** 2141–2144.

Kessler A, Baldwin IT, 2002. Plant responses to insect herbivory: the emerging molecular analysis. *Annual Review of Plant Biology* **53,** 299–328.

Kessler A, Halitschke R, Baldwin IT, 2004. Silencing the jasmonate cascade: induced plant defenses and insect populations. *Science* **305,** 665–668.

Lam E, Kato N, Lawton M, 2001. Programmed cell death, mitochondria and the plant hypersensitive response. *Nature* **411,** 848–853.

Laue G, Preston CA, Baldwin IT, 2000. Fast track to the trichome: induction of *N*-acyl nornicotines precedes nicotine induction in *Nicotiana repanda*. *Planta* **210,** 510–514.

Lawrence PK, Koundal KR, 2002. Plant protease inhibitors in control of phytophagous insects. *Electronic Journal of Biotechnology* **5,** 93–109.

Li C, Williams MM, Loh Y-T, Lee GI, Howe GA, 2002. Resistance of cultivated tomato to cell content feeding herbivores is regulated by the octadecanoid-signaling pathway. *Plant Physiology* **130,** 494–503.

Li X, Schuler MA, Berenbaum MR, 2002. Jasmonate and salicylate induce expression of herbivore cytochrome P450 genes. *Nature* **419,** 712–715.

Martel JW, Malcolm SB, 2004. Density-dependent reduction and induction of milkweed cardenolides by a sucking insect herbivore. *Journal of Chemical Ecology* **30,** 545–561.

Matsumura M, Suzuki Y, 2003. Direct feeding-induced interactions between two rice planthoppers, *Sogatella furcifera* and *Nilaparvata lugens*: effects on dispersal capability and performance. *Ecological Entomology* **28,** 174–182.

Mattiacci L, Rudelli S, Rocca BA, Genini S, Dorn S, 2001a. Systemically-induced response of cabbage plants against a specialist herbivore, *Pieris brassicae*. *Chemoecology* **11,** 167–173.

Mattiacci L, Rocca BA, Scascighini N, D'Alessandro M, Hern A, Dorn S, 2001b. Systemically induced plant volatiles emitted at the time of 'danger'. *Journal of Chemical Ecology* **27,** 2233–2252.

McConn M, Creelman RA, Bell E, Mullet JE, Browse J, 1997. Jasmonate is essential for insect defense in Arabidopsis. *Proceedings of the National Academy of Sciences USA* **94,** 5473–5477.

Moraes MCB, Laumann R, Sujii ER, Pires C, Borges M, 2005. Induced volatiles in soybean and pigeon pea plants artificially infested with the neotropical brown stink bug, *Euschistus hero*, and their effect on the egg parasitoid, *Telenomus podisi*. *Entomologia Experimentalis et Applicata* **115,** 227–237.

Moran PJ, Cheng Y, Cassell JL, Thompson GA, 2002. Gene expression profiling of *Arabidopsis thaliana* in compatible plant aphid interactions. *Archives of Insect Biochemistry and Physiology* **51,** 182–203.

Musser RO, Cipollini DF, Hum-Musser SM, Williams SA, Brown JK, Felton GW, 2005. Evidence that the caterpillar salivary enzyme glucose oxidase provides herbivore offense in solanaceous plants. *Archives of Insect Biochemistry and Physiology* **58,** 128–137.

Ollerstam O, Rohfritsch O, Höglund S, Larsson S, 2002. A rapid hypersensitive response associated with resistance in the willow *Salix viminalis* against the gall midge *Dasineura marginemtorquens*. *Entomologia Experimentalis et Applicata* **102,** 153–162.

Orozco-Cardenas M, McGurl B, Ryan CA, 1993. Expression of an antisense prosystemin gene in tomato plants reduces resistance toward *Manduca sexta* larvae. *Proceedings of the National Academy of Sciences USA* **90,** 8273–8276.

Paré PW, Lewis WJ, Tumlinson JH, 1999. Induced plant volatiles: biochemistry and effects on parasitoids. In: Agrawal AA, Tuzun S, Bent E, eds. *Induced Plant Defenses against Pathogens and Herbivores*. St. Paul, MN: American Phytopathological Society Press.

Pechan T, Cohen A, Williams WP, Luthe DS, 2002. Insect feeding mobilizes a unique plant defense protease that disrupts the peritrophic matrix of caterpillars. *Proceedings of the National Academy of Sciences USA* **99,** 13319–13323.

Pohlon E, Baldwin IT, 2001. Artificial diets capture the dynamics of jasmonate-induced defenses in plants. *Entomologia Experimentalis et Applicata* **100,** 127–130.

Roda AL, Baldwin IT, 2003. Molecular technology reveals how the induced defenses of plants work. *Basic and Applied Ecology* **4,** 15–26.

Rodriguez-Saona C, Thaler JS, 2005. The jasmonate pathway alters herbivore feeding behaviours: consequences for plant defences. *Entomologia Experimentalis et Applicata* **115,** 125–134.

Silkstone BE, 1987. The consequences of leaf damage for subsequent insect grazing on birch (*Betula* spp.). *Oecologia* **74,** 149–152.

Stout MJ, Thaler JS, Thomma BPHJ, 2006. Plant-mediated interaction between pathogenic microorganisms and herbivorous arthropods. *Annual Review of Entomology* **51,** 663–689.

Stout MJ, Workman KV, Bostock RM, Duffey SS, 1998. Stimulation and attenuation of induced resistance by elicitors and inhibitors of chemical induction in tomato (*Lycopersicon esculentum*) foliage. *Entomologia Experimentalis et Applicata* **86,** 267–279.

Stowe KA, Marquis RJ, Hochwender CG, Stowe EL, 2000. The evolutionary ecology of tolerance to consumer damage. *Annual Review of Ecology and Systematics* **31,** 565–595.

Thaler JS, 1999. Jasmonate-inducible plant defences cause increased parasitism of herbivores. *Nature* **399,** 686–688.

Thaler JS, Farag MA, Paré PW, Dicke M, 2002. Jasmonate-deficient plants have reduced direct and indirect defences against herbivores. *Ecology Letters* **5,** 764–774.

Thomma BPHJ, Tierens KFM, Penninckx IAMA, Mauch-Mani B, Broekaert WF, Cammue BPA, 2001. Different micro-organisms differentially induce Arabidopsis disease response pathways. *Plant Physiology and Biochemistry* **39,** 673–680.

Tiffin P, 2000. Mechanisms of tolerance to herbivore damages: what do we know? *Evolutionary Ecology* **14,** 523–536.

Traw MB, Dawson TE, 2002. Reduced performance of two specialist herbivores (Lepidoptera: Pieridae, Coleoptera: Chrysomelidae) on new leaves of damaged black mustard plants. *Environmental Entomology* **32,** 714–722.

Tscharntke T, Thiessen S, Dolch R, Boland W, 2001. Herbivory, induced resistance, and interplant signal transfer in *Alnus glutinosa. Biochemical Systematics and Ecology* **29,** 1025–1047.

Tumlinson JH, Lait CG, 2005. Biosynthesis of fatty acids amide elicitors of plant volatiles by insect herbivores. *Archives of Insect Biochemistry and Physiology* **58,** 54–68.

Turlings TCJ, Ton J, 2006. Exploiting scents of distress: the prospect of manipulating herbivore-induced plant odours to enhance control of agricultural pests. *Current Opinion in Plant Biology* **9,** 421–427.

Vallad EV, Goodman RM, 2004. Systemic acquired resistance and induced systemic resistance in conventional agriculture. *Crop Science* **44,** 1920–1934.

Voelckel C, Krügel T, Gase K, Heidrich N, van Dam NM, Winz R, Baldwin IT, 2001. Anti-sense expression of putrescine *N*-methyltransferase confirms defensive role of nicotine in *Nicotiana sylvestris* against *Manduca sexta. Chemoecology* **11,** 121–126.

Walling LL, 2000. The myriad plant responses to herbivores. *Journal of Plant Growth Regulation* **19,** 195–216.

Wink M, Theile V, 2002. Alkaloid tolerance in *Manduca sexta* and phylogenetically related sphingids (Lepidoptera: Sphingidae). *Chemoecology* **12,** 29–46.

Zalucki MP, Malcolm SB, Paine TD, Hanlon CC, Brower LP, Clark AR, 2001. It's the first bites that counts: survival of first-instar monarchs on milkweeds. *Austral Ecology* **26,** 547–555.

Zangerl AR, Arntz AM, Berenbaum MR, 1997. Physiological price of an induced chemical defense: photosynthesis, respiration, biosynthesis and growth. *Oecologia* **109,** 433–441.

Chapter 6
Mechanisms of defence to pathogens: biochemistry and physiology

Christophe Garcion, Olivier Lamotte and Jean-Pierre Métraux

Département de Biologie, Université de Fribourg, Switzerland

6.1 Introduction

Plants represent an interesting source of food for many micro-organisms and herbivores. They display the remarkable ability to defend themselves from various invaders. An old, but still valid, distinction is usually made between pre-existing barriers and defences induced by the plant upon perception of the pathogen leading to the concept of preformed and induced resistance. Plants perceive pathogens via various elicitors (see section 6.2) or pathogen-associated molecular patterns, and activate a variety of defence mechanisms leading to a basal level of induced resistance (Ton *et al.*, 2002). A previous exposure of the plant to specific biotic or abiotic stimuli can further amplify this basal resistance; this is referred to as induced resistance. Resistance can be induced locally at the site of infection but also systemically in uninfected parts; this is termed systemic acquired resistance (SAR) or systemic induced resistance (Hammerschmidt *et al.*, 2001; Chapter 1). Plants exhibit a large degree of specificities in regard to the recognition of pathogens. In its most extreme form, this leads to the so-called gene-for-gene resistance, whereby a plant cultivar carrying a given resistance gene is resistant to a pathogen race carrying a specific avirulence gene. The following section will review the major defence reactions that are activated in plants upon attack by pathogens. It attempts to summarize the more recent advances and references and past reviews on the various topics are indicated where appropriate.

6.2 Structural barriers

Many plant pathogens penetrate through cell walls, and a wealth of data have accumulated on the role played by the cuticle and various forms of cell-wall strengthening (for reviews, see Aist, 1976; Vance *et al.*, 1980; Kolattukudy, 1985; Nicholson & Hammerschmidt, 1992; Mendgen *et al.*, 1996; Vorwerk *et al.*, 2004).

6.2.1 Cell wall appositions

Recently, several studies have added to our knowledge on the role of appositions in resistance to pathogens. Such cell-wall alterations are rich in phenolic compounds, appear conspicuously at the site of infection and are often associated with resistance (Nicholson & Hammerschmidt,

1992; Thordal-Christensen *et al.*, 1997; McLusky *et al.*, 1999). The discovery of *pen* mutants of Arabidopsis exhibiting increased penetration by powdery mildew provided new insights on the importance of appositions (Collins *et al.*, 2003). *PEN1* encodes a protein belonging to the family of syntaxins that are characterized by a 60–70 amino acid 'SNARE' motif and function in vesicle trafficking. A similar syntaxin-dependent resistance was also observed in barley *ror2* mutants, and the Arabidopsis *PEN1* can complement this mutation (Collins *et al.*, 2003). Another protein with SNARE domains, HvSNAP34, forms a complex with ROR2 and is required for resistance to penetration, further supporting the importance of targeted vesicle transport (Collins *et al.*, 2003). Furthermore, SYP122 is a close homologue of PEN1 that may have general functions in secretion and cell wall deposition partly overlapping with those of PEN1 (Assaad *et al.*, 2004). Presumably, vesicles might contain materials for wall deposition and/or fungitoxic substances. For example, in onion epidermis cells inoculated with *Botrytis alli,* a localized deposition of feruloyl-3′-methoxytyramine and feruloyl tyramine was observed in what appeared as membrane-bound granules (McLusky *et al.*, 1999). In resistant interactions of barley with powdery mildew (*Blumeria graminis* f. sp. *hordei*), hydrogen peroxide was observed at the site of apposition formation as well as in 2 μm vesicles near the papillae (Hückelhoven *et al.*, 1999). The localized assembly and secretion of appositions are associated with major cellular rearrangements involving elements of the cytoskeleton. For example, anti-microfilament drugs prevent the localized deposition of phenolics in bean leaves inoculated with soybean rust (Perumalla & Heath, 1991) or of autofluorescent compounds in the barley – *Erysiphe pisi* interaction (Kobayashi *et al.*, 1997). In the potato–*Phytophthora infestans* interaction, important cytoplasmic rearrangements could be observed by video-microscopy at the penetration site coinciding with the apposition of fluorescent materials (Freytag *et al.*, 1994). The targeted deposition of phenolic compounds at sites of attempted penetration is associated with a rearrangement of the actin cytoskeleton (McLusky *et al.*, 1999). Constitutive over-expression in single barley cells of the hypothetical actin cytoskeleton regulator CA RACB, a RAC/ROP G-protein, partly inhibited actin reorganization in cells inoculated with *B. graminis*, whereas knockdown of RACB promoted actin focusing. This provides a novel insight in the regulation of actin reorganization and its association with localized apposition of wall materials in host–pathogen interactions (Opalski *et al.*, 2005).

6.2.2 The case of callose deposits

The β-1,3-glucan callose has often been observed to be deposited at/around penetration sites as plugs or plates referred to as papillae, which have been proposed to act as mechanical barriers to penetration (Stone & Clarke, 1992). Various observations have shown callose to be deposited or produced in resistant interactions after inoculation with pathogens (Skalamera & Heath, 1996; Hückelhoven *et al.*, 1999; Soylu *et al.*, 2003, 2004) or pre-treatment with chemicals inducing or potentiating plant resistance mechanisms (Kogel *et al.*, 1994; Zimmerli *et al.*, 2000; Ton *et al.*, 2005). The Arabidopsis *AtGsl5* gene (*glucan synthase-like 5*) encodes a glucan synthase homologous to the catalytic subunit of fungal β-1,3-glucan synthases that partially complements the yeast *fks1* mutant proposed to be involved in β-1,3-glucan synthesis (Ostergaard *et al.*, 2002). A strong correlation with disease resistance to bacterial and oomycete pathogens was shown in the constitutively resistant Arabidopsis *mapk4* mutant where the wild type *MAP kinase 4* gene is transposon-inactivated. *AtGsl5* is strongly expressed in *mapk4*, the callose synthase is highly active, and callose is over-produced supporting a role for callose in resistance (Petersen *et al.*, 2000; Ostergaard *et al.*, 2002). The role of callose deposits as a barrier for pathogen invasion has been recently questioned. In Arabidopsis, another gene, *GLUCAN SYNTHASE-LIKE ISOFORM GSL5/ POWDERY MILDEW RESISTANCE PMR4* was shown to mediate callose synthesis, and mutants with a disrupted *GSL5/PMR4* gene do not produce callose and exhibit enhanced resistance to virulent powdery mildew pathogens (Jacobs *et al.*, 2003; Nishimura *et al.*, 2003). These results support a role for *GSL5/PMR4* in the colonization of the plant by the biotroph rather than in disease resistance. Possibly, callose surrounding infection sites prevents perception of pathogen-derived elicitors for defence, or otherwise seals off the invader against the action of plant antimicrobials. One is now left with two lines of evidence indicating opposing functions of callose. These paradoxical findings might perhaps be reconciled as follows. In the absence of callose, the formation of the hyphal neck sealing off the extra-haustorial matrix is non-functional, thus strongly impairing the survival of the biotroph. In the presence of callose, a proper haustorial complex can be established, but a strong callose deposit might prevent further fungal growth by sealing off the food exchange at the plant–haustorial interface. It cannot be excluded that the deposition of callose in these two processes might perhaps result from the expression of different genes.

6.2.3 Lignification

Given its chemical and mechanical properties, lignin represents a tremendous barrier against pathogens (Lewis, 1999; Humphreys & Chapple, 2002). Defence lignin refers to lignin deposited in response to pathogen invasion (Nicholson & Hammerschmidt, 1992). Such defence lignin can be deposited over the entire wall of the infected cell or group of cells, or only at the infection site. A number of extensive chapters have reviewed the abundant correlative evidence involving induced lignification to defence (Vance *et al.*, 1980; Moersbacher & Mendgen, 2000; Heitefuss, 2001). The defence lignin (or the lignin-like material) deposited in response to biotic or abiotic stress has a different composition to that deposited during development. For example, elicitor-treated suspension cultures of *Picea abies* (L.) Karst release various materials in the medium, of which 35%, w/w is phloroglucinol/HCL reactive. Thioacidolysis and Raney nickel desulfurization indicates the presence of lignin, and the high content in *p*-hydroxyphenyl units indicates a lignin of different composition than that of structural lignin (Lange *et al.*, 1995). In cucurbits, defence lignin is rich in *p*-coumaraldehyde units in contrast to the guaiacyl-syringyl lignin deposited during development (Stange *et al.*, 2001). In Arabidopsis, two genes related to lignin biosynthesis are differently regulated during development and in response to biotic stress. The cinnamoyl-CoA reductase *AtCCR2* is induced during the incompatible interaction with *Xanthomonas campestris* pv. *campestris* but is poorly expressed during development, while the related gene *AtCCR1* is strongly expressed in tissue undergoing lignification (Lauvergeat *et al.*, 2001). The control of lignin production during development and during infection might therefore be controlled by different signal transduction pathways. Interestingly, lignin deposited in response to wounding also shows a different composition (Hawkins & Boudet, 2003). There is considerable interest in the biotechnological modification of lignin content in plants, and various ways have been used to interfere with lignification (Anterola & Lewis, 2002). Interference with the process of lignification was also used to assess its importance in defence. The redirection of tryptophan biosynthesis in potato tubers by expression of a tryptophan decarboxylase caused a reduction in lignin associated with increased susceptibility to *P. infestans* (Yao *et al.*, 1995), although it is possible that other phenolics might also be involved. An increase in lignin content associated with enhanced tolerance to fungal pathogens was observed in transgenic tobacco plants expressing the 35S-*iaaM* and *iaaH* auxin biosynthesis genes from *Agrobacterium tumefaciens* (Sitbon *et al.*, 1999). Transformed tobacco with the defence-related cationic peroxidase gene *SPI2* of Norway spruce shows no overall increase in lignin, but alterations in histochemistry and structure. Intriguingly, transgenic plants showed enhanced resistance to the bacterium *Erwinia carotovora* but increased susceptibility to the oomycete *Phytophthora parasitica* (Elfstrand *et al.*, 2002). Specific modifications of the lignin content might provide further tools to test for its implication in disease resistance. Interesting test cases might be tobacco expressing anti-sense *o*-methyltransferase sequences resulting in an altered lignin composition (Atanassova *et al.*, 1995) or tobacco with reduced lignin content expressing an anti-sense construct of a lignin-specific peroxidase (Blee *et al.*, 2003).

6.3 Phytoalexins

6.3.1 The concept of phytoalexins

The first experimental evidence for the occurrence of antibiotic plant metabolites induced by pathogen challenge was provided by Bernard (1911; cited in Grayer & Kokubun, 2001) and Müller & Börger (1940). This led to the concept of phytoalexins (from Greek alexein = to defend) defined as 'low-molecular weight, antimicrobial compounds that are both synthesized by and accumulated in plants after exposure to microorganisms' (Paxton, 1981). They are distinguished from phytoanticipins, referred to 'low-molecular-weight, antimicrobial compounds that are present in plants before challenge by micro-organisms or are produced after infection solely from pre-existing constituents' (Van Etten *et al.*, 1994). Both classes of molecules have classically included secondary metabolites and not antimicrobial peptides. We will keep this distinction in this section and focus on phytoalexins, but many of the concepts reviewed here can be applied to phytoanticipins.

6.3.2 Distribution of phytoalexins among taxa and individuals

More than 300 molecules have been identified as phytoalexins from approximately 900 species representing 40 plant families (Harborne, 1999). These compounds can be grouped into structural families and related by their biosynthetic pathways. A close association exists between some structures and taxa, e.g. isoflavonoids are mainly produced by the Papilionoideae subfamily of Leguminosae, sesquiterpenes by Solanaceae, sulfur-containing indoles by Brassicaceae (Harborne, 1999; Grayer & Kokubun, 2001). On the other hand, some phytoalexins are shared by widely divergent plant species, like stilbenes that occur in peanut, grapevine and pine. A single species may produce several related and unrelated phytoalexins; for instance, in rice, 16 different phytoalexins have been isolated, although it is not known if all of these compounds are relevant for defence. Leaves and roots of Arabidopsis do not produce the same antimicrobials (Bednarek *et al.*, 2005).

6.3.3 Biosynthetic pathways and their regulation

The number of major biosynthetic pathways is small relative to the wide chemical diversity of phytoalexins. This provides a simple way to organize and classify these compounds (Figure 6.1). Combinations of pathways and subsequent modifications (hydroxylations, methylations, cyclizations, etc.) generate extensive divergence within each structural family. Many phytoalexins belong to the phenylpropanoid family, characterized by the C6C3 skeleton of phenylalanine. The entry point into this class of molecules is catalysed by phenylalanine ammonia-lyase (PAL) through the deamination of phenylalanine into *trans*-cinnamic acid. Some phytoalexins are readily formed from this compound e.g. *p*-coumarate (Daayf *et al.*, 1997; Bais *et al.*, 2005), or from dimerization and further modifications of related compounds, producing lignans, e.g. matairesinol (Lewis & Davin, 1999), or biphenyls, e.g. aucuparin (Grayer & Kokubun, 2001). *Trans*-cinnamate and related molecules can also undergo cyclization, giving rise to a coumarin skeleton, e.g. scopoletin and umbelliferone (Matern *et al.*, 1999), which can be in turn prenylated, producing furano- and pyrano-coumarins, for instance xanthotoxin (Stanjek *et al.*, 1999; Hehmann *et al.*, 2004). The C6C3 skeleton can also be extended by polyketide synthases (PKS) such as chalcone synthase (CHS) and stilbene synthase (STS), generating committed precursors of the flavonoid and stilbenoid families, respectively. Flavanones can be further processed into isoflavonoids, through activity of isoflavonoid synthase, and subsequently modified by numerous enzymes (for a comprehensive review of isoflavonoids, see Dixon *et al.*, 1995), producing pisatin, phaseollin or glyceollin, for example. In sorghum, apigeninidin and luteolinidin 3-deoxyanthocyanidin phytoalexins also stem from flavanones following catalysis by flavanone-4-reductase (FNR) (Forkmann & Heller, 1999). The resveratrol produced by STS is a phytoalexin on its own but also constitutes a precursor of other antimicrobial compounds (Jeandet *et al.*, 2002).

Besides phenylpropanoids, terpenoids also form a structural family encompassing many phytoalexins. The precursors isopentenyldiphosphate (IPP) and dimethylallyldiphosphate (DMAPP) are generated through the cytosolic mevalonate pathway or through the plastidic, 1-deoxyxylulose pathway (Eisenreich *et al.*, 2004). Assembly of the C5 chain of IPP and DMAPP, and of the resulting products by terpene synthases, yields isoprenoids of several carbon chain lengths (C10, C15, C20 or C30) that are further modified by specialized enzymes (Liang *et al.*, 2002). Some examples of terpenoid phytoalexins include 2,7-dihydroxycadalene, momilactone A or arjunolic acid. A few phytoalexins also rely on condensation of acetate units, after previous activation in the form of malonate, for the elaboration of their carbon skeleton. These reactions are mediated by certain PKS enzymes, belonging to the same superfamily as CHS and STS. For instance, wyerone arises from desaturation and cyclization of its precursor oleate, produced by the PKS fatty acid synthase (Nawar & Kuti, 2003). Another example is 6-methoxymellein, whose precursor 6-hydroxymellein is generated by a dedicated PKS (Kurosaki, 1994; Fan *et al.*, 2000). Certain phytoalexins, in particular those containing nitrogen, are produced by yet different pathways, as shown by the indole-based phytoalexins of Brassicaceae (Pedras *et al.*, 2004). Recent developments on camalexin biosynthesis in Arabidopsis can be found in Bednarek *et al.* (2005) and Hansen & Halkier (2005).

A number of genes involved in the various pathways for phytoalexin biosynthesis have been cloned. Following biotic or abiotic elicitation, rate-limiting enzymes were found to

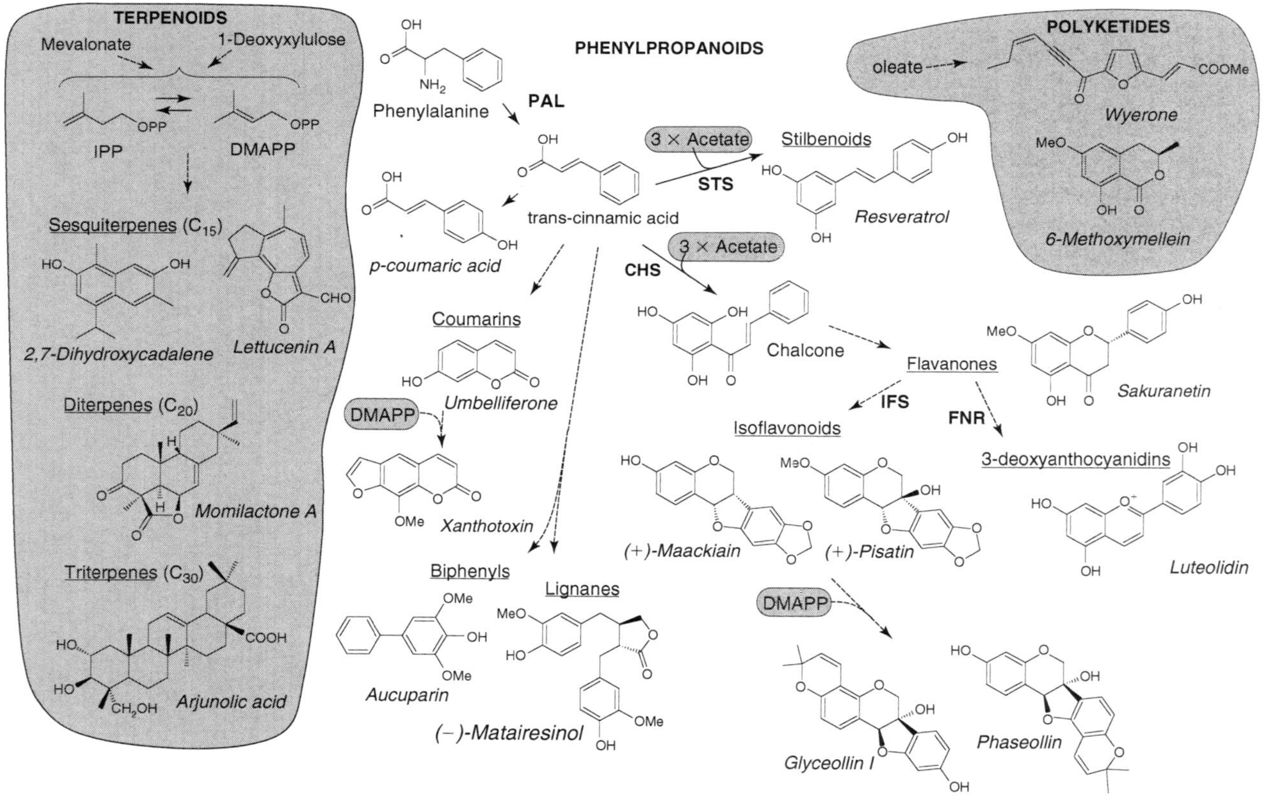

Figure 6.1 Overview on the elaboration of the carbon skeleton of terpenoid, phenylpropanoid and polyketide phytoalexins. Only selected examples are shown. Phytoalexin names are in italics, enzymes are in bold, and generic classes of compounds are underlined. Plain arrows indicate a reaction in a single step, and dashed arrows represent several consecutive enzymatic steps. Abbreviations: CHS: chalcone synthase; DMAPP: dimethylallyl pyrophosphate; FNR: flavanone reductase; IFS: isoflavone synthase; IPP: isopentenyl pyrophosphate; PAL: phenylalanine ammonia lyase; STS: stilbene synthase.

be transcriptionally controlled by dedicated transcription factors (Dixon & Paiva, 1995; Dixon *et al.*, 1995, 2002; Schmid & Amrhein, 1999; Zhao *et al.*, 2005).

Some of the genes encoding key biosynthetic enzymes belong to multigenic families (e.g. PAL, CHS, C4H) (Dixon *et al.*, 2002). The *ent*-copalyl diphosphate synthase gene is present in two copies in the rice genome, one being involved – and regulated as such – in gibberellin metabolism, and the other in the biosynthesis of the oryzalexins and phytocassanes (Prisic *et al.*, 2004). The biological significance of gene duplications as a way to independently regulate pathways sharing enzymatic activities is well illustrated in legumes, for which flavonoids serve both as phytoalexins as well as signal molecules for symbiotic micro-organisms (for reviews, refer to Aoki *et al.*, 2000; Taylor & Grotewold, 2005). The regulation of separate pathways with common enzyme activities is suggestive of metabolic channelling, which has been shown to be effective in the isoprenoid, PAL and flavonoid pathways (for reviews, refer to Winkel, 2004; Jørgensen *et al.*, 2005 and reviews therein). In this process, successive enzymes of a pathway are associated by specific interactions, allowing for rapid channelling of the substrate from one active site to another and avoiding loss or dilution into the intracellular compartment. This metabolic efficiency, together with recognition mechanisms and transcriptional activation, is probably one of the keys to the rapid production of high amounts required for phytoalexin efficiency.

6.3.4 Role of the phytoalexins in the defence response

Induced accumulation of a metabolite following pathogen infection might suggest a function for this molecule in plant defence, but nevertheless a complete demonstration would require further investigations. The criteria to examine the relevance of a phytoalexin as a defence mechanism during the plant pathogen interaction include: (1) the compound must accumulate in response to infection; (2) the compound must be inhibitory to the invading pathogen; (3) the compound must accumulate to inhibitory concentrations in the vicinity of the pathogen at the time it ceases growing in the plant; (4) variation in the rate of accumulation of the phytoalexin should cause a corresponding variation in the resistance of the plant; (5) variation in the sensitivity of the invading organism should cause a corresponding variation in its virulence. With the exception of point (1), these criteria were originally postulated to examine the importance of phytoanticipins (Wood, 1967). The various lines of evidence showing the involvement of phytoalexins in plant defence will be reviewed below, using the framework of the above criteria.

Phytoalexins are toxic towards a wide range of organisms, including bacteria, fungi, nematodes and higher animals, and even plants themselves. The EC50 (effective concentration for producing 50% of inhibition) for fungi usually ranges from 10^{-3} M to 10^{-5} M, and the MIC (minimum inhibitory concentration) for bacteria lies between 100 and 1000 $\mu g\,ml^{-1}$, classifying phytoalexins as relatively weak antifungal and antibacterial agents (Kuć, 1995; Tegos *et al.*, 2002), raising the issue of their actual concentration in the close vicinity of the pathogen. A number of studies have documented phytoalexin production at the site of pathogen attack (Yoshikawa *et al.*, 1978; Hahn *et al.*, 1985; Snyder & Nicholson, 1990; Cooper *et al.*, 1996).

Many studies established a correlation between phytoalexin accumulation and resistance to disease, although correlative evidence has to be further tested (Kuć, 1995). One of the best pieces of evidence available was provided by the transfer into different host plants of the stilbene synthase gene catalysing the one-step formation of resveratrol from the two ubiquitous plant metabolites *p*-coumarate and malonate (Figure 6.1). Introduction of this gene results in increased resistance of tobacco to *Botrytis cinerea* (Hain *et al.*, 1993), and of many crop plants against different pathogens (Zhu *et al.*, 2004 and references cited therein), although in some specific pathosystems, no effect was observed (Kobayashi *et al.*, 2000; Giorcelli *et al.*, 2004). More recently, constitutive expression of isoflavone *O*-methyltransferase, catalysing a key reaction in flavonoid biosynthesis, increased resistance of alfalfa to *Phoma medicaginis*, even if the endogenous gene was induced after infection (He & Dixon, 2000). Conversely, mutants or transgenic plants specifically affected in phytoalexin biosynthesis are more susceptible than the corresponding wild-types. The *pad3* mutant of *Arabidopsis* is defective in camalexin biosynthesis and exhibits a greater susceptibility to *Alternaria brassicicola* than the parental line (Thomma *et al.*, 1999), although the susceptibility towards other pathogens is not affected. Inhibition of the chalcone synthase in cucumber and silencing of isoflavone synthase genes in soybean lead to enhanced susceptibility to diseases, confirming that induced resistance in these species is linked with flavonoid phytoalexin accumulation (Fofana *et al.*, 2005; Subramanian *et al.*, 2005). Similarly, engineered pea plants with reduced rates of pisatin production were more susceptible than the wild-type control (Wu & Van Etten, 2004).

Virulent pathogens were generally found to be more tolerant to phytoalexins of their host than avirulent or non-pathogenic organisms, and an excellent and thoroughly studied example is the degradation of the pea phytoalexin pisatin by virulent strains of *Nectria haematococca* (Van Etten *et al.*, 2001 and references cited within). Several mechanisms can account for the resistance of bacteria and fungi to toxic compounds produced by plants (Van Etten *et al.*, 2001 and references cited therein). Many virulent pathogens have been reported to degrade phytoalexins, in some cases by several independent pathways (see Pedras & Ahiahonu, 2005 for a recent and detailed review).

Multidrug efflux pumps are emerging as a major phytoalexin tolerance mechanism in various pathogens, similar to antibiotic multi-resistance in human pathogens. Targeted mutations of different drug extrusion systems decrease virulence of *Magnaporthe grisea* on rice and barley (Urban *et al.*, 1999), of *B. cinerea* on grapevine (Schoonbeek *et al.*, 2001), of *Gibberella pulicaris* on potato (Fleissner *et al.*, 2002), of *Erwinia chrysanthemi* on witloof chicory (Barabote *et al.*, 2003) and of *Erwinia amylovora* on apple trees (Burse *et al.*, 2004). Inhibition of multidrug efflux pumps by synthetic molecules *in vitro* leads to a dramatic increase in sensitivity of several bacterial plant pathogens to plant antimicrobial metabolites (Tegos *et al.*, 2002). Exciting reports are now describing the isolation from plant tissues of inhibitors of multidrug extrusion systems (Stermitz *et al.*, 2000; Morel *et al.*, 2003; Belofsky *et al.*, 2004; Reimann & Deising, 2005). These findings expand our knowledge on plant–pathogen co-evolution, and might help in our understanding of puzzling observations (Van Etten *et al.*, 2001). Tolerance to a plant toxic compound could also be conferred by modification of the pathogen target, while preserving its fitness. For example, rifampicin-resistant strains of *Escherichia coli* have a mutation in the *rpoB* gene encoding a subunit of the RNA polymerase. But to our knowledge, such a tolerance mechanism to a phytoalexin has not yet been reported for a plant pathogen. Finally, a type III secretion system-dependent mechanism in *Pseudomonas syringae* pv. *tomato* DC3000 (*Pst* DC3000) is required for blocking synthesis or exudation of toxic compounds in *Arabidopsis* roots (Bais *et al.*, 2005). However, the relevance of this observation remains to be determined, since *Pst*

DC3000 is generally considered as a leaf pathogen. This interesting observation should now be followed up to find out if other pathogens also display a similar ability.

6.4 The hypersensitive response (HR)

6.4.1 *Cell death in plants and animals*

Programmed cell death (PCD) occurs in multicellular organisms during normal physiological processes. This genetically controlled cell suicide is observed during development (senescence, tracheary element differentiation, etc.) and in response to abiotic or biotic stresses including pathogen attack. Plants often exhibit a form of PCD, called the hypersensitive response (HR) following bacterial, viral or fungal challenge. The HR is characterized by the rapid collapse and death of the plant cells in and around the site of attempted infection. It is commonly assumed that, together with defence responses, the HR helps the plant to confine the pathogen, and prevents it spreading into healthy adjacent tissues. However, depending on the host–pathogen combination, plants can also control pathogens without the induction of an HR (Hammond-Kosack & Jones, 1996; Clough *et al.*, 2000; Torres *et al.*, 2002). Virulent, but not avirulent pathogens, can escape the *avr*-based plant recognition mechanism and are able to suppress defence responses and inhibit the HR (Nomura *et al.*, 2005 and references cited within). In the case of necrotrophic pathogens, the role of the HR in limiting disease remains questionable. Indeed, these micro-organisms are able to feed and live on dead tissues, and it has been proposed that they can induce plant cell death to their own profit (Morel & Dangl, 1997; Govrin & Levine, 2002; Lincoln *et al.*, 2002; Van Baarlen *et al.*, 2004).

Cell death during the HR displays some of the morphological features of apoptosis in animal cells, including cytoplasmic shrinkage, condensation of chromatin, cleavage of DNA, swelling of mitochondria and activation of proteases (Greenberg & Yao, 2004). But some differences remain between the two processes. Most of the structural orthologues of the key regulatory proteins of apoptosis are not encoded by the plant genome, except for BAX-INHIBITOR-1 and DEFENDER AGAINST APOPTOTIC DEATH-1. Over-expression of *Oryza sativa* BAX-INHIBITOR results in a sustainable cell survival after challenge with *M. grisea* elicitor (Matsumura *et al.*, 2003). Thus, similar strategies with functionally related molecules might have been conserved between animals and plants to control PCD (Lam, 2004). For example, expression of pro- or anti-apoptotic animal proteins in plants can impact on the development of the HR and disease resistance. Over-expression of the two animal anti-apoptotic proteins BCL-XL and CED-9 in tobacco delayed the HR induced by TMV (Mitsuhara *et al.*, 1999). In contrast, over-expression of the pro-apoptotic protein BAX in tobacco induced HR-like symptoms and defence gene activation (Lacomme & Santa Cruz, 1999).

6.4.2 *Signalling during the hypersensitive response*

The discovery in the early 1980s of maize mutants showing spontaneous development of necrotic lesions provided the first support for a genetic control of the HR by the plant (Hoisington, 1982). The analysis of these lesion-mimic mutants (LMM) has led to some of the key regulators of the HR. Initiation mutants show localized necrotic spots, whereas propagation mutants are unable to control the extent of the lesions (Lorrain *et al.*, 2003).

Based on these genetic studies, it is clear that plants are able to control the initiation of the HR and also possess the machinery to efficiently suppress its propagation when required. In addition, the use of a variety of other experimental systems (purified elicitors, suspension cells, etc.) has permitted the identification of components of the signalling pathways leading to the HR. The early events after stimulus perception necessary to trigger and control the HR include ion fluxes, oxidative burst, NO production, kinase activation, lipid signalling and also salicylic acid dependent steps (Métraux & Durner, 2004). While some of the key regulators of the HR are starting to be identified, the cascade of events and most of the genes that control PCD remain unknown. Recent microarray analyses have indicated that the expression of common sets of genes is altered in various forms of PCD indicating that a core cell death program may exist in plants (Swidzinski *et al.*, 2002; Gechev *et al.*, 2004). Proteomics approaches have been undertaken to study post-transcriptionally regulated elements involved in plant PCD. For example, this experimental strategy has identified up- or down-regulated proteins in an LMM mutant of rice compared to wild type plants suggesting their involvement in the regulation of PCD (Tsunezuka *et al.*, 2005). The same kind of result was obtained with senescence and heat-stress induced PCD (Swidzinski *et al.*, 2004). Such approaches should now be extended to the study of differential post-translational modifications (such as phosphorylation) in infected or uninfected plants. This would further our knowledge of the regulation of the HR and its signalling pathway.

Biochemical and pharmacological approaches with purified elicitors or suspension cells have shown that changes in ion fluxes and plasma membrane potential directly after stimulus perception are intimately involved in mediating HR cell death. Calcium influx from extracellular spaces and changes in free cytosolic Ca^{2+} concentration are crucial steps in the signalling cascade leading to defence responses and HR (Lecourieux *et al.*, 2002). The activation of plasma membrane anion channels can stimulate or antagonize cell death depending on the model studied. Inhibition of plasma membrane anion channel activity strongly delays HR-like symptoms in tobacco leaves infiltrated by cryptogein, a proteinaceous elicitor from *Phytophthora cryptogea* (Wendehenne *et al.*, 2002). Cell volume loss (Apoptotic Volume Decrease, AVD) is a characteristic feature of animal cells undergoing PCD (Bortner & Cidlowski, 2002). AVD is promoted by channel-mediated loss of anions, potassium and therefore water. In plants, plasma membrane anion channel-induced depolarization can participate in channel-mediated potassium efflux and subsequent water efflux (Barbier-Brygoo *et al.*, 2000). Since all these events are induced by pathogens or elicitors (Zimmermann *et al.*, 1998; Wendehenne *et al.*, 2002) it can be hypothesized that the anion channel can control HR development by such a pathway. By contrast, the harpin elicitor $HrpN_{ea}$ from *Erwinia amylovora* induces PCD through an inhibition of a CFTR-related anion channel (Reboutier *et al.*, 2005). The role of ionic fluxes in mediating HR development was confirmed genetically by the characterization of *dnd1* and *hlm1/dnd2*, two Arabidopsis mutants where the HR is compromised after pathogen challenge (Clough *et al.*, 2000; Balagué *et al.*, 2003; Jurkowski *et al.*, 2004). *DND1* and *HLM1/DND2* encode members of the *CNGC* (Cyclic Nucleotide Gated Channel) family, and the corresponding proteins are semi-selective cationic channels (Very & Sentenac, 2002).

6.4.3 The role of reactive oxygen species (ROS)

Since the first report indicating that $O_2^{\bullet-}$ production is induced in potato during its interaction with *Phytophthora infestans*, several plant tissues and suspension-cultured cells have been reported to produce ROS after pathogen infection (Doke, 1983). ROS are chemically reactive species of oxygen formed by successive one-electron reduction of molecular oxygen (O_2) and include superoxide anion ($O_2^{\bullet -}$), hydrogen peroxide (H_2O_2), hydroxyl radical ($OH^{\bullet}$) or hydroperoxyl radical ($HO_2^{\bullet}$). ROS are also generated during plant development and by a plethora of environmental factors (Laloi *et al.*, 2004). During plant pathogen interactions, ROS are mainly produced at the plasma membrane (Sagi & Fluhr, 2001 and references cited within) and ROS generation and accumulation coincide with the induction of cell death during the HR (Grant & Loake, 2000, and references cited therein). Recently, eight respiratory burst oxidase homologues (Rboh) have been identified in plants. These proteins are orthologues of the gp91^{phox} subunit of the macrophage NADPH oxidase. Single and double mutants of Arabidopsis *RbohD* and *RbohF* genes display reduced ROS production after inoculation with avirulent *P. syringae* pv. *tomato* that correlates with reduced HR (Torres *et al.*, 2002). Moreover, the silencing of *NtRboh A* and *B* in *Nicotiana benthamiana* results in reduced production of H_2O_2, a delayed and reduced HR following the infiltration of INF1, an elicitor from *P. infestans.* Although ROS seem to be a primary signal that leads to HR in some systems, they are not required for its induction in some other cases (Dorey *et al.*, 1999; Hirasawa *et al.*, 2005).

The role of ROS in triggering or executing plant cell death is still a matter of debate. ROS can have a direct cytotoxic effect in plant cells, since exogenous treatment of H_2O_2 can induce cell death (Levine *et al.*, 1996). Modulation of ROS levels *in planta* by lowering catalase or ascorbate peroxidase activity has demonstrated the role of H_2O_2 in limiting pathogen spread and suggested its involvement in cell death (Mittler *et al.*, 1999; Dat *et al.*, 2003). This direct cytotoxic effect could reflect the ability of ROS free radicals to initiate branched chain reaction, for example the Fenton reaction, which can have deleterious effects on biological structures (Grant & Loake, 2000). Hydroperoxyl radical and hydroxyl radicals can react with lipids in membranes, leading to their auto-oxidation and thus forming lipid peroxides. This reaction results in self-perpetuating lipid peroxidation, thus destroying biological membranes (Grant & Loake, 2000). In parallel, enzymatic dependent lipid peroxidation can occur in response to pathogen attack. In a first step, lipid acyl hydrolases (LAH) such as galactolipase or phospholipase A can release free fatty acids from membrane lipids which are oxidized by 9- and 13-lipoxygenase (LOX) to produce lipid hydroperoxides in a second step (La Camera *et al.*, 2004). Depending on the pathosystem studied, ROS-dependent and/or LOX-dependent lipid peroxidation may play a role during the HR (Rancé *et al.*, 1998; Rustérucci *et al.*, 1999; Jalloul *et al.*, 2002; Cacas *et al.*, 2005; Montillet *et al.*, 2005). Nevertheless, hydroperoxides formed enzymatically are at the origin of oxylipins (La Camera *et al.*, 2004), which might be involved as signalling molecules during the HR. But, it is now clear that ROS produced in response to pathogens or other stimuli are not necessarily deleterious for cells but participate in cellular signalling. For example, H_2O_2 treatment produces discrete changes in gene expression in Arabidopsis cell cultures. Altered expression occurs in genes encoding antioxidant enzymes but also signalling proteins including calmodulin, kinases or transcription factors (Desikan *et al.*, 2001). Moreover, microarray profiling experiments using catalase-silenced plants have resulted in the identification of H_2O_2-responsive genes and have outlined the pathways that are likely to participate in the cell death process (Gechev & Hille, 2005). Kinases and the MAPK module were identified among the possible signalling intermediates that may decode the ROS signals (Nakagami *et al.*, 2004; Rentel *et al.*, 2004; Gechev & Hille, 2005). Although the role of MAPK modules in mediating HR during plant pathogen interaction is established (Apel & Hirt, 2004; del Pozo *et al.*, 2004), it remains to be determined whether or not H_2O_2-mediated MAPK activation could be a part of the signal leading to HR.

6.4.4 The role of nitric oxide (NO)

NO serves as a signalling molecule in plants as in animals (Lamotte *et al.*, 2005). Its role during the HR is now established. Cell death induced by exogenous NO treatment exhibits morphological features observed during plant PCD (Zottini *et al.*, 2002; Neill *et al.*, 2003). Animal inhibitors of NO synthase reduce the extent of HR induced by avirulent *P. syringae* in Arabidopsis, and this is accompanied by more extensive bacterial growth in treated tissue, suggesting that HR prevents the pathogen spreading through an NO-dependent pathway (Delledonne *et al.*, 1998). In agreement with these observations, plants expressing the NO-scavenger haemoglobin express a reduced HR in response to tobacco necrosis virus and to avirulent *P. syringae* (Seregelyes *et al.*, 2003). Moreover, scavenging NO or inhibiting NO synthesis delayed the cell death in cryptogein-treated tobacco cell suspensions (Lamotte *et al.*, 2004). NO can contribute to cell death through different mechanisms. Peroxynitrite anion ($ONOO^-$) formed from NO and superoxide anion is a mediator of cell injury in many pathophysiological processes, but its role during the HR is controversial. Scavenging of peroxynitrite inhibits the HR induced by avirulent *P. syringae* (Alamillo & Garcia-Olmedo, 2001). For Delledonne *et al.* (2001), the balance between the level of NO and H_2O_2, produced from the dismutation of superoxide anions, is important for triggering cell death rather than the accumulation of peroxynitrite. Similarly, de Pinto *et al.* (2002) found that PCD-like cell death in tobacco cells was induced by an increase in both NO and H_2O_2. NO can also trigger cell death through a cGMP-dependent pathway (Clarke *et al.*, 2000) or through a modification of mitochondrial functionality (Saviani *et al.*, 2002; Zottini *et al.*, 2002).

6.4.5 How is cell death executed?

Although we are beginning to understand the signalling pathways leading to the HR, the final executioners of cell death and their mode of action remain largely unknown. Animal apoptosis is finely controlled by key regulators to ensure that this cell suicide is activated at the right time and in the right place, thus avoiding irreversible damage. The caspase family has a prominent role among the molecular switches of animal apoptosis (Riedl & Shi, 2004). Although orthologues of the caspase family are not found in plants (The Arabidopsis Genome, 2000), some caspase-like activities have been associated with PCD, especially during the HR (Woltering *et al.*, 2002), and caspase inhibitors can block the expansion of the HR (del Pozo & Lam, 1998). These experimental results raise the question of the involvement of caspase-like proteases in plant PCD and their role as executioners of plant PCD, as observed in animals. Many categories of proteases have been identified in plants, and interesting results have suggested the role of some of them in plant defence and in HR execution (review by van der Hoorn & Jones, 2004; Rotari *et al.*, 2005). But, up to now, the clearest results indicating the involvement of caspase-like proteases in HR comes from the study of vacuolar processing enzymes (VPEs). VPEs are cysteine proteases responsible for maturation of seed storage proteins in the vacuole of plant cells. Direct evidence indicates that VPE_γ from *Arabidopsis* has a caspase-like activity *in vivo* which is involved in HR progression and disease resistance in response to avirulent *P. syringae* or turnip mosaic virus (Rojo *et al.*, 2004). In tobacco plants carrying the N resistance gene, tobacco mosaic virus (TMV) infection results in the development of an HR. Both Caspase1 inhibitor and VPE inhibitors are able to interfere with the TMV-induced HR. Complete silencing of VPEs has resulted in a complete abrogation of the HR with a correlated TMV proliferation and also an inhibition of vacuolar collapse induced by TMV. Hatsugai *et al.* (2004) proposed that VPE is necessary to promote HR through vacuolar rupture. Vacuole collapse is an early event in all forms of PCD (Jones, 2001); it releases various hydrolases, including proteases potentially involved in executing cell death. However, proteases involved in the HR could be located in the nucleus, cytosol, chloroplast or mitochondria (Chichkova *et al.*, 2004; Rotari *et al.*, 2005; Sanmartin *et al.*, 2005). Despite the increasing interest in plant proteases, their direct role in mediating HR remains elusive, especially regarding their mode of action: are they involved in signalling during the HR or in the direct execution of HR? Both functions might be conceivable, and the identification of the molecular targets of these proteases will help us in deciphering intimate mechanisms controlling the HR.

6.5 Antifungal proteins

6.5.1 Introduction

The first observations made in the 1970s on tobacco plants reacting hypersensitively to tobacco mosaic virus showed the appearance of novel proteins accumulating in response to the infection (Datta & Muthukrishnan, 1999 and reviews therein). In the years to follow, a large number of so-called pathogenesis-related proteins (PRs) were described for various plant pathogen interactions. A major breakthrough came with the discovery that several PRs have biochemical functions that made them potentially antimicrobial. PRs were assigned to 14 distinct families in plants, and identified biochemical functions include β-1,3-glucanase (PR-2), chitinase (PR-3, PR-4, PR-8, PR-11), proteinase inhibitors (PR-6) and peroxidase (PR-9) (Van Loon &Van Strien, 1999). PRs were functionally defined as plant-encoded proteins induced in tissue infected by pathogens as well as systemically and are associated with the development of SAR (Van Loon &Van Strien, 1999).

6.5.2 PRs: current status

The advent of large-scale gene profiling has considerably broadened our knowledge on genes induced upon pathogen infection (see Chapter 3). In Arabidopsis, the number of genes induced after an incompatible interaction was found to be much larger than the repertoire of PRs hitherto recognized (Maleck *et al.*, 2000). The expression of up to 25% of the genes is modified in plants after infection based on the results obtained with an 8000 gene chip (Maleck *et al.*, 2000; Tao *et al.*, 2003). Interestingly, genes induced under situations where the plant develops subsequent resistance were found to be similar to those expressed after infection with a virulent pathogen, but they are induced faster and some of them to a higher extent, confirming previous observations on smaller sets of PRs (Tao *et al.*, 2003). Besides studies in the model system Arabidopsis (Van Wees *et al.*, 2003), other work carried out in soybean/*Phytophthora sojae* (Moy *et al.*, 2004), in cotton/*Fusarium oxysporum* (Dowd *et al.*, 2004), in alfalfa/*Colletotrichum trifolii* (Cluzet *et al.*, 2004) or in cassava/*Xanthomonas axonopodis* (Lopez *et al.*, 2005) pathosystems extended our knowledge on the gene expression changes in response to pathogens. All these studies show a larger number of induced genes than those corresponding to the classical list of PRs, and the original inventory will have to be extended considerably. These genes can be broadly assigned to the following processes: secondary metabolism, cell-wall metabolism, oxidative burst, transport, protein metabolism, antimicrobial proteins, activators of defence reactions and photosynthesis. The availability of reverse genetics in Arabidopsis with sequence-indexed populations of Arabidopsis T-DNA makes it possible to test the importance of candidate genes in disease resistance. The importance of some of the potentially antimicrobial proteins has also been assessed using plant transformation. There is considerable interest in this area, given its commercial potential.

6.5.3 Induced antifungal genes: recent evidence from transgenic plants

The importance of PRs in plant resistance has been repeatedly tested by over-expression in various plants (reviewed in Datta *et al.*, 1999). Recent examples include the over-expression of the *Xanthomonas*-induced pepper *CaPF1* gene resulting in increased resistance of Arabidopsis to *P. syringae* pv. *tomato* and tobacco to *P. syringae* pv. *tabaci.* Interestingly, expression of this regulatory protein in Arabidopsis also provides freezing tolerance. *CaPF1* encodes an ERF/AP2 transcription factor that mediates the expression of genes containing GCC or a CRT/DRE box in their promoter regions. Over-expression of *CaPF1* in Arabidopsis was found to transactivate genes that contain a GCC or a CRT/DRE box in their respective promoters such as *PRs* (plant defensin1, *PDF1.2,* and glutathione-*S*-transferase, *GST*) as well as *COR*, a cold-regulated gene associated with freezing tolerance. This result provides an interesting example of a combined regulation leading to resistance to biotic and abiotic stress (Yi *et al.*, 2004). It remains to be seen if this process can also occur naturally in Arabidopsis. A threefold increased resistance of flax to *Fusarium culmorum and F. oxysporum* was obtained by over-expression of a potato β-1,3-glucanase cDNA. The protection was shown to result from a direct effect of the enzyme against *Fusarium* growth. Metabolic profiling of transgenic plants showed a significant decrease in carbohydrate content, which could interfere with pathogen growth *in planta* (Wróbel-Kwiatkowska *et al.*, 2004). *CABPR1,* a basic *PR1* of pepper, was expressed in tobacco plants resulting in enhanced tolerance to *Phytophthora nicotianae*, *Ralstonia solanacearum* and *P. syringae* pv. *tabaci* as well as to heavy metal stress. Expression of the *CABPR1* transgene in tobacco was found to increase *PR-Q* and *GST*, but to decrease *PR-1a* and thaumatin gene expression. Presumably, this effect might be mediated by H_2O_2 as a result of a redox imbalance, as suggested by altered peroxidase activity and transcription (Sarowar *et al.*, 2005). The rice β-1,3-glucanase gene *Gns1* is induced by fungal elicitors, wounding, salicylic acid or ethylene. Transgenic plants constitutively expressing *Gns1* produce resistant-type lesions after inoculation with a virulent strain of *M. grisea*. The protection is likely to be the result of a combined action of the over-expressed *Gns* and the earlier activation of the defence-related genes *PR-1* and *PBZ1* in transgenic plants compared to control plants (Nishizawa *et al.*, 2003). The class I chitinase cDNA (*RCC2*) of rice was transformed into cucumber. The transgenic cucumbers showed varying levels of disease resistance to *B. cinerea*. Fungal growth was suppressed in the tissue, while penetration was not affected. The high expression combined with the intracellular localization of rice chitinase might explain the resistance of transgenic plants to grey mould (Kishimoto *et al.*, 2002). The effect of over-expression of a maize ribosome-inactivating protein gene, *MOD1*, and a rice basic chitinase gene, *RCH10*, in transgenic rice was tested against *Rhizoctonia solani, Bipolaris oryzae* and *M. grisea*. Significant symptom diminution was observed only to *R. solani* (Kim *et al.*, 2003). Strawberry plants which over-expressed *PCHT28* isolated from *Lycopersicon chilense* had a significantly higher resistance to *Verticillium dahliae* than did controls (Chalavi *et al.*, 2003). Transgenic tobacco plants over-expressing a thaumatin-like protein of rice show enhanced tolerance to necrotization caused by *Alternaria alternata* (Velazhahan & Muthukrishnan, 2003). Constitutive over-expression of the antifungal gene *P23* of tomato osmotin (encoding *PR*-5-like osmotin) in transgenic orange led to a significant reduction in lesion development, and a higher survival rate was observed after infection with *Phytophthora citrophthora*. This may be employed as a strategy aimed at engineering *Phytophthora* disease resistance in orange trees (Fagoaga *et al.*, 2001). Genes encoding a chitinase and a β-1,3-glucanase isolated from a *Fusarium graminearum*-infected scab-resistant wheat cultivar were transferred into a susceptible spring wheat line. One line co-expressing this gene combination showed a delay in the spread of the infection under glasshouse, but not field, conditions (Anand *et al.*, 2003). A wheat line expressing a rice thaumatin-like protein gene with moderate resistance to scab in glasshouse trials also showed no protection in the field (Anand *et al.*, 2003). This sobering result indicates the drastic difference between glasshouse evaluation and field trials.

Generally, most of these experiments have used constitutive promoters to drive the expression of the transgenes. While this approach might be appropriate to gain insight in the potential importance of a PR gene, it can be accompanied by various detrimental effects on growth and development. The use of inducible promoters responding to specific stimuli might perhaps be the answer to this problem. Success in the future use of bioengineering for practical applications might greatly depend both on inducible promoters and on a careful choice of the gene to be transformed (Gurr & Rushton, 2005a, b; see also Chapter 3).

6.6 Conclusions

Our knowledge on the resistance mechanisms of plants to pathogens has expanded considerably. While broad groups of defences have been identified early on, the details of the cascade from stimulus perception to the expression of defence are increasingly being unravelled. These studies have greatly been advanced by technical breakthroughs such as genome-wide expression studies. Without doubt, the next burst of knowledge will come from the use of proteomics and metabolomics in studies of host–pathogen interactions. From a practical point of view, these conceptual advances will eventually serve in the development of resistant plant varieties, should it be by genetic engineering or classical breeding.

6.7 References

Aist JR, 1976. Papillae and related wound plugs of plant-cells. *Annual Review of Phytopathology* **14,** 145–163.

Alamillo JM, Garcia-Olmedo F, 2001. Effects of urate, a natural inhibitor of peroxynitrite-mediated toxicity, in the response of *Arabidopsis thaliana* to the bacterial pathogen *Pseudomonas syringae*. *The Plant Journal* **25,** 529–540.

Anand A, Zhou T, Trick HN, Gill BS, Bockus WW, Muthukrishnan S, 2003. Greenhouse and field testing of transgenic wheat plants stably expressing genes for thaumatin-like protein, chitinase and glucanase against *Fusarium graminearum*. *Journal of Experimental Botany* **54,** 1101–1111.

Anterola AM, Lewis NG, 2002. Trends in lignin modification: a comprehensive analysis of the effects of genetic manipulations/mutations on lignification and vascular integrity. *Phytochemistry* **61,** 221–294.

Aoki T, Akashi T, Ayabe S, 2000. Flavonoids of leguminous plants: Structure, biological activity, and biosynthesis. *Journal of Plant Research* **113,** 475–488.

Apel K, Hirt H, 2004. Reactive oxygen species: Metabolism, oxidative stress, and signal transduction. *Annual Review of Plant Biology* **55,** 373–399.

Assaad FF, Qiu JL, Youngs H, Ehrhardt D, Zimmerli L, Kalde M, Wanner G, Peck SC, Edwards H, Ramonell K, Somerville CR, Thordal-Christensen H, 2004. The PEN1 syntaxin defines a novel cellular compartment upon fungal attack and is required for the timely assembly of papillae. *Molecular Biology of the Cell* **15,** 5118–5129.

Atanassova R, Favet N, Martz F, Chabbert B, Tollier MT, Monties B, Fritig B, Legrand M, 1995. Altered lignin composition in transgenic tobacco expressing *O*-methyltransferase sequences in sense and antisense orientation. *The Plant Journal* **8,** 465–477.

Bais HP, Prithiviraj B, Jha AK, Ausubel FM, Vivanco JM, 2005. Mediation of pathogen resistance by exudation of antimicrobials from roots. *Nature* **434,** 217.

Balagué C, Lin B, Alcon C, Flottes G, Malmström S, Köhler C, Neuhaus G, Pelletier G, Gaymard F, Roby D, 2003. HLM1, an essential signalling component in the hypersensitive response, is a member of the cyclic nucleotide-gated channel ion channel family. *Plant Cell* **15,** 365–379.

Barabote RD, Johnson OL, Zetina E, San Francisco SK, Fralick JA, San Francisco MJD, 2003. *Erwinia chrysanthemi* tolC is involved in resistance to antimicrobial plant chemicals and is essential for phytopathogenesis. *Journal of Bacteriology* **185,** 5772–5778.

Barbier-Brygoo H, Vinauger M, Colcombet J, Ephritikhine G, Frachisse J-M, Maurel C, 2000. Anion channels in higher plants: functional characterization, molecular structure and physiological role. *Biochimica et Biophysica Acta – Biomembranes* **1465,** 199–218.

Bednarek P, Schneider B, Svatos A, Oldham NJ, Hahlbrock K, 2005. Structural complexity, differential response to infection, and tissue specificity of indolic and phenylpropanoid secondary metabolism in Arabidopsis roots. *Plant Physiology* **138,** 1058–1070.

Belofsky G, Percivill D, Lewis K, Tegos GP, Ekart J, 2004. Phenolic metabolites of *Dalea versicolor* that enhance antibiotic activity against model pathogenic bacteria. *Journal of Natural Products* **67,** 481–484.

Blee KA, Choi JW, O'Connell AP, Schuch W, Lewis NG, Bolwell GP, 2003. A lignin-specific peroxidase in tobacco whose antisense suppression leads to vascular tissue modification. *Phytochemistry* **64,** 163–176.

Bortner C, Cidlowski J, 2002. Apoptotic volume decrease and the incredible shrinking cell. *Cell Death and Differentiation* **9,** 1307–1310.

Burse A, Weingart H, Ullrich MS, 2004. The phytoalexin-inducible multidrug efflux pump AcrAB contributes to virulence in the fire blight pathogen, *Erwinia amylovora*. *Molecular Plant–Microbe Interactions* **17,** 43–54.

Cacas J-L, Vailleau F, Davoine C, Ennar N, Agnel J-P, Tronchet M, Ponchet M, Blein J-P, Roby D, Triantaphylides C, Montillet J-L, 2005. The combined action of 9 lipoxygenase and galactolipase is sufficient to bring about programmed cell death during tobacco hypersensitive response. *Plant, Cell and Environment* **28,** 1367–1378.

Chalavi V, Tabaeizadeh Z, Thibodeau P, 2003. Enhanced resistance to *Verticillium dahliae* in transgenic strawberry plants expressing a *Lycopersicon chilense* chitinase gene. *Journal of the American Society for Horticultural Science* **128,** 747–753.

Chichkova NV, Kim SH, Titova ES, Kalkum M, Morozov VS, Rubtsov YP, Kalinina NO, Taliansky ME, Vartapetian AB, 2004. A plant caspase-like protease activated during the hypersensitive response. *Plant Cell* **16,** 157–171.

Clarke A, Desikan R, Hurst RD, Hancock JT, Neill SJ, 2000. NO way back: nitric oxide and programmed cell death in *Arabidopsis thaliana* suspension cultures. *The Plant Journal* **24,** 667–677.

Clough SJ, Fengler KA, Yu IC, Lippok B, Smith RK, Bent AF, 2000. The Arabidopsis dnd1 'defense, no death' gene encodes a mutated cyclic nucleotide-gated ion channel. *Proceedings of the National Academy of Sciences USA* **97,** 9323–9328.

Cluzet S, Torregrosa C, Jacquet C, Lafitte C, Fournier J, Mercier L, Salamagne S, Briand X, Esquerré-Tugayé MT, Dumas B, 2004. Gene expression profiling and protection of *Medicago truncatula* against a fungal infection in response to an elicitor from green algae *Ulva* spp. *Plant, Cell and Environment* **27,** 917–928.

Collins NC, Thordal-Christensen H, Lipka V, Bau S, Kombrink E, Qiu JL, Huckelhoven R, Stein M, Freialdenhoven A, Somerville SC, Schulze-Lefert P, 2003. SNARE-protein-mediated disease resistance at the plant cell wall. *Nature* **425,** 973–977.

Cooper RM, Resende MLV, Flood J, Rowan MG, Beale MH, Potter U, 1996. Detection and cellular localization of elemental sulphur in disease-resistant genotypes of *Theobroma cacao*. *Nature* **379,** 159–162.

Daayf F, Schmitt A, Bélanger RR, 1997. Evidence of phytoalexins in cucumber leaves infected with powdery mildew following treatment with leaf extracts of *Reynoutria sachalinensis*. *Plant Physiology* **113,** 719–727.

Dat JF, Pellinen R, Beeckman T, Van De Cotte B, Langebartels C, Kangasjärvi J, Inzé D, Van Breusegem V, 2003. Changes in hydrogen peroxide homeostasis trigger an active cell death process in tobacco. *Plant Journal* **33,** 621–632.

Datta K, Muthukrishnan S, Datta SK, 1999. Expression and function of PR-protein genes in transgenic plants. In: S.K. Datta & S. Muthukrishnan, eds. *Pathogenesis-Related Proteins in Plants*. Boca Raton, FL: CRC Press, 261–277.

de Pinto MC, Tommasi F, De Gara L, 2002. Changes in the antioxidant systems as part of the signalling pathway responsible for the programmed cell death activated by nitric oxide and reactive oxygen species in tobacco Bright-Yellow 2 cells. *Plant Physiology* **130,** 698–708.

del Pozo O, Lam E, 1998. Caspases and programmed cell death in the hypersensitive response of plants to pathogens. *Current Biology* **8,** 1129–1132.

del Pozo O, Pedley KF, Martin GB, 2004. MAPKKK is a positive regulator of cell death associated with both plant immunity and disease. *EMBO Journal* **23,** 3072–3082.

Delledonne M, Xia YJ, Dixon RA, Lamb C, 1998. Nitric oxide functions as a signal in plant disease resistance. *Nature* **394,** 585–588.

Delledonne M, Zeier J, Marocco A, Lamb C, 2001. Signal interactions between nitric oxide and reactive oxygen intermediates in the plant hypersensitive disease resistance response. *Proceedings of the National Academy of Sciences USA* **98,** 13454–13459.

Desikan R, A-H-Mackerness S, Hancock JT, Neill SJ, 2001. Regulation of the Arabidopsis transcriptome by oxidative stress. *Plant Physiology* **127,** 159–172.

Dixon RA, Achnine L, Kota P, Liu CJ, Reddy MSS, Wang LJ, 2002. The phenylpropanoid pathway and plant defense – a genomics perspective. *Molecular Plant Pathology* **3,** 371–390.

Dixon RA, Harrison MJ, Paiva NL, 1995. The isoflavonoid phytoalexin pathway: From enzymes to genes to transcription factors. *Physiologia Plantarum* **93,** 385–392.

Dixon RA, Paiva NL, 1995. Stress-induced phenylpropanoid metabolism. *Plant Cell* **7,** 1085–1097.

Doke N, 1983. Generation of superoxide anion by potato-tuber protoplasts during the hypersensitive response to hyphal wall components of *Phytophthora infestans* and specific-inhibition of the reaction by suppressors of hypersensitivity. *Physiological Plant Pathology* **23,** 359–367.

Dorey S, Kopp M, Geoffroy P, Fritig B, Kauffmann S, 1999. Hydrogen peroxide from the oxidative burst is neither necessary nor sufficient for hypersensitive cell death induction, phenylalanine ammonia lyase stimulation, salicylic acid accumulation, or scopoletin consumption in cultured tobacco cells treated with elicitin. *Plant Physiology* **121,** 163–171.

Dowd C, Wilson LW, McFadden H, 2004. Gene expression profile changes in cotton root and hypocotyl tissues in response to infection with *Fusarium oxysporum* f. sp. *vasinfectum*. *Molecular Plant-Microbe Interactions* **17,** 654–667.

Eisenreich W, Bacher A, Arigoni D, Rohdich F, 2004. Biosynthesis of isoprenoids via the non-mevalonate pathway. *Cellular and Molecular Life Sciences (CMLS)* **61,** 1401–1426.

Elfstrand M, Sitbon F, Lapierre C, Bottin A, von Arnold S, 2002. Altered lignin structure and resistance to pathogens in spi 2- expressing tobacco plants. *Planta* **214,** 708–716.

Fagoaga C, Rodrigo I, Conejero V, Hinarejos C, Tuset JJ, Arnau J, Pina JA, Navarro L, Peña L, 2001. Increased tolerance to *Phytophthora citrophthora* in transgenic orange plants constitutively expressing a tomato pathogenesis related protein PR-5. *Molecular Breeding* **7,** 175–185.

Fan XT, Mattheis JP, Roberts RG, 2000. Biosynthesis of phytoalexin in carrot root requires ethylene action. *Physiologia Plantarum* **110,** 450–454.

Fleissner A, Sopalla C, Weltring KM, 2002. An ATP-binding cassette multidrug-resistance transporter is necessary for tolerance of *Gibberella pulicaris* to phytoalexins and virulence on potato tubers. *Molecular Plant–Microbe Interactions* **15,** 102–108.

Fofana B, Benhamou N, McNally DJ, Labbé C, Séguin A, Bélanger RR, 2005. Suppression of induced resistance in cucumber through disruption of the flavonoid pathway. *Phytopathology* **95,** 114–123.

Forkmann G, Heller W, 1999. Biosynthesis of flavonoids. In: Sankawa U, ed. *Comprehensive Natural Products Chemistry*. Amsterdam: Elsevier, 713–748.

Freytag S, Arabatzis N, Hahlbrock K, Schmelzer E, 1994. Reversible cytoplasmic rearrangements precede wall apposition, hypersensitive cell-death and defense-related gene activation in potato/*Phytophthora infestans* interactions. *Planta* **194,** 123–135.

Gechev TS, Gadjev IZ, Hille J, 2004. An extensive microarray analysis of AAL-toxin-induced cell death in *Arabidopsis thaliana* brings new insights into the complexity of programmed cell death in plants. *Cellular and Molecular Life Sciences* **61,** 1185–1197.

Gechev TS, Hille J, 2005. Hydrogen peroxide as a signal controlling plant programmed cell death. *Journal of Cell Biology* **168,** 17–20.

Giorcelli A, Sparvoli F, Fulvio MY, Tava A, Balestrazzi A, Vrhovsek U, Calligari P, Bollini R, Confalonieri M, 2004. Expression of the stilbene synthase (StSy) gene from grapevine in transgenic white poplar results in high accumulation of the antioxidant compounds resveratrol glucosides. *Transgenic Research* **13,** 203–214.

Govrin EM, Levine A, 2002. Infection of Arabidopsis with a necrotrophic pathogen, *Botrytis cinerea*, elicits various defense responses but does not induce systemic acquired resistance (SAR). *Plant Molecular Biology* **48,** 267–276.

Grant JJ, Loake GJ, 2000. Role of reactive oxygen intermediates and cognate redox signalling in disease resistance. *Plant Physiology* **124,** 21–30.

Grayer RJ, Kokubun T, 2001. Plant–fungal interactions: the search for phytoalexins and other antifungal compounds from higher plants. *Phytochemistry* **56,** 253–263.

Greenberg JT, Yao N, 2004. The role and regulation of programmed cell death in plant–pathogen interactions. *Cellular Microbiology* **6,** 201–211.

Gurr SJ, Rushton PJ, 2005a. Engineering plants with increased disease resistance: what are we going to express? *Trends in Biotechnology* **23,** 275.

Gurr SJ, Rushton PJ, 2005b. Engineering plants with increased disease resistance: how are we going to express it? *Trends in Biotechnology* **23,** 283.

Hahn MG, Bonhoff A, Grisebach H, 1985. Quantitative localization of the phytoalexin glyceollin-I in relation to fungal hyphae in soybean roots infected with *Phytophthora megasperma* f. sp. *glycinea*. *Plant Physiology* **77,** 591–601.

Hain R, Reif HJ, Krause E, Langebartels R, Kindl H, Vornam B, Wiese W, Schmelzer E, Schreier PH, Stocker RH, Stenzel K, 1993. Disease resistance results from foreign phytoalexin expression in a novel plant. *Nature* **361,** 153–156.

Hammerschmidt R, Métraux J-P, van Loon LC, 2001. Inducing resistance: a summary of papers presented at the first international symposium on induced resistance to plant diseases, Corfu, May 2000. *European Journal of Plant Pathology* **107,** 1–6.

Hammond-Kosack KE, Jones JDG, 1996. Resistance gene-dependent plant defense responses. *The Plant Cell* **8,** 1773–1791.

Hansen BG, Halkier BA, 2005. New insight into the biosynthesis and regulation of indole compounds in *Arabidopsis thaliana*. *Planta* **221,** 603–606.

Harborne JB, 1999. The comparative biochemistry of phytoalexin induction in plants. *Biochemical Systematics and Ecology* **27,** 335–367.

Hatsugai N, Kuroyanagi M, Yamada K, Meshi T, Tsuda S, Kondo M, Nishimura M, Hara-Nishimura I, 2004. A plant vacuolar protease, VPE, mediates virus-induced hypersensitive cell death. *Science* **305,** 855–858.

Hawkins S, Boudet A, 2003. 'Defense lignin' and hydroxycinnamyl alcohol dehydrogenase activities in wounded *Eucalyptus gunnii*. *Forest Pathology* **33,** 91–104.

He XZ, Dixon RA, 2000. Genetic manipulation of isoflavone 7-*O*-methyltransferase enhances biosynthesis of 4′-*O*-methylated isoflavonoid phytoalexins and disease resistance in alfalfa. *The Plant Cell* **12,** 1689–1702.

Hehmann M, Lukacin R, Ekiert H, Matern U, 2004. Furanocoumarin biosynthesis in *Ammi majus* L.: Cloning of bergaptol *O*-methyltransferase. *European Journal of Biochemistry* **271,** 932–940.

Heitefuss R, 2001. Defense reactions of plants to fungal pathogens: principles and perspectives, using powdery mildew on cereals as an example. *Naturwissenschaften* **88,** 273–283.

Hirasawa K-I, Amano T, Shioi Y, 2005. Effects of scavengers for active oxygen species on cell death by cryptogein. *Phytochemistry* **66,** 463–468.

Hoisington D, 1982. Disease lesion mimics in maize. *Developmental Biology* **93,** 381–388.

Hückelhoven R, Fodor J, Preis C, Kogel K-H, 1999. Hypersensitive cell death and papilla formation in barley attacked by the powdery mildew fungus are associated with hydrogen peroxide but not with salicylic acid accumulation. *Plant Physiology* **119,** 1251–1260.

Humphreys JM, Chapple C, 2002. Rewriting the lignin roadmap. *Current Opinion in Plant Biology* **5,** 224–229.

Jacobs AK, Lipka V, Burton RA, Panstruga R, Strizhov N, Schulze-Lefert P, Fincher GB, 2003. An Arabidopsis callose synthase, GSL5, is required for wound and papillary callose formation. *The Plant Cell* **15,** 2503–2513.

Jalloul A, Montillet JL, Assigbetsé K, Agnel JP, Delannoy E, Triantaphylidès C, Daniel JF, Marmey P, Geiger JP, Nicole M, 2002. Lipid peroxidation in cotton: *Xanthomonas* interactions and the role of lipoxygenases during the hypersensitive reaction. *The Plant Journal* **32,** 1–12.

Jeandet P, Douillet-Breuil AC, Bessis R, Debord S, Sbaghi M, Adrian M, 2002. Phytoalexins from the Vitaceae: biosynthesis, phytoalexin gene expression in transgenic plants, antifungal activity, and metabolism. *Journal of Agricultural and Food Chemistry* **50,** 2731–2741.

Jones AM, 2001. Programmed cell death in development and defense. *Plant Physiology* **125,** 94–97.

Jørgensen K, Rasmussen AV, Morant M, Nielsen AH, Bjarnholt N, Zagrobelny M, Bak S, Møller BL, 2005. Metabolon formation and metabolic channeling in the biosynthesis of plant natural products. *Current Opinion in Plant Biology* **8,** 280–291.

Jurkowski GI, Smith RK, Yu IC, Ham JH, Sharma SB, Klessig DF, Fengler KA, Bent AF, 2004. Arabidopsis DND2, a second cyclic nucleotide-gated ion channel gene for which mutation causes the 'defense, no death' phenotype. *Molecular Plant–Microbe Interactions* **17,** 511–520.

Kim JK, Jang IC, Wu R, Zu WN, Boston RS, Lee YH, Alm P, Nahm BH, 2003. Co-expression of a modified maize ribosome-inactivating protein and a rice basic chitinase gene in transgenic rice plants confers enhanced resistance to sheath blight. *Transgenic Research* **12,** 475–484.

Kishimoto K, Nishizawa Y, Tabei Y, Hibi T, Nakajima M, Akutsu K, 2002. Detailed analysis of rice chitinase gene expression in transgenic cucumber plants showing different levels of disease resistance to grey mold (*Botrytis cinerea*). *Plant Science* **162,** 655–662.

Kobayashi S, Ding CK, Nakamura Y, Nakajima I, Matsumoto R, 2000. Kiwifruits (*Actinidia deliciosa*) transformed with a Vitis stilbene synthase gene produce piceid (resveratrol-glucoside). *Plant Cell Reports* **19,** 904–910.

Kobayashi Y, Kobayashi I, Funaki Y, Fujimoto S, Takemoto T, Kunoh H, 1997. Dynamic reorganization of microfilaments and microtubules is necessary for the expression of non-host resistance in barley coleoptile cells. *The Plant Journal* **11,** 525–537.

Kogel K-H, Beckhove U, Dreschers J, Münch S, Rommé Y, 1994. Acquired-resistance in barley – the resistance mechanism induced by 2,6-dichloroisonicotinic acid is a phenocopy of a genetically based mechanism governing race-specific powdery mildew resistance. *Plant Physiology* **106,** 1269–1277.

Kolattukudy PE, 1985. Enzymatic penetration of the plant cuticle by fungal pathogens. *Annual Review of Phytopathology* **23,** 223–250.

Kuć J, 1995. Phytoalexins, stress metabolism, and disease resistance in plants. *Annual Review of Phytopathology* **33,** 275–297.

Kurosaki F, 1994. Regulation of biosynthesis of carrot phytoalexin 6-methoxymellein. *Phytochemistry* **37,** 727–730.

La Camera S, Gouzerh G, Dhondt S, Hoffmann L, Fritig B, Legrand M, Heitz T, 2004. Metabolic reprogramming in plant innate immunity: the contributions of phenylpropanoid and oxylipin pathways. *Immunological Reviews* **198,** 267–284.

Lacomme C, Santa Cruz S, 1999. Bax-induced cell death in tobacco is similar to the hypersensitive response. *Proceedings of the National Academy of Sciences USA* **96,** 7956–7961.

Laloi C, Apel K, Danon A, 2004. Reactive oxygen signalling: the latest news. *Current Opinion in Plant Biology* **7,** 323–328.

Lam E, 2004. Controlled cell death, plant survival and development. *Nature Reviews Molecular Cell Biology* **5,** 305–315.

Lamotte O, Courtois C, Barnavon L, Pugin A, Wendehenne D, 2005. Nitric oxide in plants: the biosynthesis and cell signalling properties of a fascinating molecule. *Planta* **221,** 1–4.
Lamotte O, Gould K, Lecourieux D, Sequeira-Legrand A, Lebrun-Garcia A, Durner J, Pugin A, Wendehenne D, 2004. Analysis of nitric oxide signaling functions in tobacco cells challenged by the elicitor cryptogein. *Plant Physiology* **135,** 516–529.
Lange BM, Lapierre C, Sandermann Jr H, 1995. Elicitor-induced spruce stress lignin (structural similarity to early developmental lignins). *Plant Physiology* **108,** 1277–1287.
Lauvergeat V, Lacomme C, Lacombe E, Lasserre E, Roby D, Grima-Pettenati J, 2001. Two cinnamoyl-CoA reductase (CCR) genes from *Arabidopsis thaliana* are differentially expressed during development and in response to infection with pathogenic bacteria. *Phytochemistry* **57,** 1187.
Lecourieux D, Mazars C, Pauly N, Ranjeva R, Pugin A, 2002. Analysis and effects of cytosolic free calcium increases in response to elicitors in *Nicotiana plumbaginifolia* cells. *The Plant Cell* **14,** 2627–2641.
Levine A, Pennell RI, Alvarez ME, Palmer R, Lamb C, 1996. Calcium-mediated apoptosis in a plant hypersensitive disease resistance response. *Current Biology* **6,** 427–437.
Lewis N, Davin L, 1999. Lignans: Biosynthesis and function. In: Sankawa U, ed. *Comprehensive Natural Products Chemistry*. Amsterdam: Elsevier, 639–712.
Lewis NG, 1999. A 20th century roller coaster ride: a short account of lignification. *Current Opinion in Plant Biology* **2,** 153–162.
Liang P-H, Ko T-P, Wang AHJ, 2002. Structure, mechanism and function of prenyltransferases. *European Journal of Biochemistry* **269,** 3339–3354.
Lincoln JE, Richael C, Overduin B, Smith K, Bostock R, Gilchrist DG, 2002. Expression of the anti-apoptotic baculovirus p35 gene in tomato blocks programmed cell death and provides broad-spectrum resistance to disease. *Proceedings of the National Academy of Sciences USA* **99,** 15217–15221.
Lopez C, Soto M, Restrepo S, Piegu B, Cooke R, Delseny M, Tohme J, Verdier V, 2005. Gene expression profile in response to *Xanthomonas axonopodis* pv. *manihotis* infection in cassava using a cDNA microarray. *Plant Molecular Biology* **57,** 393–410.
Lorrain S, Vailleau F, Balague C, Roby D, 2003. Lesion mimic mutants: keys for deciphering cell death and defense pathways in plants? *Trends in Plant Science* **8,** 263–271.
Maleck K, Levine A, Eulgem T, Morgan A, Schmid J, Lawton KA, Dangl JL, Dietrich RA, 2000. The transcriptome of *Arabidopsis thaliana* during systemic acquired resistance. *Nature Genetics* **26,** 403–410.
Matern U, Luer P, Kreusch D, 1999. Biosynthesis of coumarins. In: Sankawa U, ed. *Comprehensive Natural Products Chemistry*. Amsterdam: Elsevier, 623–637.
Matsumura H, Nirasawa S, Kiba A, Urasaki N, Saitoh H, Ito M, Kawai-Yamada M, Uchimiya H, Terauchi R, 2003. Overexpression of Bax inhibitor suppresses the fungal elicitor-induced cell death in rice (*Oryza sativa* L.) cells. *The Plant Journal* **33,** 425–434.
McLusky SR, Bennett MH, Beale MH, Lewis MJ, Gaskin P, Mansfield JW, 1999. Cell wall alterations and localized accumulation of feruloyl-3′-methoxytyramine in onion epidermis at sites of attempted penetration by *Botrytis allii* are associated with actin polarisation, peroxidase activity and suppression of flavonoid biosynthesis. *The Plant Journal* **17,** 523–534.
Mendgen K, Hahn M, Deising H, 1996. Morphogenesis and mechanisms of penetration by plant pathogenic fungi. *Annual Review of Phytopathology* **34,** 367–386.
Métraux JP, Durner J, 2004. The role of salicylic acid and nitric oxide in programmed cell death and induced resistance. In: Sandermann H, ed. *Ecological Studies,* vol. 170: *Molecular Ecotoxicology of Plants*. Berlin: Springer, 111–150.
Mitsuhara I, Malik KA, Miura M, Ohashi Y, 1999. Animal cell-death suppressors Bcl-xL and Ced-9 inhibit cell death in tobacco plants. *Current Biology* **9,** 775–778.
Mittler R, Lam E, Shulaev V, Cohen M, 1999. Signals controlling the expression of cytosolic ascorbate peroxidase during pathogen-induced programmed cell death in tobacco. *Plant Molecular Biology* **39,** 1025–1035.
Moersbacher B, Mendgen K, 2000. Structural aspects of defense. In: Van Loon LC, Fraser RSS, eds. *Mechanisms of Resistance to Plant Diseases*. Dordrecht: Kluwer, 231–277.
Montillet J-L, Chamnongpol S, Rustérucci C, Dat J, van de Cotte B, Agnel J-P, Battesti C, Inzé D, Van Breusegem F, Triantaphylidès C, 2005. Fatty acid hydroperoxides and H_2O_2 in the execution of hypersensitive cell death in tobacco leaves. *Plant Physiology* **138,** 1516–1526.
Morel C, Stermitz FR, Tegos G, Lewis K, 2003. Isoflavones as potentiators of antibacterial activity. *Journal of Agricultural and Food Chemistry* **51,** 5677–5679.
Morel JB, Dangl JL, 1997. The hypersensitive response and the induction of cell death in plants. *Cell Death and Differentiation* **4,** 671–683.

Moy P, Qutob D, Chapman BP, Atkinson I, Gijzen M, 2004. Patterns of gene expression upon infection of soybean plants by *Phytophthora sojae*. *Molecular Plant–Microbe Interactions* **17,** 1051–1062.

Müller KO, Börger H, 1940. Experimentelle Untersuchungen über die Phytophthora-Resistenz der Kartoffel. (Berlin-Dahlem: Arbeiten aus der Biologischen Reichsanstalt für Land- und Forstwirtschaft), pp. 189–231.

Nakagami H, Kiegerl S, Hirt H, 2004. OMTK1, a novel MAPKKK, channels oxidative stress signaling through direct mapk interaction. *Journal of Biological Chemistry* **279,** 26959–26966.

Nawar HF, Kuti JO, 2003. Wyerone acid phytoalexin synthesis and peroxidase activity as markers for resistance of broad beans to chocolate spot disease. *Journal of Phytopathology* **151,** 564–570.

Neill SJ, Desikan R, Hancock JT, 2003. Nitric oxide signalling in plants. *New Phytologist* **159,** 11–35.

Nicholson RL, Hammerschmidt R, 1992. Phenolic compounds and their role in disease resistance. *Annual Review of Phytopathology* **30,** 369–389.

Nishimura MT, Stein M, Hou BH, Vogel JP, Edwards H, Somerville SC, 2003. Loss of a callose synthase results in salicylic acid-dependent disease resistance. *Science* **301,** 969–972.

Nishizawa Y, Saruta M, Nakazono K, Nishio Z, Soma M, Yoshida T, Nakajima E, Hibi T, 2003. Characterization of transgenic rice plants over-expressing the stress-inducible β-glucanase gene *Gns1*. *Plant Molecular Biology* **51,** 143–152.

Nomura K, Melotto M, He S-Y, 2005. Suppression of host defense in compatible plant–*Pseudomonas syringae* interactions. *Current Opinion in Plant Biology* **8,** 361–368.

Opalski KS, Schultheiss H, Kogel K-H, Hückelhoven R, 2005. The receptor-like MLO protein and the RAC/ROP family G-protein RACB modulate actin reorganization in barley attacked by the biotrophic powdery mildew fungus *Blumeria graminis* f. sp. *hordei*. *The Plant Journal* **41,** 291–303.

Ostergaard L, Petersen M, Mattsson O, Mundy J, 2002. An Arabidopsis callose synthase. *Plant Molecular Biology* **49,** 559–566.

Paxton JD, 1981. Phytoalexins – a working redefinition. *Phytopathologische Zeitschrift* **101,** 106–109.

Pedras MSC, Ahiahonu PWK, 2005. Metabolism and detoxification of phytoalexins and analogs by phytopathogenic fungi. *Phytochemistry* **66,** 391–411.

Pedras MSC, Montaut S, Suchy M, 2004. Phytoalexins from the crucifer rutabaga: Structures, syntheses, biosyntheses, and antifungal activity. *Journal of Organic Chemistry* **69,** 4471–4476.

Perumalla CJ, Heath MC, 1991. The effect of inhibitors of various cellular processes on the wall modifications induced in bean-leaves by the cowpea rust fungus. *Physiological and Molecular Plant Pathology* **38,** 293–300.

Petersen M, Brodersen P, Naested H, Andreasson E, Lindhart U, Johansen B, Nielsen HB, Lacy M, Austin MJ, Parker JE, Sharma SB, Klessig DF, Martienssen R, Mattsson O, Jensen AB, Mundy J, 2000. Arabidopsis MAP kinase 4 negatively regulates systemic acquired resistance. *Cell* **103,** 1111–1120.

Prisic S, Xu M, Wilderman PR, Peters RJ, 2004. Rice contains two disparate *ent*-copalyl diphosphate synthases with distinct metabolic functions. *Plant Physiology* **136,** 4228–4236.

Rancé I, Fournier J, Esquerré-Tugayé M-T, 1998. The incompatible interaction between *Phytophthora parasitica* var. *nicotianae* race 0 and tobacco is suppressed in transgenic plants expressing antisense lipoxygenase sequences. *Proceedings of the National Academy of Sciences USA* **95,** 6554–6559.

Reboutier D, Frankart C, Vedel R, Brault M, Duggleby RG, Rona JP, Barny MA, Bouteau F, 2005. A CFTR chloride channel activator prevents HrpNea-induced cell death in *Arabidopsis thaliana* suspension cells. *Plant Physiology and Biochemistry* **43,** 567–572.

Reimann S, Deising HB, 2005. Inhibition of efflux transporter-mediated fungicide resistance in *Pyrenophora tritici-repentis* by a derivative of 4′-hydroxyflavone and enhancement of fungicide activity. *Applied and Environmental Microbiology* **71,** 3269–3275.

Rentel MC, Lecourieux D, Ouaked F, Usher SL, Petersen L, Okamoto H, Knight H, Peck SC, Grierson CS, Hirt H, Knight MR, 2004. OXI1 kinase is necessary for oxidative burst-mediated signalling in Arabidopsis. *Nature* **427,** 858–861.

Riedl SJ, Shi YG, 2004. Molecular mechanisms of caspase regulation during apoptosis. *Nature Reviews Molecular Cell Biology* **5,** 897–907.

Rojo E, Martin R, Carter C, Zouhar J, Pan S, Plotnikova J, Jin H, Paneque M, Sánchez-Serrano JJ, Baker B, Ausubel FM, Raikhel NV, 2004. VPEγ exhibits a caspase-like activity that contributes to defense against pathogens. *Current Biology* **14,** 1897–1906.

Rotari VI, He R, Gallois P, 2005. Death by proteases in plants: whodunit. *Physiologia Plantarum* **123,** 376–385.

Rustérucci C, Montillet J-L, Agnel J-P, Battesti C, Alonso B, Knoll A, Bessoule J-J, Etienne P, Suty L, Blein J-P, Triantaphylidès C, 1999. Involvement of lipoxygenase-dependent production of fatty acid hydroperoxides in the development of the hypersensitive cell death induced by cryptogein on tobacco leaves. *Journal of Biological Chemistry* **274,** 36446–36455.

Sagi M, Fluhr R, 2001. Superoxide production by plant homologues of the gp91phox NADPH oxidase. Modulation of activity by calcium and by tobacco mosaic virus infection. *Plant Physiology* **126,** 1281–1290.

Sanmartin M, Jaroszewski L, Raikhel NV, Rojo E, 2005. Caspases. Regulating death since the origin of life. *Plant Physiology* **137,** 841–847.

Sarowar S, Kim YJ, Kim EN, Kim KD, Hwang BK, Islam R, Shin JS, 2005. Overexpression of a pepper basic pathogenesis-related protein 1 gene in tobacco plants enhances resistance to heavy metal and pathogen stresses. *The Plant Cell Reports* **24,** 216–224.

Saviani EE, Orsi CH, Oliveira JFP, Pinto-Maglio CAF, Salgado I, 2002. Participation of the mitochondrial permeability transition pore in nitric oxide-induced plant cell death. *FEBS Letters* **510,** 136.

Schmid J, Amrhein N, 1999. The shikimate pathway. In: Singh B, ed. *Plant Amino Acids – Biochemistry and Biotechnology*. New York: Marcel Dekker, 147–169.

Schoonbeek H, Del Sorbo G, De Waard MA, 2001. The ABC transporter BcatrB affects the sensitivity of *Botrytis cinerea* to the phytoalexin resveratrol and the fungicide fenpiclonil. *Molecular Plant–Microbe Interactions* **14,** 562–571.

Seregelyes C, Barna B, Hennig J, Konopka D, Pasternak TP, Lukacs N, Feher A, Horvath GV, Dudits D, 2003. Phytoglobins can interfere with nitric oxide functions during plant growth and pathogenic responses: a transgenic approach. *Plant Science* **165,** 541.

Sitbon F, Hennion S, Little CHA, Sundberg B, 1999. Enhanced ethylene production and peroxidase activity in IAA-overproducing transgenic tobacco plants is associated with increased lignin content and altered lignin composition. *Plant Science* **141,** 165–173.

Skalamera D, Heath MC, 1996. Cellular mechanisms of callose deposition in response to fungal infection or chemical damage. *Canadian Journal of Botany – Revue Canadienne de Botanique* **74,** 1236–1242.

Snyder BA, Nicholson RL, 1990. Synthesis of phytoalexins in sorghum as a site-specific response to fungal ingress. *Science* **248,** 1637–1639.

Soylu EM, Soylu S, Mansfield JW, 2004. Ultrastructural characterisation of pathogen development and host responses during compatible and incompatible interactions between *Arabidopsis thaliana* and *Peronospora parasitica*. *Physiological and Molecular Plant Pathology* **65,** 67–78.

Soylu S, Keshavarzi M, Brown I, Mansfield JW, 2003. Ultrastructural characterisation of interactions between *Arabidopsis thaliana* and *Albugo candida*. *Physiological and Molecular Plant Pathology* **63,** 201–211.

Stange Jr. RR, Ralph J, Peng JP, Sims JJ, Midland SL, McDonald RE, 2001. Acidolysis and hot water extraction provide new insights into the composition of the induced 'lignin-like' material from squash fruit. *Phytochemistry* **57,** 1005–1011.

Stanjek V, Piel J, Boland W, 1999. Synthesis of furanocoumarins: mevalonate-independent prenylation of umbelliferone in *Apium graveolens* (Apiaceae). *Phytochemistry* **50,** 1141–1145.

Stermitz FR, Lorenz P, Tawara JN, Zenewicz LA, Lewis K, 2000. Synergy in a medicinal plant: antimicrobial action of berberine potentiated by 5′-methoxyhydnocarpin, a multidrug pump inhibitor. *Proceedings of the National Academy of Sciences USA* **97,** 1433–1437.

Stone BA, Clarke AE, 1992. Chemistry and biology of (1–3)-β-D-glucan. Victoria, Australia: La Trobe University Press.

Subramanian S, Graham MY, Yu O, Graham TL, 2005. RNA interference of soybean isoflavone synthase genes leads to silencing in tissues distal to the transformation site and to enhanced susceptibility to *Phytophthora sojae*. *Plant Physiology* **137,** 1345–1353.

Swidzinski JA, Leaver CJ, Sweetlove LJ, 2004. A proteomic analysis of plant programmed cell death. *Phytochemistry* **65,** 1829–1838.

Swidzinski JA, Sweetlove LJ, Leaver CJ, 2002. A custom microarray analysis of gene expression during programmed cell death in *Arabidopsis thaliana*. *The Plant Journal* **30,** 431–446.

Tao Y, Xie Z, Chen W, Glazebrook J, Chang H-S, Han B, Zhu T, Zou G, Katagiri F, 2003. Quantitative nature of Arabidopsis responses during compatible and incompatible interactions with the bacterial pathogen *Pseudomonas syringae*. *The Plant Cell* **15,** 317–330.

Taylor LP, Grotewold E, 2005. Flavonoids as developmental regulators. *Current Opinion in Plant Biology* **8,** 317–323.

Tegos G, Stermitz FR, Lomovskaya O, Lewis K, 2002. Multidrug pump inhibitors uncover remarkable activity of plant antimicrobials. *Antimicrobial Agents and Chemotherapy* **46,** 3133–3141.

The Arabidopsis Genome Initiative, 2000. Analysis of the genome sequence of the flowering plant *Arabidopsis thaliana*. *Nature* **408,** 796–815.

Thomma B, Nelissen I, Eggermont K, Broekaert WF, 1999. Deficiency in phytoalexin production causes enhanced susceptibility of *Arabidopsis thaliana* to the fungus *Alternaria brassicicola. The Plant Journal* **19,** 163–171.

Thordal-Christensen H, Zhang Z, Wei Y, Collinge DB, 1997. Subcellular localization of H_2O_2 in plants. H_2O_2 accumulation in papillae and hypersensitive response during the barley powdery mildew interaction. *The Plant Journal* **11,** 1187–1194.

Ton J, Jakab G, Toquin V, Flors V, Iavicoli A, Maeder MN, Métraux J-P, Mauch-Mani B, 2005. Dissecting the β-aminobutyric acid-induced priming phenomenon in Arabidopsis. *The Plant Cell* **17,** 987–999.

Ton J, Van Pelt JA, Van Loon LC, Pieterse CMJ, 2002. Differential effectiveness of salicylate-dependent and jasmonate/ethylene-dependent induced resistance in Arabidopsis. *Molecular Plant–Microbe Interactions* **15,** 27–34.

Torres MA, Dangl JL, Jones JDG, 2002. Arabidopsis gp91phox homologues AtrbohD and AtrbohF are required for accumulation of reactive oxygen intermediates in the plant defense response. *Proceedings of the National Academy of Sciences USA* **99,** 517–522.

Tsunezuka H, Fujiwara M, Kawasaki T, Shimamoto K, 2005. Proteome analysis of programmed cell death and defense signalling using the rice lesion mimic mutant cdr2. *Molecular Plant–Microbe Interactions* **18,** 52–59.

Urban M, Bhargava T, Hamer JE, 1999. An ATP-driven efflux pump is a novel pathogenicity factor in rice blast disease. *EMBO Journal* **18,** 512–521.

Van Baarlen P, Staats M, Van Kan JAL, 2004. Induction of programmed cell death in lily by the fungal pathogen *Botrytis elliptica. Molecular Plant Pathology* **5,** 559–574.

Vance CP, Kirk TK, Sherwood RT, 1980. Lignification as a mechanism of disease resistance. *Annual Review of Phytopathology* **18,** 259–288.

Van der Hoorn RAL, Jones JDG, 2004. The plant proteolytic machinery and its role in defense. *Current Opinion in Plant Biology* **7,** 400–407.

Van Etten H, Temporini E, Wasmann C, 2001. Phytoalexin (and phytoanticipin) tolerance as a virulence trait: why is it not required by all pathogens? *Physiological and Molecular Plant Pathology* **59,** 83–93.

Van Etten HD, Mansfield JW, Bailey JA, Farmer EE, 1994. Two classes of plant antibiotics – phytoalexins versus phytoanticipins. *The Plant Cell* **6,** 1191–1192.

Van Loon LC, Van Strien EA, 1999. The families of pathogenesis-related proteins, their activities, and comparative analysis of PR-1 type proteins. *Physiological and Molecular Plant Pathology* **55,** 85–97.

Van Wees SCM, Chang HS, Zhu T, Glazebrook J, 2003. Characterization of the early response of Arabidopsis to *Alternaria brassicicola* infection using expression profiling. *Plant Physiology* **132,** 606–617.

Velazhahan R, Muthukrishnan S, 2003. Transgenic tobacco plants constitutively overexpressing a rice thaumatin-like protein (PR-5) show enhanced resistance to *Alternaria alternata. Biologia Plantarum* **47,** 347–354.

Very A-A, Sentenac H, 2002. Cation channels in the Arabidopsis plasma membrane. *Trends in Plant Science* **7,** 168–175.

Vorwerk S, Somerville S, Somerville C, 2004. The role of plant cell wall polysaccharide composition in disease resistance. *Trends in Plant Science* **9,** 203–209.

Wendehenne D, Lamotte O, Frachisse J-M, Barbier-Brygoo H, Pugin A, 2002. Nitrate efflux is an essential component of the cryptogein signalling pathway leading to defense responses and hypersensitive cell death in tobacco. *The Plant Cell* **14,** 1937–1951.

Winkel BS, 2004. Metabolic channeling in plants. *Annual Review of Plant Biology* **55,** 85–107.

Woltering EJ, van der Bent A, Hoeberichts FA, 2002. Do plant caspases exist? *Plant Physiology* **130,** 1764–1769.

Wood RKS, 1967. *Physiological Plant Pathology*. Oxford: Blackwell.

Wróbel-Kwiatkowska M, Lorenc-Kukula K, Starzycki M, Oszmiański J, Kepczyńska E, Szopa J, 2004. Expression of β-1,3-glucanase in flax causes increased resistance to fungi. *Physiological and Molecular Plant Pathology* **65,** 245–256.

Wu QD, Van Etten HD, 2004. Introduction of plant and fungal genes into pea (*Pisum sativum* L.) hairy roots reduces their ability to produce pisatin and affects their response to a fungal pathogen. *Molecular Plant–Microbe Interactions* **17,** 798–804.

Yao K, De Luca V, Brisson N, 1995. Creation of a metabolic sink for tryptophan alters the phenylpropanoid pathway and the susceptibility of potato to *Phytophthora infestans. The Plant Cell* **7,** 1787–1799.

Yi SY, Kim JH, Joung YH, Lee S, Kim WT, Yu SH, Choi D, 2004. The pepper transcription factor CaPF1 confers pathogen and freezing tolerance in Arabidopsis. *Plant Physiology* **136,** 2862–2874.

Yoshikawa M, Yamauchi K, Masago H, 1978. Glyceollin – its role in restricting fungal growth in resistant soybean hypocotyls infected with *Phytophthora megasperma* var. *sojae*. *Physiological Plant Pathology* **12,** 73–82.
Zhao J, Davis LC, Verpoorte R, 2005. Elicitor signal transduction leading to production of plant secondary metabolites. *Biotechnology Advances* **23,** 283–333.
Zhu YJ, Agbayani R, Jackson MC, Tang CS, Moore PH, 2004. Expression of the grapevine stilbene synthase gene VST1 in papaya provides increased resistance against diseases caused by *Phytophthora palmivora*. *Planta* **220,** 241–250.
Zimmerli L, Jakab C, Métraux J-P, Mauch-Mani B, 2000. Potentiation of pathogen-specific defense mechanisms in Arabidopsis by β-aminobutyric acid. *Proceedings of the National Academy of Sciences USA* **97,** 12920–12925.
Zimmermann S, Frachisse J-M, Thomine S, Barbier-Brygoo H, Guern J, 1998. Elicitor-induced chloride efflux and anion channels in tobacco cell suspensions. *Plant Physiology and Biochemistry* **36,** 665–674.
Zottini M, Formentin E, Scattolin M, Carimi F, Lo Schiavo F, Terzi M, 2002. Nitric oxide affects plant mitochondrial functionality in vivo. *FEBS Letters* **515,** 75–83.

Chapter 7
Induced resistance in natural ecosystems and pathogen population biology: exploiting interactions

Adrian Newton[1] *and Jörn Pons-Kühnemann*[2]

[1]Scottish Crop Research Institute, Dundee, UK
[2]Biometry and Population Genetics, Giessen University, Germany

7.1 Introduction

Classically trained plant pathologists tend to focus on host–causal agent interactions as, in epidemic situations, these tend to dominate. However, with a sustainability focus, control through intervention with highly targeted pesticides increasingly gives way to a systems approach where the contribution of any and every component of the biotic and abiotic environment to the host–pathogen interaction can be utilized to bring about increased stability in favour of the host or host harvest index in agriculture. First, we must recognize what the components of the plant environment are which may contribute to the interaction, and how they might be manipulated before we can determine and bias towards the best combination of these environmental factors.

7.2 Environmental variability

There is a popular conception that the 'best' plants for controlled experimentation, particularly for molecular biology, should be produced in a growth chamber under highly controlled conditions. While it may be easier to obtain reproducible conditions between experiments in this way, a 'normal' growth chamber will not produce 'normal' plants, and thus gene expression data, for example, may be very different from that in the field. The usual concession to reproducing a normal plant environment is a daily light cycle, 16 hours light, 8 hours dark, for example. Experimental work is now constrained to sampling plants at defined points in the light cycle, as light can induce gene expression but also has secondary consequences through, for example, assimilate levels in tissues. The consequences for resistance induction are demonstrated by mildew infection on barley where 30 genotypes inoculated straight out of a 12 hour dark period were about 13% more resistant than those inoculated after 12 hours of light, although there were also genotype interactions (A.C. Newton, unpublished results).

Daylight levels and temperature variation in the field have a much greater range than that normally reproduced in controlled environments. Again, attempting to reproduce this by ramping the temperature up and down on a daily cycle demonstrated differences in resistance expression (Newton *et al.*, 2003). However, this also demonstrated that host genotypes respond differentially to even such crude reproduction of variable environments. Environmental extremes and irregularity are likely to amplify such differences. Falkhof *et al.* (1988) demonstrated that their *Bacillus subtilis* culture filtrate resistance inducer was ineffective when applied to plants grown under constant temperature, light and humidity conditions, whereas greenhouse- or outdoor-grown plants which would have been subject to frequent changes in light, temperature and humidity responded strongly, substantially reducing infection following elicitor treatment. It has been noted that under real environmental/field conditions plants respond differently at different times to chemical elicitor treatments such as benzothiadiazoles (BTH) (Navarre & Mayo, 2004). For example, in a field experiment under natural conditions on the effect of nitrogen fertilization, fungicides and resistance induction on Fusarium head blight (FHB) and related mycotoxin accumulation in wheat (Heier *et al.*, 2005), two 'plant strengtheners', BTH and a compound based on the biomass of the cyanobacterium *Spirulina platensis* were used. The results indicated that less intensive fungicide strategies, including plant strengtheners, are no worse than common fungicide strategies under conditions of low FHB severity and mycotoxin accumulation. Immoderate N-fertilization, however, can increase mycotoxin levels significantly, even under conditions unfavourable for *Fusarium* spp. Indeed, under field conditions, there may be a general high level expression of defence genes compared with glasshouse grown plants (Pasquer *et al.*, 2005). Thus, the ability or requirement to respond further to pathogen attack may be unnecessary or reduced, but there will be an energetic and therefore probable yield cost to the plant (Smedegaard-Petersen & Tolstrup, 1985).

7.3 Ecology of the plant environment

The leaf surface in the field is obviously often a complex community, and each organism is affected by each other and the environment, each factor being affected differentially. The host may also respond to these factors differentially, including induction of resistance or susceptibility. Thus, the response to one organism may affect the interaction of another organism, such as inducing resistance against an otherwise pathogenic and virulent individual, or vice versa. Organisms may also interact directly with each other in various ways such as synergistically, parasitically or antagonistically. This may induce yet more differential interactions with the host plant, even allowing non-pathogens to become part of the pathogen complex. For example, Dewey *et al.* (1999) found that non-pathogen bacteria isolated from leaf surfaces with a pathogen, when re-inoculated with the pathogen induced greater symptom expression. Dewey suggested that this may have been attributable to complementary efficiency of cell wall degrading enzymes from the fungal pathogen and the bacterium. Newton *et al.* (2004) went on to demonstrate that bacterial inoculum from a previous crop can affect fungal disease development on subsequent non-host crops in practice. This has consequences for agronomy including rotations. While the synergistic enzyme theory may explain the effects on hemi-biotrophic pathogens in their necrotrophic phase, mildew on wheat was also considerably more severe on the heavy bacterial inoculum plots. The mechanism may not be attributable to direct interaction between the bacteria and *Blumeria graminis* inoculum. It could be due to induced susceptibility or even simply a nutritional effect mediated through the plant as increased nitrogen can have vary large effects on increasing susceptibility to mildew in cereals. For example, Newton *et al.* (2000) noted that high nitrogen level increased powdery mildew by 4.58 times compared with a low level under high inoculum pressure conditions. Whatever the explanation, expression of resistance was affected by other organisms in the environment directly or, more likely, indirectly.

Induced systemic resistance (ISR) by bacteria such as Pseudomonads in the rhizoplane is discussed in Chapter 8, and likewise the effects of fungi including mycorrhizae. These effects may be expressed more in natural and semi-natural ecosystems where there is less soil disturbance and therefore more time for such interactions to develop and be selected and expressed. Few experiments have been carried out to determine their effects, so little can be concluded about the importance of induced resistance per se as opposed to nutritional effects, or the interaction with other phylloplane organisms similarly affected. For example, in a study of mycorrhizal effects on the development of early blight in the pathosystem *A. solani–Solanum lycopersicum*, mycorrhizal tomato plants had significantly fewer *A. solani* symptoms than non-mycorrhizal plants. An increased P supply had no effect on disease severity in non-mycorrhizal plants but led to a higher disease severity in mycorrhizal plants. This was parallel to a P-supply-induced reduction in mycorrhiza formation. Fritz *et al.* (2006) found that the protective effect of mycorrhizas towards development of *A. solani* has some parallels to induced systemic resistance, mediated by rhizobacteria: both biocontrol agents are root-associated organisms, and both are effective against necrotrophic pathogens.

We must consider not just the organisms we can identify from the leaf surface or roots, nor just the non-culturable organisms, but pathogens in symptomless phases as it is likely that they are interacting with their host. Pathogens such as *Ramularia collo-cygni* and *Rhynchosporium secalis* are present on barley plants in significant numbers of infections long before any symptoms are expressed. *R. collo-cygni* requires a developmental trigger around 10 days after anthesis for symptom expression (Salamati *et al.*, 2003), possibly previously existing as an endophyte throughout the life of the plant from seed-borne inoculum. This implies that resistance induction is development stage regulated, being down-regulated when it becomes energetically non-beneficial to fitness or fecundity. In agricultural situations, this is too soon, as grain quality for the end user can be compromised. However, in a natural ecosystem, the consequences may be in a different balance.

Pathogens in symptomless phases may have active interactions with their host to suppress recognition. Whether this is directly costly to the host, or indirectly as the putative pathogen has to derive its resources from the host somehow, is not known. Here, we must consider not only parts of the host capable of expressing disease symptoms but also organs where they have not been observed to express any phenotypic reaction. Again, *R. secalis* is a good example where recently it has been detected in roots by PCR techniques (Fountaine, 2005), although the nature of its interaction with the host in the roots is unknown. It is conceivable that such infections, be they bacterial, fungal or any other organism including allelopathic effects of other plants, might not only induce ISR but also induce susceptibility, and this could be developmentally related. Induced susceptibility may be ecologically important for maintaining diverse populations of organisms not under direct selection, and therefore the pathogen complex diversity reservoir may be considerably more extensive or enduring than might be anticipated.

7.4 Environmental parameters

Both light and humidity have been demonstrated to affect expression of defence-related genes, including those involved in the production of reactive oxygen species (ROS) (Mateo *et al.*, 2004; Zhou *et al.*, 2004). Light will clearly directly affect the chloroplasts and thereby the redox status of cells, and this has been shown to directly affect SAR (Fobert & Després, 2005), thus partially explaining the effects of light on mildew infection of barley noted above, for example. Humidity also affects response to mildew and response to resistance elicitors, and the interaction between them. For example, chitin was found to be effective as an elicitor only at high humidity, and at low humidity, glucan was effective on some genotypes of barley but not others, indicating genetic differences in a plant's ability to induce resistance (Newton & Dashwood, 1998).

Light affects resistance expression not only due to its intensity and duration, but also due its quality or wavelength. Yalpani *et al.* (1994) report that UV-C light, and ozone, stimulated salicylic acid up to nine-fold which correlated with subsequently enhanced pathogen (virus) resistance. Long & Jenkins (1998) found similar UV-A and UV-B induction of defence-related pathways and again redox activity. In wheat, a yellow rust resistance gene which is expressed better under high light intensity is known (Ash & Rees, 1994), and in broad bean, yellow (590 nm) and red (650 nm) light stimulated resistance and stimulated anti-fungal substance production (Islam *et al.*, 1998).

What is clearly required is detailed expression profiling of some key genes. As we do not know whether these are necessarily defence-related genes per se, initially a microarray approach is required, followed by time-course multiple treatment quantitative PCR analysis of pathway-specific genes identified. The treatments should include real field environments.

7.5 Plant and pathogen population genetics

Genotypes of plants respond differentially to the environment with respect to induction of resistance in response to pathogen challenge as noted above. Therefore, we would expect them to respond in similar differential ways to the resistance elicitors derived using the principles of pathogen recognition following challenge. Indeed, there is some evidence of genotype interaction with such elicitors in wild plants (Agrawal, 1999) and even from the narrow genetic range of modern cultivars or breeding lines (Newton & Dashwood, 1998; Newton *et al.*, 2003). Thus, there may be potential to introduce more 'inducible resistance' genes from wild sources into modern cultivars.

It is argued that as induced resistance operates through the plant's own multiple defence pathways, pathogens cannot evade single gene-product triggers through mutation. Selection pressure for pathotypes able do this will be low, and only the selection pressure for enhanced pathogenicity, fitness or aggressiveness common to any non-specific resistance source will be operating. This could lead to an erosion of efficacy of induced resistance agents (McDonald & Linde, 2002), but this is unlikely to be sudden or problematic.

For induction of resistance by chemicals, Ruess *et al.* (1996) argued in the same way that because they act on the plant and not directly on the pathogen, induced resistance in the host plant was postulated to be at less risk of a rapid breakdown. As a prerequisite for this hypothesis, it is to be supposed that induced host resistance will not apply any selection pressure on the pathogen population. Therefore, the development of pathogen strains able to overcome disease management measures, as has been determined for race-specific resistance or fungicide application, is unlikely. For example, in the case of barley powdery mildew, Ruess *et al.* (1996) stated that the inducer BTH was unlikely to cause selection for resistance due to its particular mode of action. On the other hand, it has been supposed that induced resistance will have an effect on the pathogen population similar to that of horizontal resistance (Tuzun, 2001). Therefore, the effectiveness of induced resistance has the potential to 'erode' over time as the pathogen or parasite population evolves (McDonald & Linde, 2002). Also, Sticher *et al.* (1997) and Van Loon *et al.* (1998) expected a quantitative nature of induced resistance because of the cumulative effects of numerous plant defence mechanisms involved. Nevertheless, this hypothesis has rarely been tested experimentally, and there is a need for research evaluating the effects of induced resistance on the composition of pathogen or parasite populations (Vallad & Goodman, 2004).

One major study of selection within powdery mildew populations in response to a resistance elicitor has been reported. Bousset & Pons-Kühnemann (2003) subjected a population produced from 30 isolates of *Blumeria graminis* f. sp. *hordei* to selection on BTH, a fungicide or an untreated control. Using fungicide and virulence markers, they found no shift in the population attributable to BTH alone. However, together with the fungicide, a significant shift was observed, more than that attributable to the fungicide or BTH alone. The explanation for this is unclear, so careful consideration and experimentation need to be carried out to determine the likely effects of resistance elicitor deployment in integrated pest and disease management programmes. Several fungicides, e.g. F500, are thought to possess resistance induction properties (Herms *et al.*, 2002), but even if particular fungicides or fungicide–elicitor combinations do cause directional selection, diversification in host and crop protection method deployment should ensure no long term effects. Using both molecular and virulence markers, Newton *et al.* (1998) examined mildew isolates from untreated and 78% effective yeast cell wall-based elicitor-treated plots of barley and found no differences in diversity indices attributable to the treatments. However, populations should be sampled from more effective elicitors over replicated large areas before more robust conclusions can safely be made.

In contrast to the study of Bousset & Pons-Kühnemann (2003), who treated the impact of induced host resistance on pathogen population as a quantitative effect in a compatible plant–pathogen system, Romero & Ritchie (2004) examined the impact of induced systemic defence response in an incompatible (avirulent) situation. Using the pepper-bacterial spot (causal agent, *Xanthomonas axonopodis* pv. *vesicatoria*) pathosystem, they examined the effect of SAR in reducing the occurrence of race-change mutants that defeat resistance (R) genes. Pepper plants carrying one or more R genes were sprayed with the plant defence activator ASM (BTH) and challenged with incompatible strains of the pathogen. In field experiments, they found a delay in the detection of race-change mutants and a reduction in disease severity. Decreased disease severity was associated with a reduction in the number of race-change mutants and the suppression of disease caused by the race-change mutants. This suggests a possible mechanism related to a decrease in the pathogen population size, which subsequently reduces the number of race-change mutants for the selection pressure of R genes. The authors concluded that inducers of SAR are potentially useful for increasing the durability of genotype-specific resistance conferred by major R genes.

7.6 Consequences of resistance induction

Induction of resistance is costly and therefore normally only triggered upon actual pathogen recognition (see Chapter 9). Challenge of barley with a non-host and an avirulent pathogen was calculated to have significant cost in terms of grain yield (−7%), kernel weight (−4%), grain protein (−11%), straw mass (−3%) and straw length (−5%) (Smedegaard-Petersen & Støjlen, 1981). In *Arabidopsis thaliana*, induction of resistance caused reduced growth initially followed by enhanced compensatory growth (Dietrich *et al.*, 2005). This is indicative of the enhanced disease tolerance often found following elicitor treatment, where reduction in seed production as measured in crop yield in cereals, for example, is less than that expected when disease is present (Kehlenbeck *et al.*, 1994; Reglinski *et al.*, 1994). In the absence of elicitor applications, disease tolerance is best expressed under conditions of high inoculum pressure where resistance is presumably being induced by the pathogen to a high level (Newton *et al.*, 2000), but whether more disease tolerant genotypes have a greater response to elicitors has not been tested. Furthermore, Dietrich *et al.* (2005) found that fitness in terms of seed production was dependent on combinations of environmental factors, only one of which was resistance elicitation.

It has long been recognized that resistance genes and matching virulence may have a cost in terms of fitness to the pathogen, resulting in stabilizing selection (Vanderplank, 1968), and this has also recently been demonstrated experimentally in a host, *A. thaliana* (Tian *et al.*, 2003). In cultivar mixtures, induced resistance is one of the three main contributors to the mixtures effect in reducing disease, the others being the barrier effect and dilution of susceptibles (Chin & Wolfe, 1984). Calonnec *et al.* (1996) determined by experimental means, using wheat and yellow rust infection, that it contributed between 44 and 57% to disease reduction using pure stands of cvs Clement and Austerlitz respectively. As cultivar mixtures are an agricultural implementation of the heterogeneity found in natural and semi-natural ecosystems, we might expect similar cross-protection to be taking place between genotypes of the same species, and perhaps between species here, too. When resistance is characterized in non-crop systems such as *Senecio vulgaris*, many major genes for resistance to *Erysiphe fischeri* are found, indicating that such resistance induction is likely to operate (Bevan *et al.*, 1993a), albeit in the context of a range of resistance strategies including partial-, age- and temperature-dependent resistances (Bevan *et al.*, 1993b). Given that major resistance genes are likely to have a cost, they are likely to be maintained in many natural ecosystems in this diversity rather than becoming fixed in single genotypes, thus contributing also to selection for host population heterogeneity which is necessary for such stability.

In general, diseased plants will produce inferior seed quality and quantity, i.e. reduced fitness (Jarosz *et al.*, 1989). However, there is a report that, following herbivore attack, defences induced in maternal plants may be transmitted to the progeny, resulting in enhanced resistance expression (Agrawal *et al.*, 1999). While this is not heritable in a Mendelian manner, it would allow greater survival and more time for selection of truly resistant types to occur.

Constitutive expression of SAR was also shown to reduce a plant's fitness, but equally, expression of mutants unable to express SAR reduced a plant's fitness in the field when such mechanisms were required (Heidel *et al.*, 2004). However, perception of potential threats by a plant can act to 'prime' its defence recognition mechanisms for faster induction

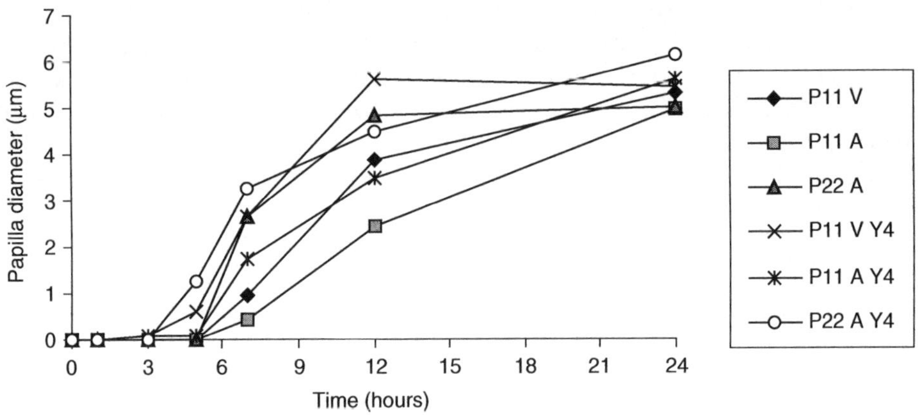

Figure 7.1 Effect of elicitor (Y4) mimicking the faster/larger papilla response of *mlo* resistance (P22) genotypes. P11 is a non-*mlo* genotype (*Mla13*). V: virulent; A: avirulent (B Hughes & AC Newton, unpublished data).

should an actual pathogen be recognized. This 'priming' can be achieved by contact with cell wall components, culture filtrates, various other elicitors or non-pathogenic organisms (Lyon *et al.*, 1995; Conrath *et al.*, 2001, 2002; Ton & Mauch-Mani, 2004). Thus, a plant's normal environment in a natural or semi-natural ecosystem will have such stimuli enabling a plant to expend energy more effectively if a pathogenic challenge is relatively likely. The highly effective *mlo* resistance gene in barley against powdery mildew leads effectively to a permanently 'primed' plant, as it responds to mildew as fast as elicitor-treated plants (Newton & Andrivon, 1995). This can be seen in the time of papilla induction (Figure 7.1), as it is speed of response which distinguishes between resistant and susceptible genotypes in barley against powdery mildew.

Both pathogens and elicitors not only affect plant fitness but can even increase somatic recombination (Lucht *et al.*, 2002). The reason is presumably to give enhanced genetic flexibility only when a stressful environment is present, while at other times the adapted genotype is more protected.

7.7 Conclusions

Clearly, induced resistance is an essential component of normal plant defence strategies. As such, it cannot be fully understood or manipulated in isolation from the plant's normal environment. It has potential for manipulation both in its expression in plants and as an applied crop protectant. However, in contrast to, for example, fungicides which act directly against their target organisms, it is entirely dependent upon the plant's extant resistance mechanisms for its efficacy. Therefore, it is dependent upon the effects of environmental factors, including all biotic and abiotic stressors, on the plant, the target pathogen(s) and all other organisms interacting with the plant and the pathogens. A systems biology approach must therefore be taken to strategically enhance induced resistance for it to be efficacious in target applications. To achieve durable resistance in practice, this is a philosophically much more satisfactory approach as it exploits the often undefined but inbuilt checks and balances of a co-evolved host–pathogen interaction. In contrast, the highly targeted, reductionist approaches of single gene expression or single mode-of-action pesticides are both fundamentally and intellectually non-durable.

As more knowledge of gene expression in induced resistance is gained, further experimentation with real environment pest and pathogen challenges should help us understand the key mechanisms responsible for induced resistance and how they might be manipulated genetically or agronomically for greater efficacy and durability of resistance.

7.8 Acknowledgements

We thank Dr Bleddyn Hughes for Figure 7.1 and SEERAD for funding.

7.9 References

Agrawal A, 1999. Induced responses to herbivory in wild radish: effects on several herbivores and plant fitness. *Ecology* **80,** 1713–1723.

Agrawal A, Laforsch C, Tollrian R, 1999. Transgenerational induction of defences in animals and plants. *Nature* **401,** 61–63.

Ash GJ, Rees RG, 1994. Effect of post-inoculation temperature and light-intensity on expression of resistance to stripe rust in some Australian wheat cultivars. *Australian Journal of Agricultural Research* **45,** 1379–1386.

Bevan JR, Clarke DD, Crute IR, 1993a. Resistance to *Erysiphe fischeri* in 2 populations of *Senecio vulgaris*. *Plant Pathology* **42,** 636–646.

Bevan JR, Crute IR, Clarke DD, 1993b. Diversity and variation in expression of resistance to *Erysiphe fischeri* in *Senecio vulgaris*. *Plant Pathology* **42,** 647–653.

Bousset L, Pons-Kühnemann J, 2003. Effects of acibenzolar-*S*-methyl and ethirimol on the composition of a laboratory population of barley powdery mildew. *Phytopathology* **93,** 305–315.

Calonnec A, Goyeau H, de Vallavieille-Pope C, 1996. Effects of induced resistance on infection efficiency and sporulation of *Puccinia striiformis* on seedlings in varietal mixtures and on field epidemics in pure stands. *European Journal of Plant Pathology* **102,** 733–741.

Chin KM, Wolfe MS, 1984. The spread of *Erysiphe graminis* f. sp. *hordei* in mixtures of barley varieties. *Plant Pathology* **33,** 89–100.

Conrath U, Pieterse CMJ, Mauch-Mani B, 2002. Priming in plant–pathogen interactions. *Trends in Plant Science* **7,** 210–216.

Conrath U, Thulke O, Katz V, Schwindling S, Kohler A, 2001. Priming as a mechanism in induced systemic resistance of plants. *European Journal of Plant Pathology* **107,** 113–119.

Dewey FM, Wong Y, Seery R, Hollins TW, Gurr SJ, 1999. Bacteria associated with *Stagonospora (Septoria) nodorum* increase pathogenicity of the fungus. *New Phytologist* **144,** 489–497.

Dietrich R, Ploss K, Heil M, 2005. Growth response and fitness costs after induction of pathogen resistance depend on environmental conditions. *Plant, Cell and Environment* **28,** 211–222.

Falkhof A-G, Dehne H-W, Schönbeck F, 1988. Dependence of the effectiveness of induced resistance on environmental conditions. *Journal of Phytopathology* **123,** 311–321.

Fobert PR, Després C, 2005. Redox control of systemic acquired resistance. *Current Opinion in Plant Biology* **8,** 378–382.

Fountaine J, 2005. Epidemiological studies of *Rhynchosporium secalis* (leaf blotch of barley). PhD thesis, University of Reading.

Fritz M, Jakobsen I, Lyngkjær MF, Thordal-Christensen H, Pons-Kühnemann J, 2006. Arbuscular mycorrhiza reduces susceptibility of tomato to *Alternaria solani*. *Mycorrhiza* **16,** 413–419.

Heidel AJ, Clarke JD, Antonovics J, Dong X, 2004. Fitness costs of mutations affecting the systemic acquired resistance pathway in *Arabidopsis thaliana*. *Genetics* **168,** 2197–2206.

Heier T, Jain SK, Kogel K-H, Pons-Kühnemann J, 2005. Influence of N-fertilization and fungicide strategies on Fusarium head blight severity and mycotoxin content in winter wheat. *Journal of Phytopathology* **153,** 551–557.

Herms S, Seehaus K, Koehle H, Conrath U, 2002. A strobilurin fungicide enhances the resistance of tobacco against tobacco mosaic virus and *Pseudomonas syringae* pv *tabaci*. *Plant Physiology* **130,** 120–127.

Islam SZ, Honda Y, Arase S, 1998. Light-induced resistance of broad bean against *Botrytis cinerea*. *Journal of Phytopathology* **146,** 479–485.
Jarosz AM, Burdon JJ, Muller WJ, 1989. Long-term effects of disease epidemics. *Journal of Applied Ecology* **26,** 725–733.
Kehlenbeck H, Krone C, Oerke E-C, Schönbeck F, 1994. The effectiveness of induced resistance on yield of mildewed barley. *Journal of Plant Disease and Protection* **101,** 11–21.
Long JC, Jenkins GI, 1998. Involvement of plasma membrane redox activity and calcium homeostasis in the UV-B and UV-A blue light induction of gene expression in Arabidopsis. *The Plant Cell* **10,** 2077–2086.
Lucht JM, Mauch-Mani B, Steiner H-Y, Métreaux J-P, Ryals J, Hohn B, 2002. Pathogen stress increases somatic recombination frequency in Arabidopsis. *Nature Genetics* **30,** 311–314.
Lyon GD, Reglinski T, Newton AC, 1995. Novel disease control chemicals: the potential to 'immunize' plants against infection. *Plant Pathology* **44,** 407–427.
Mateo A, Mühlenbock P, Rustérucci C, Chang CC, Miszalski Z, Karpinska B, Parker JE, Mullineaux PM, Karpinski S, 2004. LESION SIMULATING DISEASE 1 is required for acclimation to conditions that promote excess excitation energy. *Plant Physiology* **136,** 2818–2830.
McDonald BA, Linde C, 2002. Pathogen population genetics, evolutionary potential, and durable resistance. *Annual Review of Phytopathology* **40,** 349–379.
Navarre DA, Mayo D, 2004. Differential characteristics of salicylic acid-mediated signaling in potato. *Physiological and Molecular Plant Pathology* **64,** 179–188.
Newton AC, Andrivon D, 1995. Assumptions and implications of current gene-for-gene hypotheses. *Plant Pathology* **44,** 607–618.
Newton AC, Dashwood EP, 1998. The interaction of humidity and resistance elicitors on expression of polygenic resistance of barley to mildew. *Journal of Phytopathology* **146,** 123–130.
Newton AC, Hackett CA, Guy DC, 1998. Diversity and complexity of *Erysiphe graminis* f. sp. *hordei* collected from barley cultivar mixtures or barley plots treated with a resistance elicitor. European *Journal of Plant Pathology* **104,** 925–931.
Newton AC, Guy DC, Campbell E, Thomas WTB, 2003. The effect of variable environment on inducible powdery mildew resistance expression in barley. *Journal of Plant Diseases and Protection* **110,** 113–119.
Newton AC, Guy DC, Gaunt RE, Thomas WTB, 2000. The effect of powdery mildew inoculum pressure and fertiliser level on disease tolerance in spring barley. *Journal of Plant Diseases and Protection* **107,** 67–73.
Newton AC, Toth IK, Neave P, Hyman L, 2004. Bacterial inoculum from a previous crop affects fungal disease development on subsequent non-host crops. *New Phytologist* **163,** 133–138.
Pasquer F, Isidore E, Zarn J, Keller B, 2005. Specific patterns of changes in wheat gene expression after treatment with three antifungal compounds. *Plant Molecular Biology* **57,** 693–707.
Reglinski T, Newton AC, Lyon GD, 1994. Assessment of the ability of yeast-derived elicitors to control barley powdery mildew in the field. *Journal of Plant Diseases and Protection* **101,** 1–10.
Romero AM, Ritchie DF, 2004. Systemic acquired resistance delays race shifts to major resistance genes in bell pepper. *Phytopathology* **94,** 1376–1382.
Ruess W, Mueller K, Knauf-Beiter G, Kunk W, Staub T, 1996. Plant activator CGA 245704: an innovative approach for disease control in cereals and tobacco. *Proceedings 1996, British Crop Protection Conference – Pests and Diseases*, 53–60.
Salamati S, Reitan L, Flatcher KF, 2003. Occurrence of *Ramularia collo-cigni* in Norway. In: Yahyaoui AH, Brader L, Tekauz A, Wallwork H, Steffenson B, eds. *Meeting the Challenges of Barley Blights.* Proceedings of the Second International Workshop on Barley Leaf Blights, 7–11 April 2002, (ICARDA) Aleppo, Syria, 355–359. [http://www.barleyblights.net/workshops/barleyblights2.pdf]
Smedegaard-Petersen V, Stølen O, 1981. Effect of energy-requiring defence reactions on yield and grain quality in a powdery mildew-resistant barley cultivar. *Phytopathology* **71,** 396–399.
Smedegaard-Petersen V, Tolstrup K, 1985. The limiting effect of disease resistance on yield. *Annual Review of Phytopathology* **23,** 475–490.
Sticher L, Mauch-Mani B, Métraux J-P, 1997. Systemic acquired resistance. *Annual Review of Phytopathology* **35,** 235–270.
Tian D, Traw MB, Chen JQ, Kreitman M, Bergelson J, 2003. Fitness cost of R-gene-mediated resistance in *Arabidopsis thaliana. Nature* **423,** 74–77.
Ton J, Mauch-Mani B, 2004. β-Amino-butyric acid-induced resistance against necrotrophic pathogens is based on ABA-dependent priming for callose. *Plant Journal* **38,** 119–130.

Tuzun S, 2001. Relationship between pathogen-induced systemic resistance and horizontal resistance. *European Journal of Plant Pathology* **107,** 85–93.

Vallad GE, Goodman RM, 2004. Systemic acquired resistance and induced systemic resistance in conventional agriculture. *Crop Science* **44,** 1920–1934.

Vanderplank JE, 1968. *Disease Resistance in Plants*. New York: Academic Press.

Van Loon LC, Bakker PAHM, Pieterse CMJ, 1998. Systemic resistance induced by rhizosphere bacteria. *Annual Review of Phytopathology* **36,** 453–483.

Yalpani N, Enyedi AJ, Leon J, Raskin I, 1994. Ultraviolet-light and ozone stimulate accumulation of salicylic-acid, pathogenesis-related proteins and virus resistance in tobacco. *Planta* **193,** 372–376.

Zhou F, Menke FL, Yoshioka K, Moder W, Shirano Y, Klessig DF, 2004. High humidity suppresses *ssi4*-mediated cell death and disease resistance upstream of MAP kinase activation, H_2O_2 production and defence gene expression. *The Plant Journal* **39,** 920–932.

Chapter 8
Microbial induction of resistance to pathogens

Dale Walters[1] *and Tim Daniell*[2]

[1]Scottish Agricultural College, Edinburgh, UK
[2]Scottish Crop Research Institute, Dundee, UK

8.1 Introduction

It is well documented that prior inoculation with pathogens can induce resistance in plants to subsequent infection. Thus, resistance can be induced by prior inoculation with viral, bacterial and fungal pathogens (see Kuć, 1982). Although most reports involve the use of pathogens which cause necrosis (e.g. Uknes *et al.*, 1993), resistance to subsequent infection can also be induced by pathogens which do not cause necrosis, e.g. biotrophic fungal pathogens like rusts (Murray & Walters, 1992) and powdery mildews (Cho & Smedegaard-Petersen, 1986). Non-pathogens can also induce resistance to pathogen infection. For example, in some recent work, Nelson (2005) showed that drench inoculation of undisturbed roots of barley with *Fusarium oxysporum* f. sp. *radicis-lycopersici*, a non-pathogen of barley, induced systemic resistance to the powdery mildew pathogen *Blumeria graminis* f. sp. *hordei*. But resistance can also be induced by micro-organisms involved in non-pathogenic associations with plants, including symbiotic and endophytic associations. This chapter will focus on resistance induced by mycorrhizal fungi, plant growth-promoting rhizobacteria and a fungal endophyte, as well as by biological control agents.

8.2 Resistance induced by plant growth promoting rhizobacteria

As mentioned in Chapter 4, some bacteria in the rhizosphere actively colonise plant roots in the presence of the existing native microflora. These are known as rhizobacteria, and those which exert a beneficial effect on plant growth are called plant growth-promoting rhizobacteria (PGPR) (Zehnder *et al.*, 2001). Studies on the mechanisms underlying these beneficial effects indicated that PGPR increased growth indirectly by altering the microbial balance in the rhizosphere (Zehnder *et al.*, 1999). Iron-chelating siderophores, antibiotics and hydrogen cyanide are produced by some PGPR and have been implicated in reductions in plant pathogens and harmful rhizobacteria in the soil, with corresponding improvement in plant growth (Zehnder *et al.*, 1999). Rhizobacteria were first used commercially for biological control of plant pathogens in 1985, while in China, PGPR have been used commercially on some 20 million ha of crops for more than 20 years (Zehnder *et al.*, 1999).

In 1988, PGPR strains which increased rapeseed growth in field trials were studied for biological control activity (Kloepper *et al.*, 1988). Results from this work suggested that PGPR strains exhibiting biological control activity fell into two groups, those that control disease by antagonism to the pathogen and those that control disease by mechanisms that do not involve production of toxic compounds (Kloepper *et al.*, 1988). Later studies provided evidence that PGPR can in fact induce resistance in plants to pathogen infection, a phenomenon known as induced systemic resistance (ISR; Zehnder *et al.*, 2001; Bakker *et al.*, 2003). Although, phenotypically, ISR is similar to systemic acquired resistance (SAR), the latter requires salicylic acid (SA) accumulation, while, in a number of cases, ISR is mediated by jasmonic acid (JA) and ethylene (ET) (Bakker *et al.*, 2003). Signalling in ISR has already been dealt with in detail in Chapter 4. The present chapter will focus therefore on the use of PGPR-mediated ISR for disease control.

8.2.1 Glasshouse and controlled environment experiments

Data supporting the conclusion that PGPR can induce resistance to plant pathogens came from three studies in 1991 (Alström, 1991; Van Peer *et al.*, 1991; Wei *et al.*, 1991). Working on carnation, Van Peer *et al.* (1991) inoculated plants by pouring a suspension of PGPR onto roots of cuttings in rock wool and inoculated stems one week later with *Fusarium oxysporum* f. sp. *dianthi*. PGPR-treated plants had a significantly lower incidence of *Fusarium* wilt. In this study, antagonism and competition were ruled out as possible mechanisms for disease control, because of the spatial separation of the PGPR and the pathogen (Van Peer *et al.*, 1991). Using cucumber, Wei *et al.* (1991) screened 94 PGPR strains for ISR against the foliar pathogen *Colletotrichum orbiculare* and found that six PGPR strains provided significant disease control. Since the PGPR strains colonized the roots of the cucumber plants and were not present on leaves, competition and antagonism were ruled out as mechanisms responsible for the disease control observed (Wei *et al.*, 1991).

In order for the disease suppression observed with PGPR to be the result of ISR, it is necessary to show that the disease control is plant mediated and that it extends to parts of the plant not in contact with the inducing PGPR. Work on ISR in a number of plants has shown that the inducing PGPR were not recoverable from sites where plants were challenged with the pathogen (Van Peer *et al.*, 1991; Leeman *et al.*, 1995a; Pieterse *et al.*, 1996). In further work on these plant species, bacterial lipopolysaccharide preparations were used to induce ISR, thus ruling out any protective effects resulting from bacterial metabolism (Van Peer & Schippers, 1992; Leeman *et al.*, 1995b; Van Wees *et al.*, 1996). Further evidence comes from some elegant work using a split root system, where Liu *et al.* (1995) applied a bioluminescent PGPR strain (89B-27) to one part of the cucumber root and inoculated the other part of the root with *F. oxysporum* f. sp. *cucumerinum*. Application of the PGPR strain led to protection against the pathogen without any movement of the bioluminescent PGPR strain from its application site to the part of the root system inoculated with the pathogen (Liu *et al.*, 1995).

Further research since 1991 has demonstrated that many strains of PGPR can trigger ISR against Oomycete, bacterial and fungal pathogens. For example, Yan *et al.* (2002) found that two strains of PGPR, *Bacillus pumilus* SE34 and *Pseudomonas fluorescens* 891361, elicited ISR against late blight on tomato, caused by the Oomycete pathogen *Phytophthora infestans*, and reduced disease severity by a level equivalent to SAR induced by BABA. The authors found that the ISR elicited by both PGPR strains was SA-independent, but dependent on JA and ET (Yan *et al.*, 2002). Another Oomycete pathogen, *Peronospora tabacina*, the cause of blue mould of tobacco, was controlled by a number of strains of PGPR (Zhang *et al.*, 2004). Two strains in particular (*Serratia marcescens* 90–166 and *B. pumilus* SE34) increased plant growth and reduced disease severity. In work on bacterial wilt of tomato caused by *Ralstonia solanacearum*, the PGPR strain *P. putida* 89B61 was shown to reduce the incidence of bacterial wilt significantly when applied to transplants at the time of seeding and one week prior to inoculation with the pathogen (Anith *et al.*, 2004). In this study, BioYield, a formulated PGPR containing two *Bacillus* strains, also decreased bacterial wilt significantly. The non-pathogenic bacterium *Pseudomonas putida* BTP1 was shown to induce systemic resistance in bean to the fungal pathogen *Botrytis cinerea* (Ongena *et al.*, 2004). BTP1-treated plants exhibited increased levels of linoleic and linolenic acids, together with increased activities of the enzymes lipoxygenase and hydroperoxide lyase and elevated concentrations of the fungitoxic compound Z-3-hexenal. These results suggest an association of the oxylipin pathway in the systemic resistance induced by BTP1 (Ongena *et al.*, 2004). Subsequent studies by these workers revealed a major determinant of the ISR elicited by BTP1 to be an *n*-alkylated benzylamine derivative (Ongena *et al.*, 2005).

PGPR mediated ISR has also been reported in *A. thaliana*. In some interesting work by Ryu *et al.* (2003), the level of systemic protection elicited in *A. thaliana* by PGPR was shown to be dependent on the PGPR strain and the challenge pathogen. Six of nine PGPR strains used in this study reduced the severity of *P. syringae* pv. *tomato*, while seven of the PGPR strains reduced the severity of *P. syringae* pv. *maculicola*. These workers found that one PGPR strain differed in the ability to elicit ISR against the two pathovars of *P. syringae*. Since ISR is thought to provide broad spectrum resistance to many pathogens, this result demonstrates an unexpected specificity in defence responses elicited by this PGPR strain (Ryu *et al.*, 2003). Subsequent work by Ryu *et al.* (2004) suggested a role for volatile organic compounds (VOCs) in ISR elicited by some PGPR strains. These workers found that *B. subtilis* GB03, *B. amyloliquefaciens* IN937a, *S. marcescens* 90–166 and *B. pumilus* T4 were all capable of eliciting ISR in *A. thaliana* by emission of VOCs. However, in these studies, not all of the PGPR strains used worked in this way; four strains that elicited ISR when inoculated onto seeds failed to do so when physically separated from the plants (Ryu *et al.*, 2004). Chemical analysis of the bacterial volatile emissions revealed the release of low molecular weight hydrocarbons, including the VOC 2,3-butanediol. When seedlings were exposed to a racemic mixture of 2,3-butanediol, ISR was triggered, while transgenic lines of the PGPR *B. subtilis* that emitted reduced levels of 2,3-butanediol elicited less ISR than wild-type strains of the bacterium (Ryu *et al.*, 2004).

A number of studies have demonstrated that PGPR-mediated ISR is also effective against viruses. For example, Raupach *et al.* (1996) showed that two strains of PGPR induced resistance to cucumber mosaic virus (CMV) in cucumber and tomato. PGPR-mediated ISR has also been reported for tobacco necrosis virus (TNV) and tobacco mosaic virus (TMV) in tobacco (Maurhofer *et al.*, 1994; De Meyer *et al.*, 1999). In the experiments of Maurhofer *et al.* (1994), the resistance induced by *P. fluorescens* strain CHA0 against TNV resulted in reduced lesion numbers and size, while ISR mediated by treatment with *P. aeruginosa* strain 7NSK reduced the size of lesions caused by TMV infection (De Meyer *et al.*, 1999). However, not all studies demonstrated protective effects of PGPR treatment against virus infection. Thus, Ton *et al.* (2002a, b) found no effect of treatment with PGPR against infection of *A. thaliana* with turnip crinkle virus.

More recently, Murphy *et al.* (2003) examined the effect of combinations of two PGPR strains with chitosan, on protection of tomato against CMV. Treatment with chitosan alone is known to lead to low levels of disease resistance (Benhamou *et al.*, 1998) and enhancement of soil microflora (Rodriguez-Kabana *et al.*, 1987). Therefore, combining chitosan with PGPR strains increases the likelihood of obtaining both induced resistance and increased plant growth under variable growth conditions (Kloepper *et al.*, 2004). When tomato plants were treated with a combination of two PGPR strains and chitosan, plant height, fresh weight and flower and fruit numbers were significantly greater than controls. Moreover, treated plants also exhibited significantly lower CMV disease severity than controls (Murphy *et al.*, 2003).

8.2.2 Field experiments

The first successful field evaluations of PGPR-mediated ISR were carried out in the early to mid-1990s on cucumber. This work showed that application of PGPR as a seed treatment followed by soil drench application led to a reduction in the severity of bacterial wilt (Wei *et al.*, 1995) and control of bacterial angular leaf spot and anthracnose (Wei *et al.*, 1996). Subsequent work demonstrated that treatment of cucumber seed with PGPR resulted in increased plant growth and control of angular leaf spot and anthracnose, both in the presence and absence of methyl bromide (Raupach & Kloepper, 2000). These data indicated that use of PGPR to control these pathogens should help to compensate for the reductions in plant growth often observed in the absence of methyl bromide fumigation (Raupach & Kloepper, 2000).

The bacterial pathogen *Erwinia tracheiphila* causes bacterial wilt of cucumber. It is transmitted by spotted and striped cucumber beetles, whose feeding behaviour is strongly influenced by a group of triterpenoid metabolites found in cucumber and known as cucurbitacins. In fact, bacterial wilt is controlled by targeting the beetles with insecticides (Zehnder *et al.*, 2001). Wei *et al.* (1995) found that PGPR protected cucumber against bacterial wilt in the presence of large numbers of cucumber beetles. Subsequent work showed that PGPR-treated plants contained significantly less cucurbitacin C (the primary cucurbitacin in cucumber) than control plants, leading the workers to suggest that PGPR-mediated ISR protects cucumber against bacterial wilt in two ways (Zehnder *et al.*, 2001). The authors suggested that first, reduced levels of cucurbitacin C in PGPR-treated plants make them less palatable to the beetles, resulting in a smaller number of beetles acquiring and transmitting the bacterium; and second, treatment with PGPR may induce plant defences once the pathogen has been introduced into the plant (Zehnder *et al.*, 2001).

The PGPR-mediated ISR against CMV under glasshouse conditions mentioned above, was found to be repeatable under field conditions. Thus, Zehnder *et al.* (2001) demonstrated that although results varied from year to year, field-grown tomatoes treated with PGPR exhibited a reduction in symptoms of infection by CMV and tomato mottle virus (ToMoV), together with increased tomato yield. The authors suggested that the variation in disease control observed with CMV from one year to the next could have resulted from differences in the levels of natural infection. So, the poorer level of disease control observed with CMV on tomato in 1997 may have been due to a higher level of natural infection, with the consequence that the PGPR-induced plant defences may have been unable to compensate for the greater viral load (Zehnder *et al.*, 2001). In more recent work, strains of *P. fluorescens* were found to provide control of tomato spotted wilt virus on tomato in field and glasshouse experiments (Kandan *et al.*, 2005). In this work, disease control was accompanied by increased plant growth and yield compared to controls, and plants treated with the PGPR strains exhibited a reduction in viral antigen concentrations compared to controls (Kandan *et al.*, 2005).

In field experiments conducted in Thailand in 2001 and 2002, Jetiyanon *et al.* (2003) examined the effects of PGPR, used alone or as mixtures, on disease control in a number of crops. This work focused on southern blight of tomato caused by *Sclerotium rolfsii*, anthracnose of long cayenne pepper caused by *Colletotrichum gloeosporioides* and mosaic disease of cucumber caused by CMV. PGPR mixtures (all *Bacillus* spp.) were found to suppress disease more consistently than the PGPR strain used alone (*B. pumilus* IN937b). Indeed, one particular PGPR mixture (*B. amyloliquefaciens* IN937a + *B. pumilus* IN937b) provided significant protection against all diseases in both seasons (Jetiyanon *et al.*, 2003). In this work, the PGPR-mediated ISR was associated with increased plant growth in most cases and sometimes with enhanced total yield, but treatments which gave the best disease control were not always those which most enhanced plant growth and yield (Jetiyanon *et al.*, 2003).

Combining the use of PGPR with reduced fungicide application may be useful in cases where obtaining effective disease control is difficult. Such an approach would also help to reduce fungicide use by cutting down the number of sprays applied in a season. For example, Silva *et al.* (2004) found that combined treatments of PGPR and the fungicide chlorothalonil provided effective control of *Alternaria solani, P. infestans* and *Septoria lycopersici* on tomato under field conditions. Here, the PGPR treatment was used together with 10 fungicide sprays, compared to the 20 fungicide sprays used in practice (Silva *et al.*, 2004).

While there are many pathogens on a range of crop plants which are amenable to control by strains of PGPR, there are also examples where PGPR have failed to provide disease control. For example, in studies of late leaf spot of peanut caused by *Cercosporidium personatum*, although some PGPR strains elicited ISR in a glasshouse assay, treatment with PGPR did not provide disease control in the field (Zhang *et al.*, 2001). Moreover, a number of chemical inducers of resistance were also used in this work, including BTH and BABA. None of the chemical inducers could provide significant and consistent disease control under field conditions (Zhang *et al.*, 2001). It would appear therefore that for some crop diseases, induced resistance, irrespective of how it is elicited, is not an option for disease control.

In their review of induced resistance in conventional agriculture, Vallad & Goodman (2004) highlighted 60 examples where PGPR were used to control crop diseases. Although particularly high levels of disease control were achieved in some cases, e.g. disease control in cucumber provided by the PGPR *B. pumilis* INR-7 and *S. marcescens* 90–166 (Zehnder *et al.*, 2001), reductions in disease severity of less than 80% were obtained in 57 of these studies (Vallad & Goodman, 2004). Since ISR is a plant response, it is likely to be influenced by many factors, including genotype and environment (Vallad & Goodman, 2004; Walters *et al.*, 2005). For example, in *A. thaliana*, the PGPR strain *P. fluorescens* WCS417r was capable of eliciting an ISR response in most, but not all, ecotypes (e.g. Van Wees *et al.*, 1997). Other factors, e.g. disease pressure, can also influence the efficacy of ISR. Thus, Murphy *et al.* (2000) attributed the inconsistency of PGPR-mediated

ISR against tomato mottle virus in field trials to increased disease pressure. Maximizing the efficacy of ISR is likely to depend therefore on a sound understanding of the effects of these factors on the expression of ISR.

8.3 Induction of resistance by biological control agents

Control of plant pathogens by biological control agents (BCAs) can involve both direct and indirect mechanisms. Direct modes of action include mycoparasitism and production of inhibitory compounds, while indirect mechanisms can include competition for nutrients and space. However, data from some studies showed that some BCAs can also affect the host plant. Thus, cellulose from *Trichoderma viride* was found to induce plant defence responses in grapevine cell cultures (Calderón *et al.*, 1993), while control of *Phytophthora parasitica* var. *nicotianae* on tobacco by *T. longibrachiatum* was linked to the induction of plant defences (Chang *et al.*, 1997). Subsequent work on control of *B. cinerea* on a number of plant species using the BCA *T. harzianum* T39 provided further evidence for the involvement of induced plant defences (De Meyer *et al.*, 1998). These workers found that in tomato, lettuce, pepper, bean and tobacco, application of the BCA at sites spatially separated from sites of inoculation with *B. cinerea* led to significant disease control. Disease control under conditions where the BCA and the pathogen are spatially separated suggests the induction of systemic resistance (De Meyer *et al.*, 1998).

In some detailed work using sugar beet, Bargabus *et al.* (2002) showed that the non-pathogenic, phyllosphere-inhabiting bacterium *Bacillus mycoides* (isolate Bac J) reduced *Cercospora* leaf spot by 38–91%. They found that disease control was achieved even when the bacterium and pathogen were spatially separated and that following treatment with *B. mycoides*, sugar beet plants exhibited increased activities of chitinase, β-1,3-glucanase and peroxidase (Bargabus *et al.*, 2002). Indeed, in plants treated with *B. mycoides*, new isoforms of these three enzymes were detected. Interestingly, the same isoforms were also detected in sugar beet plants treated with BTH (Bargabus *et al.*, 2002). Based on this evidence, these workers suggested that control of Cercospora leaf spot on sugar beet using *B. mycoides* involves the induction of systemic resistance (Bargabus *et al.*, 2002). Further evidence for the involvement of induced resistance in disease control provided by a BCA came from work by Kilic-Ekici & Yuen (2003). These researchers examined the control of leaf spot of tall fescue caused by *Bipolaris sorokiniana* using the BCA *Lysobacter enzymogenes* strain C3. They found that while application of live or heat-killed BCA cells to tall fescue leaves resulted in localized resistance confined to the treated leaf, treatment of roots with the BCA led to the expression of systemic resistance in leaves (Kilic-Ekici & Yuen, 2003). The induced resistance observed was long lasting and was not pathogen or host specific, with *L. enzymogenes* controlling *B. sorokiniana* on wheat, as well as *Rhizoctonia solani* on tall fescue. Moreover, treatment of tall fescue leaves or roots with *L. enzymogenes* resulted in significantly increased peroxidase activities compared to controls (Kilic-Ekici & Yuen, 2003). Induction of defence related enzymes was also observed in disease control provided by *B. subtilis* strain AUBS1 (Jayaraj *et al.*, 2004). Here, control of sheath blight of rice, caused by *Rhizoctonia solani*, with foliar application of *B. subtilis* AUBS1, was accompanied by increased activities of PAL and peroxidase. Application of *B. subtilis* also led to accumulation of two isoforms of β-1,3-glucanase (Jayaraj *et al.*, 2004). These authors suggest that the co-ordinate up-regulation of these defences in plants treated with *B. subtilis* AUBS1 may be responsible for the disease control observed (Jayaraj *et al.*, 2004).

8.4 Resistance induced by composts

Compost is the final product of the aerobic biodegradation of organic matter. Applied to soils or container media, it has been shown to suppress the severity of diseases caused by soil-borne plant pathogens, especially those caused by *Pythium* and *Phytophthora* spp. (Hoitink & Boehm, 1999). Although microbiostasis and parasitism appear to be the key mechanisms by which these root rots are suppressed (Chen *et al.*, 1988; Mandelbaum & Hadar, 1990; Boehm *et al.*, 1997), systemic induced resistance can also play a role in the biological control provided by compost amendments (Zhang *et al.*, 1996; Pharand *et al.*, 2002). For example, Pharand *et al.* (2002) showed that incorporation of composted paper mill sludge into a peat-based potting mix induced the formation of physical barriers at infection sites in tomato, thus limiting colonization by *F. oxysporum* f. sp. *radicis-lycopersici*.

Although compost-amended media usually suppress root rots caused by *Pythium* and *Phytophthora* spp. within a few days of their formulation (Hoitink & Boehm, 1999), composts are highly variable in their suppressive effects against foliar pathogens (Zhang *et al.*, 1996; Krause *et al.*, 2003). However, inoculation of compost-amended potting mixes with micro-organisms that are capable of triggering systemic resistance (see Section 8.3 above) can enhance systemic protection (Zhang *et al.*, 1998; Pharand *et al.*, 2002). Isolates of several *Trichoderma* spp. have been reported to induce systemic resistance (De Meyer *et al.*, 1998; Yedidia *et al.*, 1999; Sid Ahmed *et al.*, 2000), and since populations of *T. hamatum* and *T. harzianum* are often abundant in composts, they may be good candidates for inoculation of composts (Hoitink *et al.*, 2006). Indeed, when radish, lettuce and tomato plants were grown in composted pine bark fortified with *T. hamatum* 382, they were less severely infected with the bacterial leaf spot pathogen, *Xanthomonas campestris*, than plants grown in commercial peat mix or vermiculite (Aidahmani *et al.*, 2005). In other work, Hoitink *et al.* (2006) showed that Phytophthora dieback of *Rhododendron* cv. *roseum elegans* and Botryosphaeria dieback of *Myrica pennsylvanica* were suppressed in a compost-amended medium containing *T. hamatum* 382. They found however, that Phytophthora dieback was not suppressed in the *Rhododendron* cvs. Aglo and PJM Elite, both of which are very susceptible to *Phytophthora*. The authors suggest that if the systemic protective effect of *T. hamatum* 382 cannot be activated in these cultivars, presumably because they lack resistance to the pathogen, this could limit the application of systemic induced resistance in some nursery crops (Hoitink *et al.*, 2006).

8.5 Disease control provided by an endophytic fungus

Ascomycete endophytes have often been reported to protect plants against attack by pathogens and pests. For example, there is a long history of associations, ranging from mutualism to antagonism, between grasses and fungi belonging to the Clavicipitaceae (Schardl *et al.*, 2004). These fungi exhibit narrow host ranges, are confined to aerial plant parts and grow intercellularly. Moreover, it would appear from the published literature that the protection afforded by these fungal endophytes to their hosts is the result of direct effects on pathogens and pests by fungally produced alkaloids (Schardl *et al.*, 2004).

However, recent work on the association between the root colonizing basidiomycete endophyte *Piriformospora indica* and barley indicates that induced resistance may be involved in protection against pathogen infection (Waller *et al.*, 2005). Infestation of barley roots with *P. indica* resulted in reduced infection of leaves by the powdery mildew fungus *Blumeria graminis* f. sp. *hordei*. This control of barley powdery mildew was associated with an increased frequency of the hypersensitive response (leading to host cell death) and cell wall associated defence, leading to reduced penetration success (Waller *et al.*, 2005). Interestingly, root colonization by *P. indica* also increased the tolerance of barley to salt stress. In fact, plants with roots colonized by this endophyte exhibited an elevated antioxidative capacity due to activation of the glutathione ascorbate cycle (Waller *et al.*, 2005). Despite the re-programmed metabolic state of *P. indica* infested barley, grain yield is not negatively affected. The authors suggest that since *P. indica* can be cultured axenically, it could be cultured on a large scale, thus offering potential for use in sustainable agriculture (Waller *et al.*, 2005).

8.6 Mycorrhizal symbiosis and induced resistance

The majority of effort relating to the suppression of disease by mycorrhizal fungi has concentrated on arbuscular mycorrhiza (AM). This is primarily due to the widespread nature of this symbiosis, with the majority of plant species, including most crop plants, capable of forming AM relationships with glomalean fungi. The effect of the symbiosis on interaction between pathogens and their hosts is complex with positive, negative and neutral outcomes reported (Dehne, 1982). The interaction between mycorrhizal plants and a wide range of pathogen types (bacterial, fungal, oomycete and nematode) has been examined, and a number of reviews are available, including general reviews (e.g. Dehne, 1982; Borowicz, 2001), reviews dealing with specific pathogen types such as nematodes (e.g. Gera Hol & Cook, 2005) and others dealing with the possible application of AM fungi for the biocontrol of root pathogens (e.g. Whipps, 2004). Five main hypotheses have been proposed to explain the effects of the tri-partite interaction between plant, mycorrhizal fungus and pathogen. These hypotheses have been eloquently discussed in a recent paper by Bennett *et al.* (2006). Briefly, the five hypotheses are: (1) the nutritional quality hypothesis, where AM fungi improve the nutritional balance of the plant, increasing its value as a resource for pathogens, thus leading to greater levels of infection; (2) an extension of hypothesis 1, where there is an indirect interaction between the AM fungus and the pathogen resulting in an increased infection rate; (3) a tolerance hypothesis where the fungus increases plant tolerance to the pathogen, indirectly increasing plant fitness; (4) an interference response where the fungus directly interferes with the ability of the pathogen to colonize the plant efficiently; (5) a defence hypothesis where the plant defence response is altered in the presence of the fungus, leading to an increased resistance to the pathogen. This is again an indirect effect but clearly is the major route for systemic induction of resistance, thus forming the main focus of this chapter.

It is, unsurprisingly, difficult to dissect which hypothesis is correct, and indeed resistance may be due to a combination of effects or may be different depending on the combination of plant, AM fungus and pathogen. It has become accepted that these fungi are both multi-functional and functionally diverse, and this further complicates the analysis of the mode of action of any resistance mechanism. The complexity of these interactions can be observed in experiments examining the effect of AM fungal colonization on pathogenicity. For example, Garmendia *et al.* (2004) examined the effectiveness of three AM fungi, *Glomus mosseae*, *G. intraradices* and *G. deserticola* for the protection of *Capsicum* against verticillium wilt. The three fungi displayed differential effectiveness against wilt, as measured by plant growth and pepper yield, with *G. intraradices* increasing pathogenicity, *G. mosseae* increasing plant growth but not reducing pathogenicity and *G. deserticola* improving performance on both scores. This study contrasts with other examples demonstrating suppression of *Rhizoctonia* root rot (Berta *et al.*, 2005) and *Phytophthora* infection (Pozo *et al.*, 2002) in tomato when colonized by *G. mosseae*, but no suppression of *Rhizoctonia* with *G. intraradices* in bean (Guillon *et al.*, 2002).

These examples show that the combination of plant, AM fungus, pathogen and conditions is important in determining the outcome of any interaction. It also appears that the timing of colonization is important, with studies showing that pre-inoculation of plants with AM fungi reduces subsequent pathogenicity. Examples of such results include the examination of the interaction between *Ammophila arenaria* and root-feeding nematodes where de la Peña *et al.* (2006) showed that pre-inoculation of dune grass increased resistance to attack by root feeding nematodes. This effect is of particular interest in this plant due to the link between constant sand burial and fitness, but the importance of pre-inoculation for effective protection is also evident in other systems such as coffee, where Vaast *et al.* (1998) observed this effect in the relationship with a root feeding nematode. In nature, this is unlikely to be an issue due to the almost ubiquitous habit of the symbiosis, and the patchy distribution of pathogens in both space and time, meaning that plants targeted by pathogens are likely to have pre-existing mycorrhizal partners.

The majority of experiments examining the relationship between mycorrhizal plants and pathogens have been performed in single pot experiments where it has been difficult, using traditional techniques, to dissect the likely mode of action of any resistance mechanism. Possible mechanisms for systemic resistance in mycorrhizal plants have been known for some time. For example, it is known that levels of plant hormones (Allen *et al.*, 1980, 1982) and plant defence genes (Lambais & Mehdy, 1995) can be elevated in mycorrhizal plants. There is now an increasing literature where authors have utilized a range of methods to assess possible systemic resistance associated with mycorrhizal colonization. The classic method for avoiding the limitation of simple single pot experiments is to utilize split root experiments where the root system of a single plant is maintained in two pots, only one of which contains a fungal inoculum. These systems thus allow the separation of local and systemic effects, although it is still difficult to make allowance for increased resistance due to improved plant health. Using such a system, Pozo *et al.* (2002) demonstrated that colonization by *G. mosseae* gave some protection of distal tomato roots against infection by *Phytophthora parasitica*, and this study and previous work (Cordier *et al.*, 1998) demonstrated that this effect was mediated by changes in the expression of defence related genes and that both local and systemic mechanisms were important for resistance. Other work utilizing split root systems has demonstrated that the relationship between tomato and *G. versiforme* showing resistance to the bacterial pathogen *Ralstonia solanacearum* was mediated both systemically and locally with an increase in the phenolic content of tissues (Zhu & Yao, 2004). Recently, the application of a proteomics approach has demonstrated that colonization of a highly susceptible line of *Medicago truncatula* with *G. intraradices* induced selected proteins to a similar level to that found in resistant lines, with a concomitant increase in resistance to the oomycete *Aphanomyces euteiches* (Colditz *et al.*, 2005). A transcriptomics approach identified the importance of a chitinase gene, expressed throughout the root system, in the resistance of the grape vine *Vitis amurensis* to the nematode *Meloidogyne incognita*, a result which was confirmed by the observed resistance of a tobacco line expressing this gene (Li *et al.*, 2006).

Interestingly, there is little or no evidence of foliar pathogen resistance enabled through a systemic mechanism, and indeed a review by Dehne (1982) suggests that any resistance mechanism may rely on better general performance. This summary is supported by later work which suggests a higher level of necrotic lesions, with delayed expression of pathogen related proteins in mycorrhizal (*G. intraradices*) tobacco plants (Shaul *et al.*, 1999). Also, it appears that mycorrhizal colonization has effects on foliar feeding by insects, again with mixed results depending on the feeding behaviour and specialism of the insect (Gehring & Whitham, 2002).

Overall, there is a scarcity of data relating to effects of AM fungi on plant pathogens, especially under field conditions, with only a few papers published, including work by Newsham *et al.* (1995) which demonstrated enhanced disease resistance in the field. There is a realization that the number of AM fungal types has been grossly underestimated for a number of years (e.g. Clapp *et al.*, 1995), and in addition, there is evidence for multifunctionality in mycorrhizal relations (e.g. van de Heijden *et al.*, 1998), including negative AM fungal–plant interactions (e.g. Modjo & Hendrix, 1986), and observed preference in the relationship (e.g. Vandenkoornhuyse *et al.*, 2002). This suggests that simple tri-partite experiments, using essentially random combinations of AM fungi and plant, may do little to explain the possible benefits of mycorrhizal colonization and explain the common occurrence of negative conclusions. Targeted experiments utilizing fungi known to associate with specific plants are required to assess likely benefits in the field. In this light, the observed low diversity of AM fungi in arable systems (e.g. Helgason *et al.*, 1998) is of concern if a shift to low input systems drives a greater interest in this symbiosis for plant defence. A greater understanding of the biology and ecology of this relationship is therefore needed to assess the utility of the symbiosis to aid plant health.

8.7 Acknowledgements

The Scottish Agricultural College and the Scottish Crop Research Institute receive financial support from the Scottish Executive, Environment and Rural Affairs Department.

8.8 References

Aidahmani JH, Abbasi PA, Sahin F, Hoitink HAJ, Miller SA, 2005. Reduction of bacterial leaf spot severity on radish, lettuce, and tomato plants grown in compost-amended potting mixes. *Canadian Journal of Plant Pathology* **27,** 186–193.

Allen MF, Moore Jr TS, Christensen M, 1980. Phytohormone changes in *Bouteloua gracilis* infected by vesicular-arbuscular mycorrhizae. I. Cytokinin increases in the host plant. *Canadian Journal of Botany* **58,** 371–374.

Allen MF, Moore Jr TS, Christensen M, 1982. Phytohormone changes in *Bouteloua gracilis* infected by vesicular-arbuscular mycorrhizae. II. Altered levels of gibberellin-like substances and abscisic acid in the host plant. *Canadian Journal of Botany* **60,** 468–471.

Alström S, 1991. Induction of disease resistance in common bean susceptible to halo blight bacterial pathogen after seed bacterisation with rhizosphere pseudomonads. *Journal of Genetic and Applied Microbiology* **37,** 495–501.

Anith KN, Momol MT, Kloepper JW, Marois JJ, Olson SM, Jones JB, 2004. Efficacy of plant growth promoting rhizobacteria, acibenzolar-*S*-methyl, and soil amendment for integrated management of bacterial wilt on tomato. *Plant Disease* **88,** 669–673.

Bakker PAHM, Ran LX, Pieterse CMJ, van Loon LC, 2003. Understanding the involvement of rhizobacteria-mediated induction of systemic resistance in biocontrol of plant diseases. *Canadian Journal of Plant Pathology* **25,** 5–9.

Bargabus RL, Zidack NK, Sherwood JE, Jacobsen BJ, 2002. Characterisation of systemic resistance in sugar beet elicited by a non-pathogenic, phyllosphere-colonising *Bacillus mycoides*, biological control agent. *Physiological and Molecular Plant Pathology* **61,** 289–298.

Benhamou N, Kloepper JW, Tuzun S, 1998. Induction of resistance against Fusarium wilt of tomato by combination of chitosan with an endophytic bacterial strain: ultrastructure and cytochemistry of the host response. *Planta* **204,** 153–168.

Bennett AE, Alers-Garcia J, Bever JD, 2006. Three way interactions among mutualistic mycorrhizal fungi, plants and plant enemies: hypothesis and synthesis. *The American Naturalist* **167,** 141–152.

Berta G, Sampo S, Gamalero E, Massa N, Lemanceau P, 2005. Suppression of *Rhizoctonia* root-rot of tomato by *Glomus mosseae* BEG 12 and *Pseudomonas fluorescens* A6RI is associated with their effect on the pathogen growth and on the root morphogenesis. *European Journal of Plant Pathology* **111,** 279–288.

Boehm MJ, Wu T, Stone AG, Kraakman B, Ianotti DA, Wilson GE, Madden LV, Hoitink HAJ, 1997. Cross-polarized magic-angle spinning ^{13}C nuclear magnetic resonance spectroscopic characterization of soil organic matter relative to culturable bacterial species composition and sustained biological control of *Pythium* root rot. *Applied and Environmental Microbiology* **63,** 162–168.

Borowicz VA, 2001. Do arbuscular mycorrhizal fungi alter plant–pathogen relations? *Ecology* **82,** 3057–3068.

Calderón AA, Zapata JM, Muñoz R, Pedreño MA, Ros Barceló A, 1993. Resveratrol production as a part of the hypersensitive-like response of grapevine cells to an elicitor from *Trichoderma viride. New Phytologist* **124,** 455–463.

Chang P-FL, Xu Y, Narasimhan ML, Cheah KT, D'Urzo MP, Damsz B, Kononowicz AK, Abad L, Hasegawa PM, Bressan RA, 1997. Induction of pathogen resistance and pathogenesis-related genes in tobacco by a heat stable *Trichoderma* mycelial extract and plant signal messengers. *Physiologia Plantarum* **100,** 341–352.

Chen W, Hoitink HAJ, Madden LV, 1988. Microbial activity and biomass in container media for predicting suppressiveness to damping-off caused by *Pythium ultimum. Phytopathology* **78,** 1447–1450.

Cho BH, Smedegaard-Petersen V, 1986. Induction of resistance to *Erysiphe graminis* f. sp. *hordei* in near isogenic barley lines. *Phytopathology* **76,** 301–305.

Clapp JP, Young JPW, Merryweather JW, Fitter AH, 1995. Diversity of fungal symbionts in arbuscular mycorrhizas from a natural community. *New Phytologist* **130,** 259–265.

Colditz F, Braun H-P, Jacquet C, Niehaus, K, Krajinski F, 2005. Proteomic profiling unravels insights into the molecular background underlying increased *Aphanomyces euteiches*-tolerance of *Medicago truncatula. Plant Molecular Biology* **59,** 387–406.

Cordier C, Pozo MJ, Barea JM, Gianinazzi S, Gianinazzi-Pearson V, 1998. Cell defense responses associated with localized and systemic resistance to *Phytophthora parasitica* induced in tomato by an arbuscular mycorrhizal fungus. *Molecular Plant–Microbe Interactions* **11,** 1017–1028.

Dehne HW, 1982. Interaction between vesicular-arbuscular mycorrhizal fungi and plant pathogens. *Phytopathology* **72,** 1115–1119.

de la Peña E, Rodríguez Echevarría S, Van Der Putten WH, Freitas H, Moens M, 2006. Mechanism of control of root-feeding nematodes by mycorrhizal fungi in the dune grass *Ammophila arenaria. New Phytologist* **169,** 829–840.

De Meyer G, Bigirimana J, Elad Y, Hofte M, 1998. Induced systemic resistance in *Trichoderma harzianum* T39 biocontrol of *Botrytis cinerea. European Journal of Plant Pathology* **104,** 279–286.

De Meyer G, Audenaert K, Hofte M, 1999. *Pseudomonas aeruginosa* 7NSK2-induced resistance in tobacco depends on in planta salicylic acid accumulation but is not associated with PR1a expression. *European Journal of Plant Pathology* **105,** 513–517.

Garmendia I, Goicoechea N, Aguirreola J, 2004. Effectiveness of three *Glomus* species in protecting pepper (*Capsicum annuum* L.) against verticillium wilt. *Biological Control* **31,** 296–305.

Gehring CA, Whitham TG, 2002. Mycorrhizae–herbivore interactions: Population and community consequences. In: van de Heijden MGA, Sanders IR, eds. *Mycorrhizal Ecology.* Berlin: Springer, 295–320.

Gera Hol WH, Cook R, 2005. An overview of arbuscular mycorrhizal fungi-nematode interactions. *Basic and Applied Ecology* **6,** 489–503.

Guillon C, St-Armand M, Hamel C, Jabaji-Hare SH, 2002. Differential and systemic alteration of defence related gene transcript levels in mycorrhizal bean plants infected with *Rhizoctonia solani*. *Canadian Journal of Botany* **80,** 305–315.

Helgason T, Daniell TJ, Husband R, Fitter AH, Young JPW, 1998. Ploughing up the wood-wide web? *Nature* **394,** 431.

Hoitink HAJ, Boehm MJ, 1999. Biocontrol within the context of soil microbial communities: a substrate-dependent phenomenon. *Annual Review of Phytopathology* **37,** 427–446.

Hoitink HAJ, Madden LV, Dorrance AE, 2006. Systemic resistance induced by *Trichoderma:* interactions between the host, the pathogen, the biocontrol agent and soil organic matter quality. *Phytopathology* **96,** 186–189.

Jayaraj J, Yi H, Liang GH, Muthukrishnan S, Velazhahan R, 2004. Foliar application of *Bacillus subtilis* AUBS1 reduces sheath blight and triggers defense mechanisms in rice. *Journal of Plant Diseases and Protection* **111,** 115–125.

Jetiyanon K, Fowler, WD, Kloepper JW, 2003. Broad-spectrum protection against several pathogens by PGPR mixtures under field conditions in Thailand. *Plant Disease* **87,** 1390–1394.

Kandan A, Ramiah M, Vasanthi VJ, Radjacommare R, Nandakumar R, Ramanathan A, Samiyappan R, 2005. Use of *Pseudomonas fluorescens*-based formulations for management of tomato spotted wilt virus (TSWV) and enhanced yield in tomato. *Biocontrol Science and Technology* **15,** 553–569.

Kilic-Ekici O, Yuen GY, 2003. Induced resistance as a mechanism of biological control by *Lysobacter enzymogenes* strain C3. *Phytopathology* **93,** 1103–1110.

Kloepper JW, Hume DJ, Scher FM, Singleton C, Tipping B, Laliberte M, Frauley K, Kutchaw T, Simonson C, Lifshitz R, Zaleska I, Lee L, 1988. Plant growth promoting rhizobacteria (PGPR) on canola (rapeseed). *Plant Disease* **72,** 42–46.

Kloepper JW, Reddy MS, Kenney DS, Kokalis-Burelle N, Martinez-Ochoa N, Vavrina CS, 2004. Theory and applications of rhizobacteria for transplant production and yield enhancement. *Acta Horticulturae* **631,** 219–229.

Krause MS, De Ceuster TJJ, Tiquia SM, Michel FC, Madden LV, Hoitink HAJ, 2003. Isolation and characterization of rhizobacteria from composts that suppress the severity of bacterial leaf spot of radish. *Phytopathology* **93,** 1292–1300.

Kuć J, 1982. Induced systemic protection in plants. *Trends in Biotechnology* **3,** 125–128.

Lambais MR, Mehdy MC, 1995. Differential expression of defense-related genes in arbuscular mycorrhiza. *Canadian Journal of Botany* **73,** S533–S540.

Leeman M, Van Pelt JA, Den Ouden FM, Heinsbroek M, Bakker PAHM, Schippers B, 1995a. Induction of systemic resistance by *Pseudomonas fluorescens* in radish cultivars differing in susceptibility to Fusarium wilt, using a novel bioassay. *European Journal of Plant Pathology* **101,** 655–664.

Leeman M, Van Pelt JA, Den Ouden FM, Heinsbroek M, Bakker PAHM, Schippers B, 1995b. Induction of systemic resistance against Fusarium wilt of radish by liposaccharides of *Pseudomonas fluorescens*. *Phytopathology* **85,** 1021–1027.

Li H-Y, Yang G-D, Shu H-R, Yang Y-T, Ye B-X, Nishada I, Zheng C-C, 2006. Colonization by the arbuscular mycorrhizal fungus *Glomus versiforme* induces a defense response against the root-knot nematode *Meloidogyne incognita* in the grapevine (*Vitis amurensis* Rupr.) which includes transcriptional activation of the class III chitinase gene *VCH3*. *Plant Cell Physiology* **47,** 154–163.

Liu L, Kloepper JW, Tuzun S, 1995. Induction of systemic resistance in cucumber against Fusarium wilt by plant growth promoting rhizobacteria. *Phytopathology* **85,** 695–698.

Mandelbaum R, Hadar Y, 1990. Effects of available carbon source on microbial activity and suppression of *Pythium aphanidermatum* in compost and peat container media. *Phytopathology* **80,** 794–804.

Maurhofer M, Hase C, Meuwly P, Métraux J-P, Defago G, 1994. Induction of systemic resistance of tobacco to tobacco necrosis virus by the root colonising *Pseudomonas fluorescens* strain CHA0: influence of the *gacA* gene and of pyoverdine production. *Phytopathology* **84,** 139–146.

Modjo HS, Hendrix JW, 1986. The mycorrhizal fungus *Glomus macrocarpum* as a cause of tobacco stunt disease. *Phytopathology* **76,** 688–691.

Murphy JF, Zehnder GW, Schuster DJ, Sikora EJ, Polston JE, Kloepper JW, 2000. Plant growth promoting rhizobacteria mediated protection in tomato against tomato mottle virus. *Plant Disease* **84,** 779–784.

Murphy JF, Reddy MS, Ryu C-M, Kloepper JW, Li R, 2003. Rhizobacteria-mediated growth promotion of tomato leads to protection against cucumber mosaic virus. *Phytopathology* **93,** 1301–1307.

Murray DC, Walters DR, 1992. Increased photosynthesis and resistance to rust infection in upper, uninfected leaves of rusted broad bean (*Vicia faba* L.). *New Phytologist* **120,** 235–242.

Nelson HE, 2005. *Fusarium oxysporum* f. sp. *radicis-lycopersici* can induce systemic resistance in barley against powdery mildew. *Journal of Phytopathology* **153,** 366–370.

Newsham KK, Fitter AH, Watkinson AR, 1995. Arbuscular mycorrhizas protect an annual grass from root pathogenic fungi in the field. *Journal of Ecology* **83,** 991–1000.

Ongena M, Duby F, Rossignol F, Fauconnier ML, Dommes J, Thonart P, 2004. Stimulation of the lipoxygenase pathway is associated with systemic resistance induced in bean by a non-pathogenic *Pseudomonas* strain. *Molecular Plant–Microbe Interactions* **17,** 1009–1018.

Ongena M, Jourdan E, Schafer M, Kech C, Budzikiewicz H, Luxen A, Thonart P, 2005. Isolation of an n-alkylated benzylamine derivative from *Pseudomonas putida* BTP1 as elicitor of induced systemic resistance in bean. *Molecular Plant–Microbe Interactions* **18,** 562–569.

Pharand B, Carisse O, Benhamou N, 2002. Cytological aspects of compost-mediated induced resistance against Fusarium crown and root rot in tomato. *Phytopathology* **92,** 424–438.

Pieterse CMJ, Van Wees SCM, Hoffland E, Van Pelt JA, Van Loon LC, 1996. Systemic resistance in Arabidopsis induced by biocontrol bacteria is independent of salicylic acid accumulation and pathogenesis related gene expression. *The Plant Cell* **8,** 1225–1237.

Pozo MJ, Cordier C, Dumas-Gaudot E, Gianinazzi S, Barea JM, Azcon-Aguilar C, 2002. Localised versus systematic effect of arbuscular mycorrhizal fungi on defense responses to *Phytophthora* infection in tomato plants. *Journal of Experimental Botany* **53,** 525–534.

Raupach GS, Kloepper JW, 2000. Biocontrol of cucumber diseases in the field by plant growth promoting rhizobacteria with and without methyl bromide fumigation. *Plant Disease* **84,** 1073–1075.

Raupach GS, Liu L, Murphy JF, Tuzun S, Kloepper JW, 1996. Induced systemic resistance in cucumber and tomato against cucumber mosaic cucumovirus using plant growth promoting rhizobacteria (PGPR). *Plant Disease* **80,** 891–894.

Rodriguez-Kabana R, Morgan-Jones G, Chet I, 1987. Biological control of nematodes: soil amendments and microbial antagonists. *Plant and Soil* **100,** 237–247.

Ryu C-M, Hu C-H, Reddy MS, Kloepper JW, 2003. Different signaling pathways of induced resistance by rhizobacteria in *Arabidopsis thaliana* against two pathovars of *Pseudomonas syringae*. *New Phytologist* **160,** 413–420.

Ryu C-M, Murphy JF, Mysore KS, Kloepper JW, 2004. Plant growth promoting rhizobacteria systemically protect *Arabidopsis thaliana* against cucumber mosaic virus by a salicylic acid and NPR1-independent and jasmonic acid-dependent signaling pathway. *The Plant Journal* **39,** 381–392.

Schardl CL, Leutchmann A, Spiering MJ, 2004. Symbioses of grasses with seedborne fungal endophytes. *Annual Review of Plant Biology* **55,** 315–340.

Shaul O, Galili S, Volpin H, Ginzberg I, Elad Y, Chet I, Kapulnik Y, 1999. Mycorrhiza-induced changes in disease severity and PR protein expression in tobacco leaves. *Molecular Plant–Microbe Interactions* **12,** 1000–1007.

Sid Ahmed A, Sanchez CP, Candela ME, 2000. Evaluation of induction of systemic resistance in pepper plants (*Capsicum annuum*) to *Phytophthora capsici* using *Trichoderma harzianum* and its relation to capsidiol accumulation. *European Journal of Plant Pathology* **106,** 817–824.

Silva HAS, Romeiro RS, Carrer R, Pereira JLA, Mizubuti ESG, Mounteer A, 2004. Induction of systemic resistance by *Bacillus cereus* against tomato foliar diseases under field conditions. *Journal of Phytopathology* **152,** 371–375.

Ton J, Van Pelt JA, Van Loon LC, Pieterse CMJ, 2002a. The *Arabidopsis ISR1* locus is required for rhizobacteria-mediated induced systemic resistance against different pathogens. *Plant Biology* **4,** 224–227.

Ton J, Van Pelt JA, Van Loon LC, Pieterse CMJ, 2002b. Differential effectiveness of salicylate-dependent and jasmonate/ethylene-dependent induced resistance in *Arabidopsis*. *Molecular Plant–Microbe Interactions* **15,** 27–34.

Uknes S, Winter AM, Delaney T, Vernooij B, Morse A, Friedrich L, Nye G, Potter S, Ward E, Ryals J, 1993. Biological induction of systemic acquired resistance in *Arabidopsis*. *Molecular Plant–Microbe Interactions* **6,** 692–698.

Vaast Ph, Caswell-Chen EP, Zasoski RJ, 1998. Influences of a root lesion nematode, *Pratylenchus coffeae*, and two arbuscular mycorrhizal fungi, *Acaulospora mellea* and *Glomus clarum* on coffee (*Coffea arabica* L.). *Biology and Fertility of Soils* **26,** 130–135.

Vallad GE, Goodman RM, 2004. Systemic acquired resistance and induced systemic resistance in conventional agriculture. *Crop Science* **44,** 1920–1934.

van de Heijden MGA, Klironomos JN, Ursic M, Moutoglis P, Streitwolf-Engel R, Boller T, Wiemken A, Sanders IR, 1998. Mycorrhizal fungal diversity determines plant biodiversity, ecosystem variability and productivity. *Nature* **396,** 69–72.

Vandenkoornhuyse P, Husband R, Daniell TJ, Watson IJ, Duck JM, Fitter AH, Young JPW, 2002. Arbuscular mycorrhizal community composition associated with two plant species in a grassland system. *Molecular Ecology* **11,** 1555–1564.

Van Peer R, Niemann GJ, Schippers B, 1991. Induced resistance and phytoalexin accumulation in biological control of Fusarium wilt of carnation by *Pseudomonas* sp. strain WCS417r. *Phytopathology* **81,** 728–734.

Van Peer R, Schippers B, 1992. Liposaccharides of plant growth promoting *Pseudomonas* spp. Strain WCS417r induce resistance in carnation to Fusarium wilt. *Netherlands Journal of Plant Pathology* **98,** 129–139.

Van Wees SCM, Pieterse CMJ, Trissenaar A, Van't Westende YAM, Hartog F, Van Loon LC, 1997. Differential induction of systemic resistance in Arabidopsis by biocontrol bacteria. *Molecular Plant Microbe Interactions* **10,** 716–724.

Van Wees SCM, Pieterse CMJ, Van't Westende YAM, Van Loon LC, 1996. *Arabidopsis* as a model to study systemic resistance induced by biocontrol bacteria. *Abstracts of the 8th International Congress of Molecular Plant Microbe Interactions, University of Tennessee, Knoxville* **D-17**.

Waller F, Achatz B, Baltruschat H, Fodor J, Becker K, Fischer M, Heier T, Hückelhoven R, Neumann C, von Wettstein D, Franken P, Kogel K-H, 2005. The endophytic fungus *Piriformospora indica* reprograms barley to salt stress tolerance, disease resistance, and higher yield. *Proceedings of the National Academy of Sciences USA* **102,** 13386–13391.

Walters D, Walsh D, Newton A, Lyon G, 2005. Induced resistance for plant disease control: maximising the efficacy of resistance elicitors. *Phytopathology* **95,** 1368–1373.

Wei G, Kloepper JW, Tuzun S, 1991. Induction of systemic resistance of cucumber to *Colletotrichum orbiculare* by select strains of plant growth promoting rhizobacteria. *Phytopathology* **81,** 1508–1512.

Wei G, Kloepper JE, Tuzun S, 1996. Induced systemic resistance to cucumber diseases and increased plant growth by plant growth promoting rhizobacteria under field conditions. *Phytopathology* **86,** 221–224.

Wei G, Yao GW, Zehnder S, Tuzun S, Kloepper JW, 1995. Induced systemic resistance by select plant growth promoting rhizobacteria against bacterial wilt of cucumber and the beetle vectors. *Phytopathology* **85,** 1154.

Whipps JM, 2004. Prospects and limitations for mycorrhizas in biocontrol of root pathogens. *Canadian Journal of Botany* **82,** 1198–1227.

Yan Z, Reddy MS, Ryu C-M, McInroy JA, Wilson M, Kloepper JW, 2002. Induced systemic protection against tomato late blight elicited by plant growth promoting rhizobacteria. *Phytopathology* **92,** 1329–1333.

Yedidia I, Benhamou N, Chet I, 1999. Induction of defense responses in cucumber plants (*Cucumis sativus* L.) by the biocontrol agent *Trichoderma harzianum. Applied and Environmental Microbiology* **65,** 1061–1070.

Zehnder GW, Murphy JF, Sikora EJ, Kloepper JW, 2001. Application of rhizobacteria for induced resistance. *European Journal of Plant Pathology* **107,** 39–50.

Zehnder GW, Yao C, Murphy JF, Sikora ER, Kloepper JW, Schuster DJ, Polston JE, 1999. Microbe-induced resistance against pathogens and herbivores: evidence of effectiveness in agriculture. In: Agrawal AA, Tuzun S, Bent A, eds. *Induced Plant Defences Against Pathogens and Herbivores: Biochemistry, Ecology and Agriculture.* St Paul, MN: APS Press, 335–355.

Zhang S, Reddy MS, Kokalis-Burelle N, Wells LW, Nightengale SP, Kloepper JW, 2001. Lack of induced resistance in peanut to late leaf spot disease by plant growth promoting rhizobacteria and chemical elicitors. *Plant Disease* **85,** 879–884.

Zhang W, Dick WA, Hoitink HAJ, 1996. Compost-induced systemic acquired resistance in cucumber to *Pythium* root rot and anthracnose. *Phytopathology* **86,** 1066–1070.

Zhang W, Han DY, Dick WA, Davis KR, Hoitink HAJ, 1998. Compost and compost water extract-induced systemic acquired resistance in cucumber and Arabidopsis. *Phytopathology* **88,** 450–455.

Zhang S, Reddy MS, Kloepper JW, 2004. Tobacco growth enhancement and blue mold disease protection by rhizobacteria: relationship between plant growth promotion and systemic disease protection by PGPR strain 90-166. *Plant and Soil* **262,** 277–288.

Zhu HH, Yao Q, 2004. Localized and systemic increase of phenols in tomato roots induced by *Glomus versiforme* inhibits *Ralstonia solanacearum. Journal of Phytopathology* **152,** 537–542.

Chapter 9
Trade-offs associated with induced resistance

Martin Heil

Department of General Botany–Plant Ecology,
University of Duisburg-Essen, Germany

9.1 Introduction

Plants can respond to a local infection with a broad-spectrum resistance to subsequent infections. The so-called systemic acquired resistance (SAR) is also active in the yet uninfected plant parts and is directed against diseases caused by the infecting pathogen as well as against other, taxonomically unrelated pathogens (Kuć, 1982; Ryals *et al.*, 1994; Hunt *et al.*, 1996; Sticher *et al.*, 1997; Hammerschmidt & Smith-Becker, 1999; Durrant & Dong, 2004; Bostock, 2005). SAR seems to be in general most active against fungi, less effective against bacteria and least effective against viruses (Kuć, 2001). It is a widespread and phylogenetically conserved trait, since the phenomenon is known from monocotyledonous and dicotyledonous plants.

9.1.1 *SAR in crop protection*

Cultivated plants in general suffer from much higher infection rates than their wild relatives. Crop protection from pests and diseases thus requires the use of increasing amounts of pesticides, with accompanying problems for the environment and consumers. The search for alternatives has therefore been an important driving force in research conducted in this area (Campbell *et al.*, 2002).

The high susceptibility of crops to pathogens results mainly from (1) their high apparency, i.e. the probability that they will be detected by their enemies (Feeny, 1976), since many conspecific and genetically similar (or even identical) plants are growing in the same area; and (2) a reduction in the plants' own defensive systems, which results from domestication and breeding for high growth rates and yield. Since contemporary cultivation methods do not allow for a reduction in crop plant apparency, sustainable protection strategies will rely fully on plant resistance mechanisms. The need to remove pesticides from the market, along with the problems arising from increasing crop resistance by means of genetic engineering (low acceptance by consumers, evolution of counter-adaptations, ecological risks resulting from 'jumping genes', etc.), have led to a general interest in the search for alternative strategies. Induced plant defences rely on a plant's own resistance mechanisms and thus seem to form a promising alternative to existing strategies (Walling, 2001; Gozzo, 2003; Vallad & Goodman, 2004).

9.1.2 Signalling and biochemical changes associated with SAR

Recent work has focused mainly on the signalling events leading to resistance expression, while less attention has been paid to the biochemical and physiological basis of phenotypic disease resistance. SAR in general is activated by pathogens that cause necrosis, either as part of successful disease development or as part of a hypersensitive response (HR). The latter represents a resistance mechanism itself, since it leads to programmed cell death and thus traps infecting pathogens in a ring of dead cells (Heath, 1998; Beers & McDowell, 2001; Dong, 2001). Salicylic acid (SA) is involved in the signal transduction pathway (Raskin, 1992; Mauch-Mani & Métraux, 1998; Cameron, 2000; Métraux, 2001; Shah, 2003; see also Chapter 4), and the SA-dependent NPR1 (non-expressor of pathogenesis-related genes 1) plays a central role as a key regulatory protein (Pieterse & Van Loon, 2004). Although it is not the transported signal, endogenous levels of SA increase locally and systemically in response to infection, SA levels increase in the phloem before SAR occurs, and exogenous application of SA can induce expression of many of the genes that respond to natural pathogen infection (Malamy *et al.*, 1990; Métraux *et al.*, 1990; Rasmussen *et al.*, 1991; Reymond & Farmer, 1998; Hammerschmidt & Smith-Becker, 1999; Maleck *et al.*, 2000). Transgenic plants expressing a bacterial gene, *NahG*, that encodes salicylate hydroxylase, are unable to accumulate SA in response to infection and are unable to express a full SAR response (Delaney *et al.*, 1994; Gaffney *et al.*, 1994). For further information concerning the signalling pathway, see Hunt & Ryals (1996), Gozzo (2003), Durrant & Dong (2004), Bostock (2005) and Chapter 4.

Based on current knowledge of the biochemistry of resistance, it can be concluded that SAR results from the expression of several parameters, including changes in cell wall composition and *de novo* synthesis of phytoalexins (Hammerschmidt, 1999a, b) and PR (pathogenesis related) proteins. Changes in cell wall composition such as increased cross-linkage among cell wall constituents and increased lignification and callose formation are important defensive mechanisms, which frequently occur in cells around those exhibiting programmed cell death (Greenberg, 1997). These responses inhibit penetration by pathogens (Hammerschmidt, 1999a; Hammerschmidt & Nicholson, 1999) that have been able to 'escape' from their HR-expressing – and therefore dying – host cell. Moreover, the local *de novo* synthesis of phytoalexins is often related to the induced resistance stage. Phytoalexins are secondary plant compounds induced by and active against microbial pathogens (Bailey & Mansfield, 1982; Dixon, 1986, 2001; Kuć, 1995; Hammerschmidt, 1999b).

Among the physiological changes associated with SAR, *de novo* synthesis of PR proteins is the most intensively studied. Originally, PR proteins were detected and defined as being absent in healthy plants but accumulating in large amounts after infection; they have now been found in more than 40 species and are assigned to at least 13 families (van Loon, 1985, 1997; Van Loon & van Strien, 1999). Some PR proteins exhibit β-1,3-glucanase (EC 3.2.1.39) or chitinase (EC 3.2.1.14) activity, i.e. they catalyse the degradation of β-1,3-glucan and chitin, which are major cell wall components of many pathogens (Boller, 1992). Another enzyme regularly induced during SAR is phenylalanine ammonia-lyase (PAL; EC 4.3.1.5), since it catalyses the first step in the phenylpropanoid pathway, products of which comprise many compounds involved in pathogen resistance such as flavonoids, condensed tannins, coumarin, lignin and also the signalling molecule SA (Dixon *et al.*, 2002).

In order to establish resistance, plants thus have to cope with metabolic efforts that, depending on plant growth stage and resource availability, can cause considerable costs. The concept of fitness costs assumes that resistant plants have lower reproduction than less resistant plants when compared under 'enemy-free' conditions, which prevent the resistance from having beneficial effects (Simms & Fritz, 1990). If such costs occur – and there now is convincing evidence that this is indeed the case – they would lead to important limitations. Thus, so-called allocation costs might lead to reduced growth and yield as soon as resources limit overall plant growth, and ecological costs in the form of trade-offs among SAR and other plant resistance mechanisms might severely compromise the resistance of plants to, for example, insect herbivores, as soon as these plants are induced to express SAR. The main focus of this chapter is to summarise the current knowledge on such trade-offs and to point to their potential effects when SAR is to be used as a crop protection strategy.

9.1.3 Definition of SAR

Usually, the expression of PR genes in general, and of *PR-1* in particular, is used as a molecular marker for a successful induction of SAR, although their role in phenotypic resistance in many cases is unclear (Durrant & Dong, 2004). In a study on BTH (benzothiadiazole, an artificial resistance elicitor; see below and Chapters 2, 10 and 11) elicited resistance of pear to fire blight, Sparla *et al.* (2004) found a successful resistance induction in terms of both disease incidence and severity, which was, however, not associated with an increased expression of *PR-1*. BTH treatment of melon seeds elicited resistance to fungal pathogens and induced chitinases and peroxidases. A similar response both on the level of biological resistance and in the patterns of isoenzymes induced was, however, elicited by methyl jasmonate rather than by salicylic acid (Buzi *et al.*, 2004). Similarly, a novel rice PR10 protein has been described that is induced in response to abiotic stress and fungal infection through the jasmonic acid signalling pathway rather than via SA signalling (Hashimoto *et al.*, 2004). The *dth9* (detachment 9) mutant of Arabidopsis fails to develop SAR in response to pathogen infection or SA treatment, although the expression of *PR-1* and *PR-2* under these conditions is unaltered (Mayda *et al.*, 2000). Focusing the definition on one or a few particular PR genes as markers might thus be too narrow in order to completely describe a broad-spectrum resistance phenomenon such as SAR. For the purpose of this chapter, every induced resistance to pathogens that is not an R-gene mediated specific response and not induced systemic resistance (ISR) elicited by plant growth promoting rhizobacteria (Bloemberg & Lugtenberg, 2001; Pieterse *et al.*, 2001; Ramamoorthy *et al.*, 2001; Zehnder *et al.*, 2001; Pozo *et al.*, 2005) (see also Chapter 8), but rather an unspecific, induced systemic resistance of plants to pathogens, is called SAR.

9.2 Artificial resistance inducers

It was the discovery of artificial resistance inducers that prompted a strong interest in SAR as a strategy for crop protection (Vallad & Goodman, 2004). Exogenous application of SA can elicit SAR (see above), and the same, or at least similar, responses occur also in plants treated with SA mimics such as 2,6-dichloroisonicotinic acid (INA), benzo(1,2,3) thiadiazole-7-carbothioic acid-*S*-methyl ester, CGA-245 704 (or acibenzolar-*S*-methyl, also known as BTH or ASM) (Schurter *et al.*, 1987; Oostendorp *et al.*, 2001) and Tiadinil (*N*-[3-chloro-4-methylphenyl]-4-methyl-1,2,3-thiadiazole-5-carboxamide; Yasuda *et al.*, 2004). BTH induces resistance to pathogens in wheat and several other plant species (Görlach *et al.*, 1996; Lawton *et al.*, 1996; Molina *et al.*, 1999) and because it does not possess *in vitro* antimicrobial activity, locally applied BTH can elicit systemic resistance to fungi, bacteria and viruses (Sticher *et al.*, 1997; Tally *et al.*, 1999; see also Chapter 10). Since biotrophic pathogens in general induce SA-dependent resistance, while necrotrophic pathogens induce JA-mediated responses, similar differences are also likely to occur in these pathogens' responses to BTH treatment (D.F. Cipollini, Wright State University, Dayton, OH, personal communication). BTH inhibits catalase and ascorbate peroxidase and therefore functions as an analogue of SA. The structurally related compound Tiadinil induces the expression of several PR genes and induced resistance of rice to viral and bacterial diseases (Yasuda *et al.*, 2004). In tobacco, CMPA (3-chloro-1-methyl-1H-pyrazole-5-carboxylic acid) was reported to elicit PR gene expression and pathogen resistance (Yasuda *et al.*, 2003).

Various studies now have demonstrated that BTH can successfully induce resistance to various pathogens (reviewed in Vallad & Goodman, 2004; see also Chapter 10). In most cases, this effect was associated with an increase in PR proteins such as peroxidase, chitinase or β-1,3-glucanase. BTH induces *de novo* synthesis of PR proteins in species such as Arabidopsis (Dietrich *et al.*, 2004, 2005), cauliflower (Ziadi *et al.*, 2001), melon (Buzi *et al.*, 2004), cocoa (Resende *et al.*, 2002), potato (Bokshi *et al.*, 2003), papaya (Zhu *et al.*, 2003) and rose (Suo & Leung, 2001). However, sets of genes induced by BTH application are not necessarily identical to defence response genes induced by pathogens (Yu & Muehlbauer, 2001). In the majority of studies, BTH treatment decreased disease incidence and/or severity measured, for example, as number of plants affected, number and size of necrotic lesions visible on leaves, leaf area affected by pathogens and so on (Table 9.1 and references therein; see also Chapter 10).

Although BTH can thus elicit successful disease resistance in many plant–pathogen combinations, there are reports where BTH either did not induce resistance or the level of resistance induced was poor (see Chapters 10 and 11). For example, BTH treatment had no significant effect on disease incidence or yield in a field study conducted on sclerotinia blight (causal agent: *Sclerotinia minor*) of peanut (Lemay *et al.*, 2002) and also failed to elicit significant resistance to late leaf spot disease (causal agent *Cercosporidium personatum*) in glasshouse and field studies conducted with the same crop plant (Zhang *et al.*, 2001). BTH treatment applied additionally to traditional copper-based fungicides did not significantly reduce citrus canker incidence on foliage or fruit drop compared to Cu alone (Graham & Leite, 2004), and BTH also failed to induce resistance to *Fusarium* wilt of cucumber, although significant resistance induction in other pathosystems was found in the same study (Ishii *et al.*, 1999). Therefore, whether or not BTH can function as a promising tool for crop protection has to be tested individually for each crop–pathogen combination.

9.2.1 Priming

Although treatment with BTH can lead to significant increases in the activities of PR proteins, SAR induction per se often does not lead to marked physiological changes. However,

Table 9.1 Selected studies demonstrating successful resistance induction by benzothiadiazole (acibenzolar-*S*-methyl) in various plant–pathogen interactions (see Table 1 in Vallad & Goodman 2004, for further examples, and Table 1 in Iriti & Faoro, 2003a).

Plant	Pathogen	Conditions	Reference
Bacterial diseases			
Pepper (*Capsicum anuum*)	*Xanthomonas campestris* pv. *vesicatoria*	Glasshouse + field	Buonaurio *et al.* (2002)
Tomato (*Lycopersicum esculentum*)	*Xanthomonas axonopodis* pv. *vesicatoria, Pseudomonas syringae* pv. *tomato*	Field	Louws *et al.* (2001)
Pear (*Pyrus communis*)	*Erwinia amylovora*	Glasshouse	Sparla *et al.* (2004)
Tobacco (*Nicotiana tabacum*)	*Pseudomonas syringae* pvs *tabaci* tox + and tox −	Glasshouse + field	Cole (1999)
Bell pepper (*Capsicum annuum*)	*Xanthomonas axonopodis* pv. *vesicatoria*	Glasshouse + field	Romero *et al.* (2001)
Fungal diseases			
Cashew (*Anacardium occidentale*)	*Colletotrichum gloeosporioides*	Glasshouse + field	Lopez & Lucas (2002)
Cowpea (*Vigna unguiculata*)	*Colletotrichum destructivum*	Laboratory	Latunde-Dada & Lucas (2001)
Rice (*Oryza sativa*)	*Rhizoctonia solani*	Glasshouse	Rohilla *et al.* (2002)
Cocoa (*Theobroma cacao*)	*Crinipellis perniciosa, Verticillium dahliae*	Growth chamber	Resende *et al.* (2002)
Potato (*Solanum tuberosum*)	*Alternaria solani, Erysiphe cichoracearum, Fusarium semitectum*	Glasshouse + field	Bokshi *et al.* (2003)
Bean (*Phaseolus vulgaris*)	*Uromyces appendiculatus*	Glasshouse	Iriti & Faoro (2003b)
Japanese pear (*Pyrus pyrifolia*)	*Alternaria alternata*	Field	Ishii *et al.* (1999)
Papaya (*Carica papaya*)	*Phytophthora palmivora*	Glasshouse	Zhu *et al.* (2003)
Grapevine (*Vitis vinifera*)	*Botrytis cinerea*	Growth chamber	Ishii *et al.* (1999)
Cucumber (*Cucumis sativus*)	*Cladosporium cucumerinum, Colletotrichum lagenarium, Fusarium oxysporum*	Growth chamber	Ishii *et al.* (1999)
Sunflower (*Helianthus annuus*)	*Puccinia helianthi*	Glasshouse + field	Prats *et al.* (2002)
Viral diseases			
Tobacco (*Nicotiana tabacum*)	*Tomato spotted wilt virus* (TSWV)	Glasshouse + field[1]	Csinos *et al.* (2001)
Others			
Sunflower (*Helianthus annuus*)	*Orobanche cumana*	Growth chamber	Sauerborn *et al.* (2002)

[1]Protective effects detected in some, but not all individual experiments.

when plants are challenged thereafter by a pathogen, the physiological responses (and thus resistance development) occur much faster and to a greater extent than in untreated control plants. This phenomenon is known as 'priming' (Zimmerli *et al.*, 2000; Conrath *et al.*, 2001, 2002).

For example, BTH treatment of cucumber induced expression of an acidic peroxidase and a pathogenesis-related protein 1 homologue (*PR1-1a*), yet not of phenylalanine ammonia lyase (PAL). However, even PAL was primed by BTH and responded much faster to subsequent inoculation with a pathogenic fungus (Cools & Ishii, 2002). In another study, BTH did not induce PAL gene expression and callose formation in Arabidopsis, while both resistance responses were significantly augmented by pre-treatment with BTH in plants then infected with bacteria (Kohler *et al.*, 2002). Rose shoots did respond to BTH treatment with significant increases in chitinase and β-1,3-glucanase activities, but the differences between BTH treated plants and controls became much more pronounced in response to inoculation with a fungal pathogen (Suo & Leung, 2001). In cauliflower, BTH treatment alone led to a significant increase in activities of glucanase, yet not chitinase, while in response to inoculation with the causal agent of downy mildew, *Peronospora parasitica*, glucanase activity responded much faster, and chitinase activity was significantly higher in plants that had been pre-treated with BTH than in controls (Ziadi *et al.*, 2001). In cowpea, treatment with BTH led to marked increases in phytoalexin concentrations, as well as in the activities of PAL and chalcone isomerase (EC 5.5.1.6) as compared to untreated controls only when plants were subsequently challenged with bacteria (Latunde-Dada & Lucas, 2001). Similarly, pre-treatment with the non-protein amino acid β-aminobutyric acid primes plants to be more resistant to various pathogens without eliciting measurable physiological changes on its own (Zimmerli *et al.*, 2000; Jakab *et al.*, 2001). Treatment with these compounds thus appears to generally increase the responsiveness of plants to subsequent pathogen infection.

9.2.2 Chemical resistance elicitation and yield

BTH can successfully induce resistance in many plant–pathogen combinations and can also lead to improved yield. However, several studies did not find significant increases in resistance (see above), or yield, in response to BTH treatment. Csinos *et al.* (2001) studied the effects of BTH treatment on the development of the thrips-transmitted tomato spotted wilt virus on tobacco and found significant reductions in disease development in some, but not in other, experiments. Obviously, time of treatment, time of infection and the plants' physiological state at the time of treatment affected whether BTH treatment could elicit significant disease resistance, or otherwise hamper growth due to phytotoxic effects and/or allocation costs (Csinos *et al.*, 2001). Although BTH application on citrus had reduced lesions, the same treatment did not lead to any detectable decrease in disease incidence in orchard trials on *C. sinensis* (Graham & Leite, 2004).

At higher doses of BTH (0.5 and 2 mg ml^{-1}), light chlorosis of sunflower was observed, leading to a reduction in shoot fresh weight 14 days after treatment (Prats *et al.*, 2002). Treating tobacco plants under glasshouse conditions once with 1 g of BTH per 7000 plants or three times with 0.5 g BTH per 7000 plants resulted in significantly shorter plants and suppressed root growth compared to untreated controls, whereas lower doses did not cause any significant effects (Csinos *et al.*, 2001). BTH treatment of melon seeds delayed germination (Buzi *et al.*, 2004), cowpea seedlings pre-treated as seeds with BTH suffered from significant reduction in shoot growth and leaf enlargement when treated with concentrations >20 ppm (Latunde-Dada & Lucas, 2001), and significantly reduced growth in response to BTH treatment was also reported for seedlings of cauliflower (Ziadi *et al.*, 2001).

Using BTH to elicit resistance of tomato to bacterial spot and bacterial speck reduced disease severity in some, but not other, experiments, and it did not increase yield even in those experiments in which plants suffered significantly less from bacterial infection (Louws *et al.*, 2001). In one experiment, BTH application significantly reduced yield, in spite of successful resistance elicitation (Louws *et al.*, 2001). Similarly, BTH protected bell pepper successfully from bacterial spot disease but had a negative effect on yield when applied weekly during the entire crop season (Romero *et al.*, 2001). Under several growing condition–cultivar combinations tested, regular BTH treatment resulted in yields similar to that of untreated plants, in spite of successful disease reduction (Romero *et al.*, 2001). This is similar to the results obtained by Stadnik & Buchenauer (1999), who found that treating wheat with BTH in addition to traditional fungicides led to a reduction of disease symptoms but no increase in yield.

What are the reasons for the observation that BTH treatment can fail to increase growth and yield in spite of successfully reducing disease symptoms? Although BTH is generally regarded as not being phytotoxic, chlorosis and reduced growth have been reported after its application (Cole, 1999; Lopez & Lucas, 2002). Suppression of growth of wheat plants in response to BTH treatment was most pronounced under limiting nitrogen conditions (Heil *et al.*, 2000). To date, a final explanation is not yet possible. Additional problems in obtaining convincing explanations for these phenomena result from their high dependency on abiotic growing conditions (Heil *et al.*, 2000; Cipollini, 2002; Délano-Frier *et al.*, 2004), making comparisons among different studies very difficult. Several lines of evidence, however, point to the possibility that SAR causes relevant costs in terms of reduced fitness when expressed under pathogen-free conditions (Heil & Baldwin, 2002). In the following, the concept of fitness costs is discussed, and evidence for its applicability and relevance to SAR is reviewed. Such costs might strongly compromise the suitability of a preventive SAR elicitation in crop protection.

9.3 Costs of SAR

The concept of fitness costs currently provides the most powerful explanation for the evolution of induced resistance. Why should a broad-spectrum resistance be expressed only in response to attack, leaving the plant with a time lag between infection and phenotypic resistance expression (Heil & Baldwin, 2002)? At first glance, constitutive (all the time) mechanisms appear to be the preferred solution. Saving metabolic effort or avoiding other putative negative effects of resistance traits when resistance actually is not required appears to be a convincing reason for the evolution of an inducible 'just in time' mechanism. This concept is consistent with most, if not all, empirical findings published thus far on different aspects of induced resistance (Heil, 2001a; Heil & Baldwin, 2002). It has been widely applied to induced resistance to herbivores (Karban & Baldwin, 1997; Baldwin, 1998; Baldwin & Preston, 1999; Tollrian & Harvell, 1999; Rausher, 2001; Cipollini *et al.*, 2003 and references therein) and has now been extended to induced resistance to pathogens (Heil, 1999, 2001a).

According to this definition, relevant (i.e. fitness) costs occur as soon as expression of a resistance trait under conditions not actually requiring resistance reduces a plant's genetic contribution to subsequent generations. Causal reasons for this reduction might be manifold (Heil, 2002; Heil & Baldwin, 2002). For example, expression of resistance compounds might use up limited resources, which then cannot be used for other fitness-relevant traits such as growth and reproduction (allocation costs). Negative impacts of resistance traits on the plant producing them might also result from autotoxicity, since some resistance traits are toxic to the plant, and their constitutive expression might impose a significant metabolic burden (Baldwin & Callahan, 1993). Finally, ecological costs can result if resistance negatively affects some of the many other interactions a plant has with its environment.

9.3.1 Allocation costs

Like all organisms, plants have to make the best use possible of resources that are available. As soon as any limited resources are allocated to defence, these are not available for other fitness-relevant processes such as further growth, reproduction or other defence mechanisms. Although no linear relations among allocation of resources and negative impacts on overall plant fitness exist, such allocation processes can easily translate into ecologically and agronomically relevant fitness costs. The term 'resources' in this context can comprise exogenous factors such as supply of light, nutrients and water, but also internal parameters such as overall transcription and translation capacities and so on.

There now is convincing empirical evidence that SAR indeed results in costs, which most probably are caused by resource allocation to resistance. In order to detect these allocation costs, resistance must be activated, or active, in the absence of pathogens (Heil & Baldwin, 2002), since otherwise the beneficial (i.e. protective) effects overlay the costs of resistance. In recent years, two strategies have been followed in this context. Several studies have made use of artificial resistance elicitors such as SA, INA or BTH to induce resistance (Cole, 1999; Csinos *et al.*, 2001; Latunde-Dada & Lucas, 2001; Ziadi *et al.*, 2001; Lopez & Lucas, 2002; Prats *et al.*, 2002) and found significant negative effects on plant growth and seed production. In a study designed specifically to quantify allocation costs of SAR, wheat plants treated with BTH under enemy-free conditions showed a significant reduction in biomass gain and seed yield, and these effects were strongly dependent on nitrogen availability (Heil *et al.*, 2000). Similarly, Arabidopsis plants treated with SA produced fewer seeds than untreated controls (Cipollini, 2002). These observations, together with the fact that the intensity of resistance induction can be compromised by resource availability (Cipollini & Bergelson, 2001; Cipollini, 2002; Dietrich *et al.*, 2004), are consistent with the interpretation that limiting resources have to be allocated to resistance compounds (e.g. nitrogen has to be allocated to the *de novo* synthesis of PR proteins) and that this process causes significant allocation costs of SAR (Heil, 1999, 2001a; Heil *et al.*, 2000; Dietrich *et al.*, 2005).

Instead of applying chemical resistance elicitors, other studies followed a genetic approach and used mutants that either constitutively express at least parts of the SAR signalling pathway or are compromised in resistance expression. In general, the phenotypes of these mutants are consistent with the assumption that SAR expression causes costs (Heil & Baldwin, 2002). The constitutive expressor of SAR, *cep 1*, produced fewer seeds than wild-type plants (Cipollini, 2002). This appears to be a general phenomenon, since SAR overexpressing Arabidopsis lines are usually described as stunted or dwarfed and gain significantly less biomass than the corresponding wild type lines (Bowling *et al.*, 1994, 1997; Clarke *et al.*, 1998; Yu *et al.*, 1998; Rate *et al.*, 1999; Petersen *et al.*, 2000; Jirage *et al.*, 2001; Mauch *et al.*, 2001; see further examples in Heil & Baldwin, 2002). Loss-of-function mutants unfortunately have received much less attention. However, two mutants deficient in JA signalling were taller and produced more seeds (or larger tubers, respectively) than wild type plants (Royo *et al.*, 1999; Greenberg *et al.*, 2000). Heidel *et al.* (2004) investigated several different Arabidopsis mutants comprising both constitutive expressors of SAR (*crp1* and *cpr5*), one mutant that prevents induction (*npr1*) and one with an altered level of expression (*NPR1-H*) under both growth chamber and field conditions. The fitness effects observed were consistent with the expectation that SAR is costly, yet benefits plants that are attacked by pathogens. Thus, over-expression of SAR was costly, whereas the loss of inducibility had no significant effects under sterile growth chamber conditions, although it affected fitness negatively in the field. Since all mutants were in the same genetic background and back-crossed several times, the authors concluded that 'the loss of fitness in *cpr1* and *cpr5* is not from pleiotropy but rather from the constitutive activation of the SAR pathway itself' (Heidel *et al.*, 2004).

Although it is very difficult to separate putative phytotoxic effects of artificial resistance elicitors and putative autotoxic effects of resistance compounds from allocation costs, all the results cited above demonstrate that constitutive expression of SAR under enemy-free conditions can have deleterious effects on plant growth and reproduction (Durrant & Dong, 2004).

The physiological processes underlying these effects are not yet understood, although the diversion of limiting resources from growth to defence provides a convincing explanation. For example, reductions in transcripts and/or proteins related to photosynthesis have repeatedly been observed during resistance induction (Logemann *et al.*, 1995; Somssich & Hahlbrock, 1998; Lian *et al.*, 2000; Bailey *et al.*, 2005). This general 'shift from housekeeping to defence metabolism' (Scheideler *et al.*, 2002) might reduce the actual needs for *de novo* protein synthesis in terms of resources, yet can be a mechanism by which costs of resistance are caused. Allocation costs provide a convincing explanation for the fact that SAR has evolved as an inducible rather than a constitutive trait. Such effects, however, have the potential to lead to relevant trade-offs when SAR is used as crop protection strategy.

9.3.2 Ecological costs

9.3.2.1 Trade-offs with mutualistic plant–microbe interactions

SAR is a broad-spectrum form of defence which is active against many different pathogens. On the other hand, plants rely on mutualistic interactions with micro-organisms – legumes interact intimately with nitrogen-fixing *Rhizobia* bacteria, many plants depend on mycorrhizal associations or are infected by mutualistic fungi that contribute to their defence against herbivores, and plant-growth promoting rhizobacteria (PGPR) have the potential to improve both growth and resistance of plants. Side-effects of an unspecific resistance mechanism such as SAR on mutualistic micro-organisms therefore appear likely. Such effects on mutualists can easily be overlooked in simplified lab study systems, but they might have a strong influence on the net outcome of SAR under natural growing conditions.

Unfortunately, little empirical evidence exists for such trade-offs, but general considerations make it very likely that they indeed exist. SAR components such as chitinases and β-1,3-glucanases are involved in the establishment of root nodules and mycorrhizae, which have to be permanently stabilized at a level between infection that is too heavy and defence that is too effective (Vierheilig *et al.*, 1994; Dumas-Gaudot *et al.*, 1996; Martínez-Abarca *et al.*, 1998; Schultze & Kondorosi, 1998; Ruiz-Lozano *et al.*, 1999). For example, nod-factors (i.e. factors that enable *Rhizobium leguminosarum* bacteria to successfully infect roots of their host plant) contain a chitin side chain, and it is exactly this side chain that is hydrolysed by host chitinases in order to avoid heavy infection by the bacteria (Oldroyd, 2001). Other results demonstrated that chitinases and β-1,3-glucanases have roles in plant/microbe signal perception, for example in symbioses with arbuscular mycorrhizal (AM) fungi (Dumas-Gaudot *et al.*, 1996).

Infection by mutualistic micro-organisms activates at least parts of the SAR pathway (Dumas-Gaudot *et al.*, 1996; Cordier *et al.*, 1998; Ruiz-Lozano *et al.*, 1999). Ruiz-Lozano *et al.* (1999) investigated early events in mycorrhiza and nodule establishment in pea and reported the induction of seven defence-related genes. Could this effect function in the other direction as well, leading to an inhibition of these beneficial infections when plants are expressing SAR at a high level? Rhizobacteria have to overcome their host's resistance for successful establishment of functioning nodules (Mithöfer, 2002). Vierheilig *et al.* (1995) and Glandorf *et al.* (1997) reported negative effects on colonization of tobacco roots by *Glomus mosseae* in plants constitutively expressing β-1,3-glucanase. Herbivory and fungal infections can inhibit nodule development and N_2-fixing activity (Russin *et al.*, 1990). Other studies have demonstrated inhibitory effects of chemically induced SAR on the development of root nodules. For example, in experiments in the author's laboratory, faba beans (*Vicia faba* cv. 'Hang down') were treated with BION to elicit SAR. Plants were harvested six weeks later, and the total dry weight of root nodules was determined. Under two different nutrient treatments, BION-treated plants had developed fewer and smaller nodules than untreated controls (Heil, 2001b). Similar results were reported by Martínez-Abarca *et al.* (1998), Ramanujam *et al.* (1998) and Lian *et al.* (2000). SA treatment of *Vigna mungo* reduced both nodulation and N_2-fixing activity (Ramanujam *et al.*, 1998). When SA was exogenously applied prior to inoculation of alfalfa plants with compatible *Rhizobium* strains, significant inhibition of nodule primordial formation and a reduction of the number of emerging nodules, as well as a delay in nodule appearance, were observed (Martínez-Abarca *et al.*, 1998). A negative effect on the number and total dry weight of root nodules developed was observed when soybean (*Glycine max* cv. 'Maple Glen') seedlings received high concentrations of SA in the rooting medium (Lian *et al.*, 2000). All these results indicate that chemical elicitation of SAR can negatively affect establishment and function of root nodules.

Does this occur in nature as well, or is it a consequence of chemical elicitation rather than a 'real' effect of SAR itself? The data reported by Russin *et al.* (1990) indicate that this effect can indeed be elicited by fungal infection. It is, however, not clear whether these effects resulted from SAR elicited by the plant enemies (or, in the other cases, by chemical treatment), or rather from a reduced allocation of assimilates to nodules (the latter interpretation being given by Russin *et al.*, 1990). Are there comparative influences on other forms of plant–microbe mutualisms such as mycorrhizae or interactions with endosymbiotic fungi (Heil, 1999)? Although usually regarded as a non-toxic elicitor of plant resistance that does not interact directly with pathogens, BTH can also exhibit direct fungicidal effects (Rohilla *et al.*, 2002). Artificial SAR elicitation thus has an even greater potential of leading to unwanted side-effects on beneficial micro-organisms.

Given the apparently low specificity of SAR, these effects are likely to occur under natural conditions as well. Intensive research using both physiological methods and ecologically realistic study systems is required to give an impression of the relative importance of the possible interactions between plant defence against pathogenic micro-organisms and plant mutualisms with other micro-organisms.

9.3.2.2 Trade-offs with other plant resistance mechanisms

Signalling pathways leading to induced resistance to herbivores, pathogens or abiotic stresses form highly interconnected networks rather than independent linear chains of signals, and there is therefore a high potential for interactions among different resistance traits (Feys & Parker, 2000; Genoud *et al.*, 2001; Kunkel & Brooks, 2002; Ramonell & Somerville, 2002; Katagiri, 2004). Trade-offs among SAR and other resistance mechanisms, particularly to herbivores, have received much attention in previous years, and several excellent reviews have already dealt with the issue (Walling, 2000; Bostock *et al.*, 2001; Cipollini, 2004; Durrant & Dong, 2004; Bostock, 2005). However, it is difficult to detect consistent patterns from the information available (Heil & Bostock, 2002). Several examples of cross-resistance (insect feeding leading to induction of aspects of SAR) have been reported. For example, resistance to fungal pathogens was induced by soybean looper feeding on soybean (Padgett *et al.*, 1994), by thrips and aphids feeding on watermelon (Russo *et al.*, 1997), by beetle feeding on *Rumex obtusifolius* (Hatcher & Paul, 2000) and by plant hopper feeding on rice (Kanno *et al.*, 2005). Caterpillar feeding can induce resistance of tomato to aphids, mites and bacteria (Stout *et al.*, 1998). Genes of the octadecanoid pathway were also induced by SA in sorghum (Salzman *et al.*, 2005).

However, most studies found trade-offs, i.e. compromised resistance against insects in plants expressing SAR or vice versa (Felton & Korth, 2000; Heil & Bostock, 2002). In a study applying both SA and JA on Arabidopsis, SA inhibited expression of the JA-dependent resistance to *Spodoptera exigua* caterpillars (Cipollini *et al.*, 2004). BTH application to field-grown tomato plants compromised the JA-dependent induction of polyphenol oxidase and thus resistance to *Spodoptera* caterpillars (Thaler *et al.*, 1999). SA had a negative effect on the JA-dependent production of trichomes by Arabidopsis (Traw & Bergelson, 2003). Transcriptional profiling of sorghum revealed both one-way and mutual antagonisms between the responses to SA and JA (Salzman *et al.*, 2005), and a large scale study taking advantage of the high number of signalling mutants available from Arabidopsis also found mutual inhibition between these two pathways (Glazebrook *et al.*, 2003). Interestingly, these effects can even be ecotype specific (Traw *et al.*, 2003).

While there is now convincing evidence that SA inhibits the synthesis of JA and thus the induction of JA-dependent resistance mechanisms (usually to herbivores), much less evidence exists for SA-dependent resistance being inhibited by JA. A reason behind this could be that factors leading to JA-dependent resistances are generally associated with leaf damage (e.g. by herbivores) and thus usually lead to wounds that could also be used by infecting pathogens (Cheong *et al.*, 2002). In contrast, pathogen infection does not necessarily lead to a higher risk of being attacked by herbivores (Heil & Bostock, 2002). Moreover, those herbivores that are generally involved in induced cross-resistance phenomena (i.e. herbivores eliciting SAR or similar responses or being affected by SAR) are often thrips and aphids (Russo *et al.*, 1997) – insects that (a) affect only few cells and thus cause a damage pattern that resembles infection by necrotizing pathogens rather than 'classical' herbivore damage and (b) are used as vectors by many pathogens. Compromised JA-dependent resistance in plants expressing SAR thus might be a cost-saving strategy that could not evolve in the reverse form (SAR inhibition by JA). It is, however, a severe constraint when SAR is generally up-regulated in crop plants, since these plants are likely suffering from a much higher level of herbivore attack.

9.3.3 Evolutionary consequences of artificial SAR elicitation

Plant resistance traits represent adaptations against plant-damaging herbivores and pathogens, and these enemies of course have the potential to respond with the evolution of counter-adaptations (Gould, 1991). Most studies on co-evolution between plants and pests have been done on insect pests rather than on pathogens. However, the underlying general patterns should not differ between these two groups of organisms. The medicinal race between pharmacists and human pathogens gives many examples of pathogens that have rapidly evolved counter-resistances to toxins.

The evolution of counter-adaptations by pests to pesticides or crop resistance traits is one of the most challenging problems in crop protection (Rausher, 2001). An example illustrating this problem is the co-evolutionary race among wheat breeders and the important wheat pest, Hessian fly (*Mayetiola destructor*) (Foster *et al.*, 1991). Adaptations of plant pathogens to artificially introduced resistance traits can evolve in a similar way, and they can occur extremely rapidly. Cultivars of oat carrying single resistance genes became susceptible to the rapidly evolving crown rust within 1–3 years after onset of their widespread use. Similarly, the southern corn leaf blight overcame the resistance of corn. A small scale experiment on resistant potato plants demonstrated that several mutants of the pathogen *Phytophthora infestans* occurred within a few weeks, thus rendering the formerly avirulent pathogen virulent (all examples from Whitham *et al.*, 1984). Under glasshouse conditions, exposing a resistant wheat cultivar to the avirulent pathogen *Erysiphe graminis* f. sp. *tritici* resulted in the evolution of three different virulent races within only 440 days (Leijerstam, 1972).

SAR might in part have evolved as induced resistance in order to slow down the evolution of such counter-adaptations. SAR adds a considerable level of variability to the presence of the resistance traits. This variability is evident at temporal and spatial levels and both within and among individual plants. Variability within plants and/or populations is considered to be one of the most important factors that may slow down the evolution of adaptations by plant pests (Whitham & Slobodchikoff, 1981; Whitham *et al.*, 1984).

One important factor that determines the rate at which pests adapt to a particular resistance trait is the fitness difference between adapted and non adapted genotypes. The higher this difference, the higher the selective pressure becomes and the more rapidly adapted phenotypes will dominate subsequent generations. SAR seems to be a highly efficient resistance trait. Mutations allowing a pathogen to overcome SAR thus will cause high fitness benefits and will rapidly spread within the population. Another parameter is whether or not refuges remain, allowing reproduction of non-adapted genotypes as well (Rausher, 2001). Gould & Anderson (1991) investigated the effects of *Bacillus thuringiensis* and a particular endotoxin on fitness of *Heliothis virescens* and concluded that pests will much more rapidly adapt to these resistance types when facing no-choice situations, i.e. when only toxin-containing food sources are available. Therefore, a particular strategy has been developed for the release of resistant crop varieties (the high dose/refuge strategy) in order to slow down the evolution of such counter-adaptations (Rausher, 2001). Resistance traits in general will first appear in heterozygous individuals (where they provide only low, if any, fitness benefits). Therefore, they will be 'diluted' within the population and are less likely to become homozygous, as long as there are enough resources left where individuals not carrying the resistance trait can reproduce without difficulty.

SAR is not based on the expression of one or a few single genes but rather seems to be achieved by a whole array of resistance responses, against all of which a pathogen might hardly be able to adapt. However, it has evolved as a temporally and spatially variable resistance, leaving unprotected plants as 'refuges' (Rausher, 2001) and providing pests with 'choice' rather than 'no-choice' (Gould & Anderson, 1991) conditions. Functionally converting SAR from an induced to a constitutive trait and using it for preventive crop protection purposes will deprive the plants of a considerable part of their variability and thus of one their most important weapons in the co-evolutionary race. This strategy would ignore most of the knowledge that has already been obtained from studies on co-evolution among crops and pests, and the evolution of counter-adaptations thus might take place much faster.

9.3.4 PR proteins as allergens

Not a physiological or evolutionary trade-off, but still a factor that might heavily compromise the use of artificial SAR elicitation in crop protection is the allergenic potential of PR proteins (Malandain & Lavaud, 2004). For example, PR-10 from common bean has high sequence similarities to tree-pollen allergens and to allergens of apple (Walter *et al.*, 1996). The major allergen of celery, Api g 1, exhibits high sequence similarities to pathogenesis-related and stress-induced proteins (Breiteneder *et al.*, 1995). Jun a 3, a highly potent allergen of *Juniperus* pollen, is homologous to members of the thaumatin-like PR proteins (PR-5) (Midoro-Horiuti *et al.*, 2000). PR proteins belonging to different families can exhibit allergenic potential, since plant-derived allergens have been identified with sequence similarities to PR-protein families 2, 3, 4, 5, 8, 10 and 14 (Hoffmann-Sommergruber, 2002). Overall, about 42% of the 440 allergens of plant origin that have been characterized so far belong to one of the families of PR proteins (Malandain & Lavaud, 2004). An artificial upregulation of SAR and the resulting increased presence of these compounds in food plants will strongly increase the risk of food-derived allergies in consumers.

9.4 Conclusions

SAR provides a plant-wide broad-spectrum resistance to pathogens that consists only of internal plant traits. Research into the elicitation and regulation of SAR expression has led to a level of understanding of the underlying signalling mechanisms that now allows artificial resistance induction via the exogenous application of chemical resistance elicitors or via the control of single steps by means of genetic engineering. A preventative up-regulation of SAR thus is technically possible and appears to be a promising tool in crop protection.

The understanding of the evolution of SAR is, however, much less elaborate, and predictions concerning the long-term consequences of a constitutive SAR expression are hard to make. SAR has evolved as an inducible rather than a constitutive mechanism, and factors that prevented it from becoming established as a constitutive trait are also likely to act under agronomic field conditions. The most convincing explanation for the evolution of SAR as an inducible mechanism is that its expression causes costs, which have to be saved when resistance actually is not required. Synthesis of SAR compounds causes allocation costs that can lead to significant reductions in further growth and productivity when this occurs under a limiting supply of resources. As an unspecific resistance to a broad range of micro-organisms, SAR can negatively affect the plants' interactions with mutualistic bacteria or fungi, and its crosstalk with signalling processes involved in herbivore resistance can lead to a compromised defence to herbivores in plants expressing SAR. In the long run, the continuous presence of SAR-mediating compounds such as phytoalexins and PR-proteins will make counter-adaptations by the respective pathogens much more likely. All these trade-offs must be taken into consideration and must be carefully investigated in long-tcrm experiments under natural and agronomic field conditions before reliable estimations of the net consequences of a preventative up-regulation of SAR as a measure in crop protection can be made.

9.5 Acknowledgements

I thank Don Cipollini for valuable comments on an earlier version of the manuscript. Financial support by the DFG (Emmy-Noether program) and the Max-Planck-Gesellschaft is gratefully acknowledged.

9.6 References

Bailey BA, Strem MD, Bae HH, de Mayolo GA, Guiltinan MJ, 2005. Gene expression in leaves of *Theobroma cacao* in response to mechanical wounding, ethylene, and/or methyl jasmonate. *Plant Science* **168,** 1247–1258.

Bailey JA, Mansfield JW, 1982. *Phytoalexins*. Glasgow: Blackie.

Baldwin IT, 1998. Jasmonate-induced responses are costly but benefit plants under attack in native populations. *Proceedings of the National Academy of Sciences USA* **95,** 8113–8118.

Baldwin IT, Callahan P, 1993. Autotoxicity and chemical defense: nicotine accumulation and carbon gain in solanaceous plants. *Oecologia* **94,** 534–541.

Baldwin IT, Preston CA, 1999. The eco-physiological complexity of plant responses to insect herbivores. *Planta* **208,** 137–145.

Beers EP, McDowell JM, 2001. Regulation and execution of programmed cell death in response to pathogens, stress and developmental cues. *Current Opinion in Plant Biology* **4,** 561–567.

Bloemberg GV, Lugtenberg BJJ, 2001. Molecular basis of plant growth promotion and biocontrol by rhizobacteria. *Current Biology in Plant Biology* **4,** 343–350.

Bokshi AI, Morris SC, Deverall BJ, 2003. Effects of benzothiadiazole and acetylsalicylic acid on beta-1,3-glucanase activity and disease resistance in potato. *Plant Pathology* **52,** 22–27.

Boller T, 1992. Biochemical analysis of chitinases and β-1,3 glucanases. In: *Practical Methods in Molecular Plant Pathology*, Bowles DJ, Gurr S, eds. Oxford: Oxford University Press, 23–30.

Bostock RM, 2005. Signal crosstalk and induced resistance: straddling the line between cost and benefit. *Annual Review of Phytopathology* **43,** 545–580.

Bostock RM, Karban R, Thaler JS, Weyman PD, Gilchrist D, 2001. Signal interactions in induced resistance to pathogens and insect herbivores. *European Journal of Plant Pathology* **107,** 103–111.
Bowling SA, Clarke JD, Liu Y, Klessig DF, Dong X, 1997. The *cpr5* mutant of *Arabidopsis* expresses both NPR1-dependent and NPR1-independent resistance. *The Plant Cell* **9,** 1573–1584.
Bowling SA, Guo A, Cao H, Gordon AS, Klessig DF, Dong X, 1994. A mutation in *Arabidopsis* that leads to constitutive expression of systemic acquired resistance. *The Plant Cell* **6,** 1845–1857.
Breiteneder H, Hoffman-Sommergruber K, O'Riordain G, Susani M, Ahorn H, Ebner C, Kraft D, Scheiner O, 1995. Molecular characterization of Api-g-1, the major allergen of celery (*Apium graveolens*), and its immunological and structural relationships to a group of 17-kDa tree pollen allergens. *European Journal of Biochemistry* **233,** 484–489.
Buonaurio R, Scarponi L, Ferrara M, Sidoti P, Bertona A, 2002. Induction of systemic acquired resistance in pepper plants by acibenzolar-S-methyl against bacterial spot disease. *European Journal of Plant Pathology* **108,** 41–49.
Buzi A, Chilosi G, De Sillo D, Magro P, 2004. Induction of resistance in melon to *Didymella bryoniae* and *Sclerotinia sclerotiorum* by seed treatments with acibenzolar-*S*-methyl and methyl jasmonate but not with salicylic acid. *Journal of Phytopathology* **152,** 34–42.
Cameron RK, 2000. Salicylic acid and its role in plant defense responses: what do we really know? *Physiological and Molecular Plant Pathology* **56,** 91–93.
Campbell MA, Fitzgerald HA, Ronald PC, 2002. Engineering pathogen resistance in crop plants. *Transgenic Research* **11,** 599–613.
Cheong YH, Chang HS, Gupta R, Wang X, Zhu T, Luan S, 2002. Transcriptional profiling reveals novel interactions between wounding, pathogen, abiotic stress, and hormonal responses in Arabidopsis. *Plant Physiology* **129,** 661–677.
Cipollini D, 2002. Does competition magnify the fitness costs of induced resistance in *Arabidopsis thaliana*? A manipulative approach. *Oecologia* **131,** 514–520.
Cipollini D, 2004. Stretching the limits of plasticity: Can a plant defend against both competitors and herbivores? *Ecology* **85,** 28–37.
Cipollini D, Enright S, Traw MB, Bergelson J, 2004. Salicylic acid inhibits jasmonic acid-induced resistance of *Arabidopsis thaliana* to *Spodoptera exigua*. *Molecular Ecology* **13,** 1643–1653.
Cipollini D, Purrington CB, Bergelson J, 2003. Costs of induced responses in plants. *Basic and Applied Ecology* **4,** 79–89.
Cipollini DF, Bergelson J, 2001. Plant density and nutrient availability constrain constitutive and wound-induced expression of trypsin inhibitors in *Brassica napus*. *Journal of Chemical Ecology* **27,** 593–610.
Clarke JD, Liu Y, Klessig DF, Dong X, 1998. Uncoupling PR gene expression from NPR1 and bacterial resistance: Characterization of the dominant *Arabidopsis cpr6-1* mutant. *The Plant Cell* **10,** 557–569.
Cole DL, 1999. The efficacy of acibenzolar-*S*-methyl, an inducer of systemic acquired resistance, against bacterial and fungal diseases of tobacco. *Crop Protection* **18,** 267–273.
Conrath U, Pieterse CMJ, Mauch-Mani B, 2002. Priming in plant–pathogen interactions. *Trends in Plant Science* **7,** 210–216.
Conrath U, Thulke O, Katz V, Schwindling S, Kohler A, 2001. Priming as mechanism in induced systemic resistance of plants. *European Journal of Plant Pathology* **107,** 113–119.
Cools HJ, Ishii H, 2002. Pre-treatment of cucumber plants with acibenzolar-*S*-methyl systemically primes a phenylalanine ammonia lyase gene (PAL1) for enhanced expression upon attack with a pathogenic fungus. *Physiological and Molecular Plant Pathology* **61,** 273–280.
Cordier C, Pozo MJ, Barea JM, Gianinazzi S, Gianinazzi-Pearson V, 1998. Cell defense responses associated with localized and systemic resistance to *Phytophthora parasitica* induced in tomato by an arbuscular mycorrhizal fungus. *Molecular Plant–Microbe Interactions* **11,** 1017–1028.
Csinos AS, Pappu HR, McPherson RM, Stephenson MG, 2001. Management of *Tomato spotted wilt virus* in flue-cured tobacco with acibenzolar-*S*-methyl and imidacloprid. *Plant Disease* **85,** 292–296.
Delaney TP, Uknes S, Vernooij B, Friedrich L, Weymann K, Negrotto D, Gaffney T, Gut-Rella M, Kessmann H, Ward E, Rylas J, 1994. A central role of salicylic acid in plant disease resistance. *Science* **266,** 1247–1249.
Délano-Frier JP, Martínez-Gallardo NA, Martínez-de la Vega O, Salas-Araiza MD, Barbosa-Jaramillo ER, Torres A, Vargas P, Borodanenko A, 2004. The effect of exogenous jasmonic acid on induced resistance and productivity in amaranth (*Amaranthus hypochondriacus*) is influenced by environmental conditions. *Journal of Chemical Ecology* **30,** 1001–1034.

Dietrich R, Ploss K, Heil M, 2004. Constitutive and induced resistance to pathogens in *Arabidopsis thaliana* depend on nitrogen supply. *Plant, Cell and Environment* **27,** 896–906.

Dietrich R, Ploss K, Heil M, 2005. Growth responses and fitness costs after induction of pathogen resistance depend on environmental conditions. *Plant Cell and Environment* **28,** 211–222.

Dixon RA, 1986. The phytoalexin response: elicitation, signalling and control of host gene expression. *Biological Reviews of the Cambridge Philosophical Society* **61,** 239–291.

Dixon RA, 2001. Natural products and plant disease resistance. *Nature* **411,** 843–847.

Dixon RA, Achnine L, Kota P, Liu CJ, Reddy MSS, Wang LJ, 2002. The phenylpropanoid pathway and plant defence – a genomics perspective. *Molecular Plant Pathology* **3,** 371–390.

Dong X, 2001. Genetic dissection of systemic acquired resistance. *Current Opinion in Plant Biology* **4,** 309–314.

Dumas-Gaudot E, Slezack S, Dassi B, Pozo MJ, Gianinazzi-Pearson V, Gianinazzi S, 1996. Plant hydrolytic enzymes (chitinases and beta-1,3-glucanases) in root reactions to pathogenic and symbiotic microorganisms. *Plant and Soil* **185,** 211–221.

Durrant WE, Dong X, 2004. Systemic acquired resistance. *Annual Review of Phytopathology* **42,** 185–209.

Feeny P, 1976. Plant apparency and chemical defense. *Recent Advances in Phytochemistry* **10,** 1–40.

Felton GF, Korth KL, 2000. Trade-offs between pathogen and herbivore resistance. *Current Opinion in Plant Biology* **3,** 309–314.

Feys BJ, Parker JE, 2000. Interplay of signaling pathways in plant disease resistance. *Trends in Genetics* **16,** 449–455.

Foster JE, Ohm HW, Patterson FL, Taylor PL, 1991. Effects of deploying single gene resistance in wheat for controlling damage by the Hessian fly (Diptera: Cecidomyiidae). *Environmental Entomology* **20,** 964–969.

Gaffney T, Friedrich L, Vernooij B, Negrotto D, Nye G, Uknes S, Ward E, Kessmann H, Ryals J, 1994. Requirement of salicylic acid for the induction of systemic acquired resistance. *Science* **261,** 754–756.

Genoud T, Santa Cruz MBT, Métraux JP, 2001. Numeric simulation of plant signaling networks. *Plant Physiology* **126,** 1430–1437.

Glandorf DCM, Bakker PAH, van Loon LC, 1997. Influence of the production of antibacterial and antifungal proteins by transgenic plant on the saprophytic soil microflora. *Acta Botanica Neerlandica* **46,** 85–104.

Glazebrook J, Chen WJ, Estes B, Chang HS, Nawrath C, Métraux JP, Zhu T, Katagiri F, 2003. Topology of the network integrating salicylate and jasmonate signal transduction derived from global expression phenotyping. *The Plant Journal* **34,** 217–228.

Görlach J, Volrath S, Knauf-Beiter G, Hengy G, Beckhove U, Kogel K-H, Oostendorp M, Staub T, Ward E, Kessmann H, Ryals J, 1996. Benzothiadiazole, a novel class of inducers of systemic acquired resistance, activates gene expression and disease resistance in wheat. *The Plant Cell* **8,** 629–643.

Gould F, 1991. The evolutionary potential of crop pests. *American Scientist* **79,** 496–507.

Gould F, Anderson A, 1991. Effects of *Bacillus thuringiensis* and HD-73 delta-endotoxin on growth, behavior, and fitness of susceptible and tocin-adapted strains of *Heliothis virescens* (Lepidoptera: Noctuidae). *Environmental Entomology* **20,** 30–38.

Gozzo F, 2003. Systemic acquired resistance in crop protection: from nature to a chemical approach. *Journal of Agricultural and Food Chemistry* **51,** 4487–4503.

Graham JH, Leite RP, 2004. Lack of control of citrus canker by induced systemic resistance compounds. *Plant Disease* **88,** 745–750.

Greenberg GT, 1997. Programmed cell death in plant–pathogen interactions. *Annual Review of Plant Physiology and Plant Molecular Biology* **48,** 525–545.

Greenberg JT, Silverman FP, Liang H, 2000. Uncoupling salicylic acid-dependent cell death and defense-related responses from disease resistance in the *Arabidopsis* mutant *acd5*. *Genetics* **156,** 341–350.

Hammerschmidt R, 1999a. Induced disease resistance: how do induced plants stop pathogens? *Physiological and Molecular Plant Pathology* **55,** 77–84.

Hammerschmidt R, 1999b. Phytoalexins: what have we learned after 60 years? *Annual Review of Phytopathology* **37,** 285–306.

Hammerschmidt R, Nicholson RL, 1999. A survey of defense responses to pathogens. In: *Induced Plant Defenses Against Pathogens and Herbivores: Biochemistry, Ecology, and Agriculture.* Agrawal AA, Tuzun S, Bent E, eds. St. Paul, MN: APS Press, 55–71.

Hammerschmidt R, Smith-Becker JA, 1999. The role of salicylic acid in disease resistance. In: *Induced Plant Defenses against Pathogens and Herbivores: Biochemistry, Ecology, and Agriculture.* Agrawal AA, Tuzun S, Bent E, eds. St. Paul, MN: APS Press, 37–53.

Hashimoto M, Kisseleva L, Sawa S, Furukawa T, Komatsu S, Koshiba T, 2004. A novel rice PR10 protein, RSOsPR10, specifically induced in roots by biotic and abiotic stresses, possibly via the jasmonic acid signaling pathway. *Plant and Cell Physiology* **45,** 550–559.

Hatcher PE, Paul ND, 2000. Beetle grazing reduces natural infection of *Rumex obtusifolius* by fungal pathogens. *New Phytologist* **146,** 325–333.

Heath MC, 1998. Apoptosis, programmed cell death and the hypersensitive response. *European Journal of Plant Pathology* **104,** 117–124.

Heidel AJ, Clarke JD, Antonovics J, Dong XN, 2004. Fitness costs of mutations affecting the systemic acquired resistance pathway in *Arabidopsis thaliana. Genetics* **168,** 2197–2206.

Heil M, 1999. Systemic acquired resistance: available information and open ecological questions. *Journal of Ecology* **87,** 341–346.

Heil M, 2001a. The ecological concept of costs of induced systemic resistance (ISR). *European Journal of Plant Pathology* **107,** 137–146.

Heil M, 2001b. Induced systemic resistance (ISR) against pathogens – a promising field for ecological research. *Perspectives in Plant Ecology, Evolution and Systematics* **4,** 65–79.

Heil M, 2002. Ecological costs of induced resistance. *Current Opinion in Plant Biology* **5,** 345–350.

Heil M, Baldwin IT, 2002. Fitness costs of induced resistance: emerging experimental support for a slippery concept. *Trends in Plant Science* **7,** 61–67.

Heil M, Bostock RM, 2002. Induced systemic resistance (ISR) in the context of induced plant defences. *Annals of Botany* **89,** 503–512.

Heil M, Hilpert A, Kaiser W, Linsenmair KE, 2000. Reduced growth and seed set following chemical induction of pathogen defence: does systemic acquired resistance (SAR) incur allocation costs? *Journal of Ecology* **88,** 645–654.

Hoffmann-Sommergruber K, 2002. Pathogenesis-related (PR)-proteins identified as allergens. *Biochemical Society Transactions* **30,** 930–935.

Hunt MD, Neuenschwander UH, Delaney TP, Weymann KB, Friedrich LB, Lawton KA, Steiner H-Y, Ryals JA, 1996. Recent advances in systemic acquired resistance research – a review. *Gene* **179,** 89–95.

Hunt MD, Ryals JA, 1996. Systemic acquired resistance signal transduction. *Critical Reviews in Plant Sciences* **15,** 583–606.

Iriti M, Faoro F, 2003a. Benzothiadiazole (BTH) induces cell-death independent resistance in *Phaseolus vulgaris* against *Uromyces appendiculatus. Journal of Phytopathology* **151,** 171–180.

Iriti M, Faoro F, 2003b. Does benzothiadiazole-induced resistance increase fitness cost in bean? *Journal of Plant Pathology* **85,** 265–270.

Ishii H, Tomita Y, Horio T, Narusaka Y, Nakazawa Y, Nishimura K, Iwamoto S, 1999. Induced resistance of acibenzolar-*S*-methyl (CGA 245704) to cucumber and Japanese pear diseases. *European Journal of Plant Pathology* **105,** 77–85.

Jakab G, Cottier V, Toquin V, Rigoli G, Zimmerli L, Métraux J-P, Mauch-Mani B, 2001. β-Aminobutyric acid-induced resistance in plants. *European Journal of Plant Pathology* **107,** 29–37.

Jirage D, Zhou N, Cooper B, Clarke JD, Dong X, Glazebrook J, 2001. Constitutive salicylic acid-dependent signaling in *cpr1* and *cpr6* mutants requires PAD4. *The Plant Journal* **26,** 395–407.

Kanno H, Satoh M, Kimura T, Fujita Y, 2005. Some aspects of induced resistance to rice blast fungus, *Magnaporthe grisea,* in rice plant infested by white-backed planthopper, *Sogatella furcifera. Applied Entomology and Zoology* **40,** 91–97.

Karban R, Baldwin IT. (1997) *Induced Responses to Herbivory.* Chicago: University of Chicago Press.

Katagiri F, 2004. A global view of defense gene expression regulation – a highly interconnected signaling network. *Current Opinion in Plant Biology* **7,** 506–511.

Kohler A, Schwindling S, Conrath U, 2002. Benzothiadiazole-induced priming for potentiated responses to pathogen infection, wounding, and infiltration of water into leaves requires the *NPR1/NIM1* gene in Arabidopsis. *Plant Physiology* **128,** 1046–1056.

Kuć J, 1982. Induced immunity to plant disease. *Bioscience* **32,** 854–860.

Kuć J, 1995. Phytoalexins, stress metabolism, and disease resistance in plants. *Annual Review of Phytopathology* **33,** 275–297.

Kuć J, 2001. Concepts and direction of induced systemic resistance in plants and its application. *European Journal of Plant Pathology* **107,** 7–12.

Kunkel BN, Brooks DM, 2002. Cross talk between signaling pathways in pathogen defense. *Current Opinion in Plant Biology* **5,** 325–331.

Latunde-Dada AO, Lucas JA, 2001. The plant defence activator acibenzolar-*S*-methyl primes cowpea [*Vigna unguiculata* (L.) Walp.] seedlings for rapid induction of resistance. *Physiological and Molecular Plant Pathology* **58,** 199–208.

Lawton KA, Friedrich L, Hunt M, Weymann K, Delaney T, Kessmann H, Staub T, Ryals JA, 1996. Benzothiadiazole induces disease resistance in Arabidopsis by activation of the systemic acquired resistance signal transduction pathway. *The Plant Journal* **10,** 71–82.

Leijerstam B, 1972. Studies in powdery mildew on wheat in Sweden. III. Variability of virulence in *Erysiphe graminis* f. sp. *tritici* due to gene recombination and mutation. *Contributions of the National Swedish Institute of Plant Protection* **15,** 231–248.

Lemay AV, Bailey JE, Shew BB, 2002. Resistance of peanut to sclerotinia blight and the effect of acibenzolar-*S*-methyl and fluazinam on disease incidence. *Plant Disease* **86,** 1315–1317.

Lian B, Zhou X, Miransari M, Smith DL, 2000. Effects of salicylic acid on the development and root nodulation of soybean seedlings. *Journal of Agronomy and Crop Sciences* **185,** 187–192.

Logemann E, Wu S-C, Schröder J, Schmelzer E, Somssich IE, Hahlbrock K, 1995. Gene activation by UV light, fungal elicitor or fungal infection in *Petroselinum crispum* is correlated with repression of cell-cycle-related genes. *The Plant Journal* **8,** 865–876.

Lopez AMQ, Lucas JA, 2002. Effects of plant defence activators on anthracnose disease of cashew. *European Journal of Plant Pathology* **108,** 409–420.

Louws FJ, Wilson M, Campbell HL, Cuppels DA, Jones JB, Shoemaker PB, Sahin F, Miller SA, 2001. Field control of bacterial spot and bacterial speck of tomato using a plant activator. *Plant Disease* **85,** 481–488.

Malamy J, Carr JP, Klessig DF, Raskin I, 1990. Salicylic acid: a likely endogenous signal in the resistance response of tobacco to viral infection. *Science* **250,** 1002–1004.

Malandain H, Lavaud F, 2004. Allergénicité de protéines de défense végétale. *Revue francaise d'allergologie et d'immunologie clinique* **44,** 469–475.

Maleck K, Levine A, Eulgem T, Morgan A, Schmid J, Lawton KA, Dangl JL, Dietrich RA, 2000. The transcriptome of *Arabidopsis thaliana* during systemic acquired resistance. *Nature Genetics* **26,** 403–410.

Martínez-Abarca F, Herrera-Cervera JA, Bueno P, Sanjuan J, Bisseling T, Olivares J, 1998. Involvement of salicylic acid in the establishment of the *Rhizobium meliloti*–alfalfa symbiosis. *Molecular Plant–Microbe Interactions* **11,** 153–155.

Mauch F, Mauch-Mani B, Gaille C, Kull B, Haas D, Reimmann C, 2001. Manipulation of salicylate content in *Arabidopsis thaliana* by the expression of an engineered bacterial salicylate synthase. *The Plant Journal* **25,** 67–77.

Mauch-Mani B, Métraux J-P, 1998. Salicylic acid and systemic acquired resistance to pathogen attack. *Annals of Botany* **82,** 535–540.

Mayda E, Mauch-Mani B, Vera P, 2000. Arabidopsis *dth9* mutation identifies a gene involved in regulating disease susceptibility without affecting salicylic acid-dependent responses. *The Plant Cell* **12,** 2119–2128.

Métraux J-P, 2001. Systemic acquired resistance and salicylic acid: current state of knowledge. *European Journal of Plant Pathology* **107,** 13–18.

Métraux J-P, Signer H, Ryals J, Ward E, Wyss-Benz M, Gaudin J, Raschdorf K, Schmid E, Blum W, Inverardi B, 1990. Increase in salicylic acid at the onset of systemic acquired resistance. *Science* **250,** 1004–1006.

Midoro-Horiuti T, Goldblum RM, Kurosky A, Wood TG, Brooks EG, 2000. Variable expression of pathogenesis-related protein allergen in mountain cedar (*Juniperus ashei*) pollen. *Journal of Immunology* **164,** 2188–2192.

Mithöfer A, 2002. Suppression of plant defence in rhizobia–legume symbiosis. *Trends in Plant Science* **10,** 440–444.

Molina A, Görlach J, Volrath S, Ryals J, 1999. Wheat genes encoding two types of PR-1 proteins are pathogen inducible, but do not respond to activators of systemic acquired resistance. *Molecular Plant–Microbe Interactions* **12,** 53–58.

Oldroyd GE, 2001. Dissecting symbiosis: development in nod factor signal transduction. *Annals of Botany* **87,** 709–718.

Oostendorp M, Kunz W, Dietrich B, Staub T, 2001. Induced resistance in plants by chemicals. *European Journal of Plant Pathology* **107,** 19–28.

Padgett GB, Russin JS, Snow JP, Boethel DJ, Berggren GT, 1994. Interactions among the soybean looper (Lepidoptera: Noctuidae), threecorned alfalfa hopper (Homoptera: Membracidae), stem canker, and red crown rot in soybean. *Journal of Entomological Science* **29,** 110–119.

Petersen M, Brodersen P, Naested H, Andreasson E, Lindhart U, Johansen B, Nielsen HB, Lacy M, Austin MJ, Parker JE, Sharma SB, Klessig DF, Martienssen R, Mattsson O, Jensen AB, Mundy J, 2000. *Arabidopsis* MAP kinase 4 negatively regulates systemic acquired resistance. *Cell* **103,** 1111–1120.

Pieterse CMJ, van Loon LC, 2004. NPR1: the spider in the web of induced resistance signaling pathways. *Current Opinion in Plant Biology* **7,** 456–464.

Pieterse CMJ, van Pelt JA, van Wees SCM, Ton J, Léon-Kloosterziel K, Keurentjes JJB, Verhagen BWM, Knoester M, van der Sluis I, Bakker PAHM, van Loon LC, 2001. Rhizobacteria-mediated induced systemic resistance: triggering, signalling and expression. *European Journal of Plant Pathology* **107,** 51–61.

Pozo MJ, van Loon LC, Pieterse CMJ, 2005. Jasmonates – Signals in plant–microbe interactions. *Journal of Plant Growth Regulation* **23,** 211–222.

Prats E, Rubiales D, Jorrin J, 2002. Acibenzolar-*S*-methyl-induced resistance to sunflower rust (*Puccinia helianthi*) is associated with an enhancement of coumarins on foliar surface. *Physiological and Molecular Plant Pathology* **60,** 155–162.

Ramamoorthy V, Viswanathan R, Raguchander T, Prakasam V, Samiyappan R, 2001. Induction of systemic resistance by plant growth promoting rhizobacteria in crop plants against pests and diseases. *Crop Protection* **20,** 1–11.

Ramanujam MP, Abdul Jaleel V, Kumara Velu G, 1998. Effect of salicylic acid on nodulation, nitrogenous compounds and related enzymes of *Vigna mungo*. *Biologia Plantarum* **41,** 307–311.

Ramonell KM, Somerville S, 2002. The genomics parade of defense responses: to infinity and beyond. *Current Opinion in Plant Biology* **5,** 291–294.

Raskin I, 1992. Role of salicylic acid in plants. *Annual Review of Plant Physiology and Plant Molecular Biology* **43,** 439–463.

Rasmussen JB, Hammerschmidt R, Zook MN, 1991. Systemic induction of salicylic acid accumulation in cucumber after inoculation with *Pseudomonas syringae* pv. *syringae*. *Plant Physiology* **97,** 1342–1347.

Rate DN, Cuenca JV, Bowman GR, Guttman DS, Greenberg JT, 1999. The gain-of-function *Arabidopsis acd6* mutant reveals novel regulation and function of the salicylic acid signaling pathway in controlling cell death, defenses, and cell growth. *The Plant Cell* **11,** 1695–1708.

Rausher MD, 2001. Co-evolution and plant resistance to natural enemies. *Nature* **411,** 857–864.

Resende MLV, Nojosa GBA, Cavalcanti LS, Aguilar MAG, Silva L, Perez JO, Andrade GCG, Carvalho GA, Castro RM, 2002. Induction of resistance in cocoa against *Crinipellis perniciosa* and VERTICILLIUM DAHLIAE by acibenzolar-*S*-methyl (ASM). *Plant Pathology* **51,** 621–628.

Reymond P, Farmer EE, 1998. Jasmonate and salicylate as global signals for defense gene expression. *Current Opinion in Plant Biology* **1,** 404–411.

Rohilla R, Singh US, Singh RL, 2002. Mode of action of acibenzolar-*S*-methyl against sheath blight of rice, caused by *Rhizoctonia solani* Kuhn. *Pest Management Science* **58,** 63–69.

Romero AM, Kousik CS, Ritchie DF, 2001. Resistance to bacterial spot in bell pepper induced by acibenzolar-S-methyl. *Plant Disease* **85,** 189–194.

Royo J, León J, Vancanneyt G, Albar JP, Rosahl S, Ortego F, Castañera P, Sánchez-Serrano JJ, 1999. Antisense-mediated depletion of a potato lipoxygenase reduces wound induction of proteinase inhibitors and increases weight gain of insect pests. *Proceedings of the National Academy of Sciences USA* **96,** 1146–1151.

Ruiz-Lozano JM, Roussel H, Gianinazzi S, Gianinazzi-Pearson V, 1999. Defense genes are differentially induced by a mycorrhizal fungus and *Rhizobium* sp. in wild-type and symbiosis-defective pea genotypes. *Molecular Plant–Microbe Interactions* **12,** 976–984.

Russin JS, Layton MB, Boethel DJ, McGawley EC, Snow JP, Berggren GT, 1990. Growth, nodule development, and N_2-fixing ability in soybean damaged by an insect–fungus–herbivore–nematode pest complex. *Journal of Economic Entomology* **83,** 247–254.

Russo VM, Russo BM, Peters M, Perkins-Veazie P, Cartwright B, 1997. Interaction of *Colletotrichum orbiculare* with thrips and aphid feeding on watermelon seedlings. *Crop Protection* **16,** 581–584.

Ryals J, Uknes S, Ward E, 1994. Systemic acquired resistance. *Plant Physiology* **104,** 1109–1112.

Salzman RA, Brady JA, Finlayson SA, Buchanan CD, Summer EJ, Sun F, Klein PE, Klein RR, Pratt LH, Cordonnier-Pratt MM, Mullet JE, 2005. Transcriptional profiling of sorghum induced by methyl jasmonate, salicylic acid, and aminocyclopropane carboxylic acid reveals cooperative regulation and novel gene responses. *Plant Physiology* **138,** 352–368.

Sauerborn J, Buschmann H, Ghiasi KG, Kogel KH, 2002. Benzothiadiazole activates resistance in sunflower (*Helianthus annuus*) to the root-parasitic weed *Orobanche cumana. Phytopathology* **92,** 59–64.

Scheideler M, Schlaich NL, Fellenberg K, Beissbarth T, Hauser NC, Vingron M, Slusarenko AJ, Hoheisel JD, 2002. Monitoring the switch from housekeeping to pathogen defense metabolism in *Arabidopsis thaliana* using cDNA arrays. *Journal of Biological Chemistry* **277,** 10555–10561.

Schultze M, Kondorosi A, 1998. Regulation of symbiotic root nodule development. *Annual Review of Genetics* **32,** 33–57.

Schurter R, Kunz W, Nyfelder R. (1987) Process and a composition for immunizing plants against diseases. In *U.S. Patent no. 4931581.*

Shah J, 2003. The salicylic acid loop in plant defense. *Current Opinion in Plant Biology* **6,** 365–371.

Simms EL, Fritz RS, 1990. The ecology and evolution of host-plant resistance to insects. *Trends in Ecology and Evolution* **5,** 356–360.

Somssich IE, Hahlbrock K, 1998. Pathogen defence in plants – a paradigm of biological complexity. *Trends in Plant Science* **3,** 86–90.

Sparla F, Rotino L, Valgimigli MC, Pupillo P, Trost P, 2004. Systemic resistance induced by benzothiadiazole in pear inoculated with the agent of fire blight (*Erwinia amylovora*). *Scientia Horticulturae* **101,** 269–279.

Stadnik MJ, Buchenauer H, 1999. Control of wheat diseases by a benzothiadiazole-derivative and modern fungicides. *Zeitschrift für Pflanzenkrankheiten und Pflanzenschutz – Journal of Plant Diseases and Protection* **106,** 466–475.

Sticher L, Mauch-Mani B, Métraux J-P, 1997. Systemic acquired resistance. *Annual Review of Phytopathology* **35,** 235–270.

Stout MJ, Workman KV, Bostock RM, Duffey SS, 1998. Specificity of induced resistance in the tomato, *Lycopersicon esculentum. Oecologia* **113,** 74–81.

Suo YY, Leung DWM, 2001. Elevation of extracellular beta-1,3-glucanase and chitinase activities in rose in response to treatment with acibenzolar-*S*-methyl and infection by *D. rosae. Journal of Plant Physiology* **158,** 971–976.

Tally A, Oostendorp M, Lawton K, Staub T, Bassi B, 1999. Commercial development of elicitors of induced resistance to pathogens. In *Induced Plant Defenses against Pathogens and Herbivores: Biochemistry, Ecology, and Agriculture.* Agrawal AA, Tuzun S, Bent E, eds. St. Paul, MN: APS Press, 357–369.

Thaler JS, Fidantsef AL, Duffey SS, Bostock RM, 1999. Trade-offs in plant defense against pathogens and herbivores: a field demonstration of chemical elicitors of induced resistance. *Journal of Chemical Ecology* **25,** 1597–1609.

Tollrian R, Harvell CD, 1999. *The Ecology and Evolution of Inducible Defenses.* Princeton, NJ: Princeton University Press.

Traw MB, Bergelson J, 2003. Interactive effects of jasmonic acid, salicylic acid, and gibberellin on induction of trichomes in Arabidopsis. *Plant Physiology* **133,** 1367–1375.

Traw MB, Kim J, Enright S, Cipollini DF, Bergelson J, 2003. Negative cross-talk between salicylate- and jasmonate-mediated pathways in the Wassilewskija ecotype of *Arabidopsis thaliana. Molecular Ecology* **12,** 1125–1135.

Vallad GE, Goodman RM, 2004. Systemic acquired resistance and induced systemic resistance in conventional agriculture. *Crop Science* **44,** 1920–1934.

van Loon LC, 1985. Pathogenesis-related proteins. *Plant Molecular Biology* **4,** 111–116.

van Loon LC, 1997. Induced resistance in plants and the role of pathogenesis-related proteins. *European Journal of Plant Pathology* **103,** 753–765.

van Loon LC, van Strien EA, 1999. The families of pathogenesis-related proteins, their activities, and comparative analysis of PR-1 type proteins. *Physiological and Molecular Plant Pathology* **55,** 85–97.

Vierheilig H, Alt M, Lange J, Gut Rella M, Wiemken A, Boller T, 1995. Colonization of transgenic tobacco constitutively expressing pathogenesis-related proteins by the vesicular-arbuscular mycorrhizal fungus *Glomus mosseae. Appplied and Environmental Microbiology* **61,** 3031–3034.

Vierheilig H, Alt M, Mohr U, Boller T, Wiemken A, 1994. Ethylene biosynthesis and activities of chitinase and β-1,3-glucanase in the roots of host and non-host plants of vesicular-arbuscular mycorrhizal fungi after inoculation with *Glomus mosseae*. *Journal of Plant Physiology* **143,** 337–343.

Walling LL, 2000. The myriad plant responses to herbivores. *Journal of Plant Growth Regulation* **19,** 195–216.

Walling LL, 2001. Induced resistance: from the basic to the applied. *Trends in Plant Science* **6,** 445–447.

Walter MH, Liu JW, Wunn J, Hess D, 1996. Bean ribonuclease-like pathogenesis-related protein genes (*Ypr10*) display complex patterns of developmental, dark-induced and exogenous-stimulus-dependent expression. *European Journal of Biochemistry* **239,** 281–293.

Whitham TG, Slobodchikoff CN, 1981. Evolution by individuals, plant–herbivore interactions, and mosaics of genetic variability: the adaptive significance of somatic mutations in plants. *Oecologia* **49,** 287–292.

Whitham TG, Williams AG, Robinson AM, 1984. The variation principle: individual plants as temporal and spatial mosaics of resistance to rapidly evolving pests. In: *A New Ecology: Novel Approaches to Interactive Systems.* Price PW, Slobodchikoff CN, Gaud WS, eds. New York: Wiley, 15–51.

Yasuda M, Nakashita H, Yoshida S, 2004. Tiadinil, a novel class of activator of systemic acquired resistance, induces defense gene expression and disease resistance in tobacco. *Journal of Pesticide Science* **29,** 46–49.

Yasuda M, Nishioka M, Nakashita H, Yamaguchi I, Yoshida S, 2003. Pyrazolecarboxylic acid derivative induces systemic acquired resistance in tobacco. *Bioscience Biotechnology and Biochemistry* **67,** 2614–2620.

Yu GY, Muehlbauer GJ, 2001. Benzothiadiazole-induced gene expression in wheat spikes does not provide resistance to *Fusarium* head blight. *Physiological and Molecular Plant Pathology* **59,** 129–136.

Yu I-C, Parker J, Bent AF, 1998. Gene-for-gene disease resistance without the hypersensitive response in *Arabidopsis dnd1* mutant. *Proceedings of the National Academy of Sciences USA* **95,** 7819–7824.

Zehnder GW, Murphy JF, Sikora EJ, Kloepper JW, 2001. Application of rhizobacteria for induced resistance. *European Journal of Plant Pathology* **107,** 39–50.

Zhang S, Reddy MS, Kokalis-Burelle N, Wells LW, Nightengale SP, Kloepper JW, 2001. Lack of induced systemic resistance in peanut to late leaf spot disease by plant growth-promoting rhizobacteria and chemical elicitors. *Plant Disease* **85,** 879–884.

Zhu YJ, Qiu XH, Moore PH, Borth W, Hu J, Ferreira S, Albert HH, 2003. Systemic acquired resistance induced by BTH in papaya. *Physiological and Molecular Plant Pathology* **63,** 237–248.

Ziadi S, Barbedette S, Godard JF, Monot C, Le Corre D, Silue D, 2001. Production of pathogenesis-related proteins in the cauliflower (*Brassica oleracea* var. *botrytis*)-downy mildew (*Peronospora parasitica*) pathosystem treated with acibenzolar-*S*-methyl. *Plant Pathology* **50,** 579–586.

Zimmerli L, Jakab G, Métraux J-P, Mauch-Mani B, 2000. Potentiation of pathogen-specific defense mechanisms in *Arabidopsis* by beta-aminobutyric acid. *Proceeding of the National Academy of Sciences USA* **97,** 12920–12925.

Chapter 10
Topical application of inducers for disease control

Philippe Reignault[1] *and Dale Walters*[2]

[1]Université du Littoral Côte d'Opale, Calais, France
[2]Scottish Agricultural College, Edinburgh, UK

10.1 Introduction

Plants possess a variety of defence mechanisms to protect themselves against microbial attack, and the co-ordination and timing of these mechanisms are crucial for effective defence. In the early stages of a plant–pathogen interaction, elicitor molecules are released. These elicitors can be of plant or fungal origin and include lipids, carbohydrate polymers, glycopeptides and glycoproteins (Walters *et al.*, 2005). The elicitor molecules are perceived by plant cells, thereby activating a signalling pathway (see Chapter 4) and leading in turn to the formation of defence mechanisms. The prospect of broad-spectrum disease control by artificially activating plant defences has resulted in great interest in the development of agents capable of mimicking natural inducers of resistance (Walters *et al.*, 2005). Activity in this area has concentrated on elicitor molecules released early in the host–pathogen interaction and on the signalling pathways responsible for triggering defences locally and systemically. This chapter deals with both the natural inducers of resistance (biotic inducers) and agents which mimic the action of these natural inducers (abiotic inducers) (see Table 10.1), concentrating on the effects of topical treatment with inducers on disease control under controlled conditions and in the field (refer to Chapter 2 for a comprehensive treatment of agents which have been reported to induce defence responses).

10.2 Biotic inducers

Because of their crucial importance in plant responses to pathogen infection, signalling molecules or elicitors have received considerable attention over the past 20 years. Strong elicitor activities have been correlated with pectic fragments released from plant cell walls through the action of pathogen-produced pectinases (Collmer & Keen, 1988) and with extracellular products of pathogen origin, like fungal cell wall oligomers (Lawton & Lamb, 1987). Pathogen-derived elicitors are thought to be the primary signals responsible for the induction of plant defence responses (Callow, 1984) and since glucan oligomers from fungal cell walls were shown to be active elicitors of phytoalexins (Albersheim & Valent, 1978), a range of fungal derived molecules, including unsaturated lipids, glycoproteins and polysaccharides, have been shown to activate plant defences (Pearce & Ride, 1982; Bostock *et al.*, 1986).

Table 10.1 Elicitors of induced resistance covered in Chapter 10.

Type of elicitor	Protected plant	Targeted pathogen	Cited reference
Abiotic inducers			
Reactive oxygen species			
Oxycom™	Tobacco	*Pseudomonas syringae* pv. *tabaci*	Yang *et al.* (2002)
Oxycom™	Tobacco		Blee *et al.* (2004)
Oxycom™	Lettuce	*Bremia lactucae*	Kim *et al.* (2001)
MSB	Banana	*Fusarium oxysporum* f. sp. *cubense*	Borges *et al.* (2004)
MSB	Oilseed rape	*Leptosphaeria maculans*	Borges *et al.* (2003)
Benzo(1,2,3)thiadiazole-7-carbothioic acid S-methyl ester (BTH)/acibenzolar-S-methyl (ASM)			
See main text for details			
Lipids			
JA & MJ	Grapefruit	*Penicillium digitatum*	Droby *et al.* (1999)
JA & MJ	Norway spruce	*Pythium ultimum*	Kozlowski *et al.* (1994)
JA & MJ	Potato	*Phytophthora infestans*	Cohen *et al.* (1993)
JA	Barley	*Blumeria graminis*	Schweizer *et al.* (1993)
MJ	Barley	*B. graminis*	Mitchell & Walters (1995)
MJ	Melon	*Didymella bryoniae, Sclerotinia sclerotiorum* and *F. oxysporum* f. sp. *melonis*	Buzi *et al.* (2004a)
Trihydroxy-oxylipins	Barley	*B. graminis*	Cowley & Walters (2005)
Minerals and ions			
Phosphate	Barley	*B. graminis*	Mitchell & Walters (2004)
Phosphate	Cucumber	*Colletotrichum lagenarium* various	Gottstein & Kuć (1989), Mucharromah & Kuć (1991)
Phosphate	Cucumber	*Sphaerotheca fuligenea*	Reuveni *et al.* (2000), Orober *et al.* (2002)
Phosphate	Grapevine		Reuveni & Reuveni (1995)
Phosphate	Pepper		Reuveni *et al.* (1998)
Phosphate	Rice	*Magnaporthe grisea*	Mandahar *et al.* (1998)
Phosphonate	Cauliflower	*Peronospora parasitica*	Bécot *et al.* (2000)
Phosphonate	Lettuce	*B. lactucae*	Pajot *et al.* (2001)
Silicon	Rice	*M. grisea, Bipolaris oryzae* and *Rhizoctonia solani*	Kim *et al.* (2002), Seebold *et al.* (2004)
Silicon	Sweet cherry	*Penicillium expansum* and *Monilinia fructigena*	Qin & Tian (2005)
Silicon	Wheat	*B. graminis*	Bélanger *et al.* (2003)
Protein, peptide and amino acid-derived inducers			
AABA	Tobacco	TMV	Siegrist *et al.* (2000)
BABA	*A. thaliana*	*Botrytis cinerea*	Zimmerli *et al.* (2001)
BABA	*A. thaliana*	*P. brassicae*	Si-Ammour *et al.* (2003)

(Continued)

Table 10.1 (*Continued*)

Type of elicitor	Protected plant	Targeted pathogen	Cited reference
BABA	*A. thaliana*	*Alternaria brassicicola* and *Plectosphaerella cucumerina*	Ton & Mauch-Mani (2004)
BABA	Cauliflower	*P. parasitica*	Silué *et al.* (2002)
BABA	Cereals	*Heterodera avenae* and *H. latipons*	Oka & Cohen (2001)
BABA	Grapevine	*Plasmopara viticola*	Cohen *et al.* (1999)
BABA	Pearl millet	*Sclerospora graminicola*	Shailasree *et al.* (2001)
Harpin	*A. thaliana*	*P. syringae* pv. *tomato* and *P. parasitica*	Dong *et al.* (1999)
Harpin	*A. thaliana*	*Erwinia carotovora* subsp. *carotovora*	Kariola *et al.* (2003)
Harpin	*A. thaliana*	*P. parasitica*	Peng *et al.* (2003)
Harpin	Tobacco	TMV	Peng *et al.* (2003)
Syringolin A	Rice	*M. grisea*	Wäspi *et al.* (1998)
Syringolin A	Wheat	*B. graminis*	Wäspi *et al.* (2001)
Salicylic acid (SA) and structurally related compounds			
BIT	*A. thaliana*	*P. syringae* pv. *tomato*	Yoshioka *et al.* (2001)
BIT	Rice	*Xanthomonas oryzae* pv. *oryzae*	Nakashita *et al.* (2002b)
BIT	Tobacco	TMV, *P. syringae* pv. *tabaci* and *Oidium lycopersici*	Nakashita *et al.* (2002b)
NCI	Rice	*M. grisea* and *X. oryzae* pv. *oryzae*	Nakashita *et al.* (2002a)
NCI	Tobacco	TMV, *P. syringae* pv. *tabaci* and *O. lycopersici*	Nakashita *et al.* (2002a)
SA	Bean	WCMV	Gális *et al.* (2004)
SA	Pear	*E. amylovora*	Sparla *et al.* (2004)
SA	Sweet cherry	*M. fructicola*	Yao and Tian (2005)
SA	Tomato	*A. solani*	Spletzer & Enyedi (1999)
SA & ASA	Chickpea	*F. oxysporum*	Saikia *et al.* (2003)
SHZ & 4HBHZ	Tomato	*F. oxysporum* f. sp. *lycopersici*	Miyazawa *et al.* (1998)
Sugars			
Trehalose	Wheat	*B. graminis*	Reignault *et al.* (2001)
Ultraviolet irradiation			
	Apple	*P. expansum*	de Capdeville *et al.* (2002)
	Cabbage	*X. campestris* pv. *campestris*	Brown *et al.* (2001)
	Citrus	*P. digitatum*	Arcas *et al.* (2000)
	Peach	*M. fructigena*	Stevens *et al.* (1998)
	Sweet potato	*Fusarium solani*	Stevens *et al.* (1999)
Biotic inducers			
Chitin			
	Groundnut	*Phaeoisariopsis personata*	Kishore *et al.* (2005)

(*Continued*)

Table 10.1 (*Continued*)

Type of elicitor	Protected plant	Targeted pathogen	Cited reference
Chitosan			
	Carrot	*S. sclerotiorum*	Molloy *et al.* (2004)
	Cucumber	*B. cinerea*	Ben-Shalom *et al.* (2003)
	Groundnut	*Puccinia arachidis*	Sathiyabama & Balasubramanian (1998)
	Pearl millet	*S. graminicola*	Sarathchandra *et al.* (2004)
	Potato	*E. carotovora*	Dutton *et al.* (1997)
	Potato	*R. solani*	Linden *et al.* (2000)
	Strawberry	*B. cinerea* and *Rhizopus stolonifer*	El Ghaouth *et al.* (1992)
	Strawberry	*Phytophthora cactorum*	Elkemo *et al.* (2003)
	Tomato	*F. oxysporum* f. sp. *radicis-lycopersici*	Benhamou & Theriault (1992)
	Wheat	*Fusarium graminearum*	Reddy *et al.* (1999)
	Wheat	*Microdochium nivale*	Hofgaard *et al.* (2005)
Fragments and extracts of fungal cell walls			
P. chrysogenum extract	Cotton	*Verticillium dahliae*	Dong *et al.* (2003)
P. oligandrum extract	Sugar beet	*R. solani*	Takenaka *et al.* (2003)
S. cerevisiae extract	Barley	*B. graminis*	Reglinski *et al.* (1994a, b)
S. cerevisiae extract	Lettuce	*B. cinerea* and *R. solani*	Reglinski *et al.* (1995)
Lipids			
Cerebroside B	Lettuce	*F. oxysporum*	Umemura *et al.* (2004)
Cerebroside B	Tomato	*F. oxysporum*	Umemura *et al.* (2004)
Cerebroside B	Melon	*F. oxysporum*	Umemura *et al.* (2004)
Cerebroside B	Sweet potato	*F. oxysporum*	Umemura *et al.* (2004)
LPS	Pepper	*X. campestris* pv. *campestris*	Newman *et al.* (2002)

10.2.1 Chitin and chitosan

Chitin, the main wall component of many filamentous fungi (Aronson, 1981), and chitosan, the deacetylated derivative of chitin, have been shown to elicit several plant defence responses including lignification (Barber *et al.*, 1989) and phytoalexin production (Kendra & Hadwiger, 1984). However, although chitin can induce defence responses, its performance in providing protection against infection is mixed. For example, the addition of 4% chitin to the growth medium led to significantly enhanced growth of cucumber plants, but although the soil population of root and stem rot pathogen *Fusarium oxysporum* f. sp. *radicis-cucumerinum* was reduced, disease severity increased compared to controls (Rose *et al.*, 2003). In a study of the effects of foliar applications of *Serratia marcescens* and *Pseudomonas aeruginosa* on control of the groundnut pathogen *Phaeoisariopsis personata,* supplementation of the treatments with chitin led to different effects, depending on the treatment applied. So, chitin supplementation had no effect on defence enzyme activation and disease control provided by *P. aeruginosa*, whereas chitin supplementation of *S. marcescens* resulted in increased activities of defence related enzymes and enhanced disease control, compared to controls (Kishore *et al.*, 2005).

Chitosan, a β-1,4-D-glucosamine polymer found in the walls of many fungi, is a polycationic polymer that may readily interfere with negatively charged molecules exposed at the cell surface. In a detailed study of the interaction between tomato and the crown and root rot pathogen *F. oxysporum* f. sp. *radicis-lycopersici,* Benhamou & Theriault (1992) demonstrated that application of chitosan as a foliar spray or a root treatment markedly reduced root infection and greatly increased the formation of physical barriers in root tissues. In fact, chitosan has been found to protect a range of hosts against important pathogens, including potato tubers against *Erwinia carotovora* and *Rhizoctonia solani* (Dutton *et al.*, 1997; Linden *et al.*, 2000), carrots against the storage pathogen *Sclerotinia sclerotiorum* (Molloy *et al.*, 2004), wheat against *F. graminearum* (Reddy *et al.*, 1999), groundnut against the rust *Puccinia arachidis* (Sathiyabama & Balasubramanian, 1998) and cucumber against the grey mould pathogen *Botrytis cinerea* (Ben-Shalom *et al.*, 2003). A commercial formulation of chitosan developed by Glycogenesys Inc. (Boston, MA), Elexa™, contains 4% chitosan as its active ingredient and has been shown to protect a range of crops against important pathogens (Agostini *et al.*, 2003). In pearl millet, Elexa™ was shown to reduce downy mildew severity by 58% when used as a seed treatment, by 75% when used as a foliar spray and by 77% when used as a combined seed treatment and foliar spray (Sarathchandra *et al.*, 2004).

However, although in many of these cases chitosan induced host defence responses (Reddy *et al.*, 1999; Ben-Shalom *et al.*, 2003), chitosan also possesses direct antifungal activity. Thus, El Ghaouth *et al.* (1992) found that coating strawberry fruits with chitosan markedly reduced decay caused by *B. cinerea* and *Rhizopus stolonifer,* but could find no evidence of increased activities of host antifungal enzymes in treated whole fruits. They did, however, find that chitosan was a very effective inhibitor of spore germination, germ tube elongation and mycelial growth of both fungi *in vitro* (El Ghaouth *et al.*, 1992). More recently, Elkemo *et al.* (2003) found that chitosan protected strawberry against the crown rot pathogen *Phytophthora cactorum,* but not against the red core pathogen *Phytophthora fragariae* var. *fragariae*. Interestingly, *in vitro* growth of both pathogens was reduced by chitosan (Elkemo *et al.*, 2003). Chitosan also induced resistance in winter wheat to the snow mould pathogen *Microdochium nivale,* although the effect was variable (Hofgaard *et al.*, 2005). Chitosan treatment of wheat led to increased chitinase gene expression, but chitosan was also found to exert a direct inhibitory effect on *in vitro* growth of *M. nivale* (Hofgaard *et al.*, 2005). It is possible therefore that both induced resistance and direct antifungal activity were responsible for the control of *M. nivale* on wheat. The authors suggested that the variability in disease control obtained may have been the result of differences in disease pressure, with much higher disease pressures leading to poorer disease control (Hofgaard *et al.*, 2005).

10.2.2 Fragments and extracts of fungal cell walls

Cell wall extracts from the yeast *Saccharomyces cerevisiae* have been shown to control plant diseases, providing up to 95% control of powdery mildew infection in barley in field trials (Reglinski *et al.*, 1994a, b). In that work, the yeast extracts were found to possess high phytoalexin elicitor activity, and in barley treated with the extracts, there was rapid stimulation of PAL activity and faster formation of papillae in response to attempted fungal penetration (Reglinski *et al.*, 1994a). The yeast cell wall extracts also controlled *B. cinerea* and

R. solani on lettuce (Reglinski *et al*., 1995). The fungus *Penicillium chrysogenum* is known to be a potent biological control agent against *B. fabae* (Jackson *et al*., 1994), and water extracts of the dry mycelium of *P. chrysogenum* were found to provide significant protection of cotton against the wilt pathogen *V. dahliae* (Dong *et al*., 2003). This protection was cultivar dependent, with better protection obtained with cultivars of *Gossypium hirsutum,* compared to cultivars of *G. barbadense* (Dong *et al*., 2003). Water extracts of *P. chrysogenum* increased POX activity and lignin deposition within 24 hours of treatment, leading the authors to suggest that induced resistance was involved (Dong *et al*., 2003). Recent work by Thuerig *et al*. (2006) showed that an aqueous extract of *P. chrysogenum* induced resistance against several pathogens in *A. thaliana*. These workers obtained strong evidence that the *P. chrysogenum* extract induced resistance against downy mildew in *A. thaliana* via a salicylic acid-dependent, but NPR1-independent, pathway. Another biological control agent, *Pythium oligandrum,* has been used in studies of induced resistance. Takenaka *et al*. (2003) isolated protein fractions from the cell walls of *P. oligandrum* and found that sugar beet seedlings treated with these fractions exhibited enhanced activities of PAL and chitinase, and in some cases there were also increased amounts of cell wall bound phenolics. Sugar beet seedlings treated with the protein fractions were significantly less infected with *R. solani* campared to controls (Takenaka *et al*., 2003).

10.2.3 Lipids

In addition to the oxylipins described below in the section on abiotic inducers, two distinct lipid molecules have also been investigated as inducers of resistance. The first of these is liposaccharides (LPS) isolated from the bacteria *Xanthomonas campestris* pv. *campestris*, *Salmonella minnesota* and *Escherichia coli*. Infiltration of pepper leaves with a $50\,\mu g\,ml^{-1}$ solution of LPS 20 hours prior to inoculation with *X. campestris* pv. *campestris* led to a 10-fold decrease in the number of viable bacteria (Newman *et al*., 2002). The second example is cerebroside B, a compound categorized as a sphingolipid and found in various strains of *F. oxysporum* and other soil-borne pathogens (Umemura *et al*., 2004). Resistance to pathogenic strains of *F. oxysporum* was induced in various plant species following treatment with cerebroside B. Thus, when cerebroside B was applied to lettuce, tomato, melon and sweet potato by dipping in solutions of concentrations ranging from 1 to $50\,\mu g\,ml^{-1}$, resistance to pathogenic strains of *F. oxysporum* was obtained under both laboratory and field conditions.

10.3 Abiotic inducers

The use of abiotic elicitors (elicitors which are not directly derived from living organisms) is a promising approach for the agricultural application of induced resistance. Several chemical inducers have now been released commercially in various countries, and there is continuing research on the mode of action of these agents and on the potential for practical application. Abiotic resistance inducers have been listed and described previously (Tuzun & Kloepper, 1995; Karban & Kuć, 1999; Lyon & Newton, 1999), but the information presented here focuses on the literature published since the release of these previous reviews, predominantly in the last eight years.

10.3.1 Benzo(1,2,3)thiadiazole-7-carbothioic acid S-methyl ester (BTH)/acibenzolar-S-methyl (ASM)

It is now almost a decade since BTH or ASM was described as a structurally related functional analogue of salicylic acid (SA) and therefore as a novel class of SAR inducer. In addition to the activation of defence gene expression, it was first reported to protect wheat against powdery mildew in growth cabinets when sprayed at a 0.3 mM final concentration and when treatments were applied between four and seven days before challenge inoculation (Görlach *et al.*, 1996). These authors also observed induced resistance to both *Puccinia recondita* and *Septoria* spp. in wheat following BTH treatment and in the field achieved protection against powdery mildew for an entire season with a single application of 30 g of BTH per hectare. Since then, the molecule has been shown to be effective against a wide range of pathogens on a range of crops and was released commercially under the trade names Bion® (in Europe) and Actigard® (in the USA). The study of resistance induced by BTH is a flourishing area of research, and so many papers have been published in the last seven years, both on different crop/pathogen combinations and at different experimental scales, that they cannot be dealt with comprehensively here. Rather, this section will provide an overview of the effects of BTH on a range of different host–pathogen interactions.

10.3.1.1 Diseases caused by leaf and stem-infecting fungi and Oomycetes

Under laboratory conditions, resistance to *Collectotrichum destructivum* was induced rapidly in cowpea seedlings following seed treatment with BTH (Latunde-Dada & Lucas, 2001). These workers found that the destructive necrotrophic phase of disease development was blocked by a hypersensitive-like response. The enhanced resistance observed was accompanied by rapid, transient increases in the activities of PAL and chalcone isomerase and accelerated accumulation of the isoflavanoid phytoalexins kievitone and phaseollidin (Latunde-Dada & Lucas, 2001). The authors suggested that BTH protected cowpea seedlings by potentiating an early defence response, rather than by altering constitutive resistance. Iriti & Faoro (2003) showed that glasshouse-grown French bean was fully protected against rust caused by *Uromyces appendiculatus* by a single foliar spray of 0.3 mM BTH applied seven days before inoculation, while subsequent work demonstrated a reduction in the severity of *B. cinerea* infection on grapevine using three applications of 0.3 mM BTH (Iriti *et al.*, 2004). The efficacy of foliar sprays of BTH has also been investigated for diseases caused by Oomycetes. Thus, infection of cauliflower seedlings by the downy mildew pathogen, *Peronospora parasitica*, was reduced by 69% using a 0.045 $mg\,ml^{-1}$ solution of BTH applied four days before inoculation (Godard *et al.*, 1999). Similar results were obtained on 30 day old plants. On tobacco, blue mould is caused by *Peronospora hyoscyami* f. sp. *tabacina*. Field experiments showed that its development was prevented by foliar applications of BTH at 10 day intervals (Perez *et al.*, 2003).

10.3.1.2 Diseases caused by bacteria

In a series of detailed experiments on tobacco, Cole (1999) examined the ability of BTH to provide protection against a number of pathogens. Under controlled conditions, 10 week old tobacco seedlings were protected against wildfire and angular leaf spot, caused by *Pseudomonas syringae* pv. *tabaci* tox+ and *Ps. syringae* pv. *tabaci* tox− respectively, by

BTH applied alone at 0.5 g a.i. l^{-1} or with the bactericide copper oxychloride. Outdoor experiments on seedlings treated two or three times with BTH at 0.05–0.1 g a.i. m^{-2} either alone or mixed with copper oxychloride confirmed the protective effect obtained on wildfire (Cole, 1999). In this work, BTH was also found to reduce the incidence of *Cercospora nicotianae* on field-grown tobacco and in the colouring phase of leaf curing (Cole, 1999). BTH has also been shown to reduce the severity of bacterial spot (*Xanthomonas axonopodis* pv. *vesicatoria*) and bacterial speck (*P. syringae* pv. *tomato*) on tomato under field conditions (Louws *et al.*, 2001) and to induce resistance to bacterial spot (*X. axonopodis* pv. *vesicatoria*) of glasshouse-grown bell pepper (Romero *et al.*, 2001). In the latter work, under field conditions, application of BTH every two weeks resulted in control levels similar to those obtained with standard fungicide treatments (Romero *et al.*, 2001). The efficacy of BTH against bacterial spot of pepper was subsequently confirmed by Buonaurio *et al.* (2002) using a 300 μM solution on plants in growth chambers and by spraying field-grown peppers six or seven times with a mixture of BTH and copper hydroxide. On apple trees, sprays of BTH at 75 mg a.i. l^{-1} at weekly intervals partially reduced fire blight caused by *Erwinia amylovora* (Maxson-Stein *et al.*, 2002). Later work by Sparla *et al.* (2004) demonstrated partial control of fire blight on two year old pear plants using BTH applied 10 days before inoculation.

10.3.1.3 Soil-borne diseases

BTH has been shown to provide control of a range of damaging soil-borne diseases. For example, BTH has been reported to control soft rot and white mould diseases caused by *S. sclerotiorum*. Buzi *et al.* (2004a) showed that soaking melon seeds in BTH for 12 hours provided effective control of *S. sclerotiorum* on seedlings. On field-grown soybean, two or four applications of BTH at 35 or 375 mg a.i. l^{-1} reduced disease severity by 20–60% (Dann *et al.*, 1998). In this work, levels of disease control depended on the soybean cultivar used, with greater levels of disease control achieved on highly susceptible cultivars (Dann *et al.*, 1998). This highlights the influence of genotype on the expression of induced resistance. There are also reports that BTH is effective against vascular wilt diseases. Thus, a foliar spray of 1.5 mM BTH provided systemic protection of tomatoes against *F. oxysporum* f. sp. *radicis-lycopersici* (Benhamou & Bélanger, 1998), while BTH also induced protection against wilt in glasshouse-grown cocoa caused by *Verticillium dahliae,* reducing disease severity by more than 50% (Resende *et al.*, 2002). Investigation of the effect of BTH on rice sheath blight caused by *R. solani* led to the conclusion that it inhibited both development and spread of the disease irrespective of the way it is applied to the plant, i.e. as either a soil drench or a foliar spray (Rohilla *et al.*, 2001). Impressively, BTH has also been shown to be effective against root rot and blight caused by *Phytophthora palmivora* on papaya (Zhu *et al.*, 2003). In this work, BTH drench treatments were performed in the greenhouse one week before inoculation of roots with the pathogen. Use of BTH at 5 μM reduced disease development, whereas plants treated with BTH at 100 and 500 μM prior to inoculation did not exhibit any disease symptoms six weeks after inoculation.

10.3.1.4 Post-harvest diseases

All of the studies mentioned above have used vegetative plant tissues. In contrast to the hundreds of reports on the use of BTH to control pathogens on growing plants, comparatively little work has been done to examine the effects of BTH against post-harvest diseases. Nevertheless, a number of studies have shown that BTH can control post-harvest disease (Terry & Joyce, 2004). In a recent study, Liu *et al.* (2005) showed that post-harvest BTH treatment induced resistance of peach fruit to infection by *Penicillium expansum*. In this study, harvested fruits that were immersed for five minutes in 200 mg l^{-1} BTH and stored in controlled conditions for 60 hours, showed reductions in lesion area and disease incidence of 64.1 and 49.5% respectively (Liu *et al.*, 2005).

10.3.2 Salicylic acid and structurally related compounds

It is now nearly 30 years since the discovery that tobacco leaves treated with SA or acetyl salicylic acid (ASA; aspirin) exhibited increased PR-protein accumulation and enhanced resistance to TMV infection (White, 1979; Antoniw & White, 1980). Since those pioneering studies, there have been many reports of the effectiveness of SA at enhancing the resistance of a range of plants to bacterial, fungal and viral pathogens (Dempsey *et al.*, 1999). Work on SA and induced resistance was rejuvenated following the reports in 1990 that SA might be an endogenous signal for the activation of defence responses (Malamy *et al.*, 1990; Métraux *et al.*, 1990). Although subsequent work suggested that SA is not the long-distance SAR signal (Dempsey *et al.*, 1999), it is now well established that SA is a major determinant of SAR (Bostock, 2005). Indeed, it is often used as a positive control treatment in experiments aimed at the characterization of novel inducers (Hammerschmidt & Smith-Becker, 1999). Thus, studies on the effects of SA on resistance to pathogen infection continue. For example, Spletzer & Enyedi (1999) showed that the addition of 200 μM SA to the roots of hydroponically grown tomatoes resulted in 83% and 77% reductions in the number of lesions per leaf and in the blighted leaf area, respectively, following inoculation with *Alternaria solani*. In more recent work, Saikia *et al.* (2003) found that addition of 80 μg ml^{-1} SA to the roots of chickpea plants in a hydroponic system, or injection of chickpea plants with 2000 μg ml^{-1} SA, gave reductions of 60% and a 56%, respectively, in infection by *F. oxysporum*. Similar levels of protection were obtained with ASA (Saikia *et al.*, 2003).

Exogenous application of SA has also been shown to provide control of diseases caused by bacterial and viral pathogens e.g. *E. amylovora* on pear (Sparla *et al.*, 2004) and white clover mosaic virus (WCLMV) on two-week-old bean plants (Gális *et al.*, 2004). In a recent study, the effects of both pre- and post-harvest applications of SA on the resistance of sweet cherry fruits to *Monilia fructicola* were investigated in storage (Yao & Tian, 2005). Here, pre-harvest sprays with 2 mM SA applied three days before the fruits were harvested led to significant reductions in both disease incidence and lesion diameter compared to water-treated fruits.

Initial laboratory screening of 39 different derivatives of SA led to the identification of salicylic hydrazide (SHZ) and 4-hydroxybenzoic hydrazide (4HBHZ) as inducers of resistance to wilt caused by *F. oxysporum* f. sp. *lycopersici* (Miyazawa *et al.*, 1998). Another structurally related compound known to induce resistance is *N*-cyanomethyl-2-chloroisonicotinamide (NCI). This compound has been shown to provide protection against a broad spectrum of pathogens, including *Magnaporthe grisea* and *Xanthomonas oryzae* pv. *oryzae* on rice and TMV, *P. syringae* pv. *tabaci* and *Oïdium lycopersici* on tobacco (Nakashita *et al.*, 2002a). Although NCI induced PR-protein encoding gene expression in tobacco, NCI activity did not require SA, leading the authors to suggest that it acts at the same level as, or downstream of, SA accumulation (Nakashita *et al.*, 2002a).

10.3.2.1 Probenazole

Another molecule with structural similarities to SA, probenazole (PBZ), is the active ingredient of the commercially available product Oryzemate®. This chemical has been used widely for two decades in the rice-growing areas of Asia, mainly to protect rice crops against rice blast caused by *M. grisea*. The action of PBZ is studied in parallel with one of its closely related active metabolites, benzisothiazole (BIT). In a study aiming at characterizing the mode of action of PBZ and BIT, the latter was shown to induce SAR in *A. thaliana*. Foliar sprays with 2 mM BIT four days before inoculation with *P. syringae* pv. *tomato* resulted in a 10-fold inhibition of bacterial growth, while a 0.2 mM BIT solution also strongly reduced downy mildew caused by *P. parasitica* (Yoshioka *et al.*, 2001). BIT has also been shown to provide protection against TMV, *P. syringae* pv. *tabaci* and *O. lycopersici* in tobacco and *X. oryzae* pv. *oryzae* in rice (Nakashita *et al.*, 2002b). Both PBZ and BIT have been shown to induce SAR in tobacco by triggering signalling upstream at a point upstream of SA accumulation (Nakashita *et al.*, 2002b).

10.3.3 Protein, peptide and amino acid-derived inducers

10.3.3.1 β-Aminobutyric acid (BABA)

The non-protein amino acid BABA occurs rarely in nature and is hardly found in plants. However, it is a potent inducer of resistance in plants with broad spectrum activity. Isomers of BABA such as DL-2-aminobutyric acid (AABA) and 4-aminobutyric acid (GABA) have also been investigated for their resistance-inducing activity. Whereas BABA exhibited both local and systemic activity against downy mildew caused by *Plasmopara viticola* on grapevines when applied to leaves as a 0.25 mM spray, both AABA and GABA showed no activity (Cohen *et al.*, 1999). Similar results were obtained in studies using 10 mM solutions sprayed onto tobacco leaves, with AABA exhibiting slight inducing activity (Siegrist *et al.*, 2000), and studies using foliar sprays and soil drenches at 8000 mg l^{-1} and 125 mg l^{-1} respectively, against the cyst and root-knot nematodes *Heterodera avenae* and *H. latipons* on cereals, where only BABA exhibited inducing activity (Oka & Cohen, 2001). BABA has also been shown to provide protection against *Sclerospora graminicola* on pearl millet (Shailasree *et al.*, 2001), *P. parasitica* on cauliflower (Silué *et al.*, 2002) and *B. cinerea* (Zimmerli et al., 2001), *Phytophthora brassicae* (Si-Ammour *et al.*, 2003), *Alternaria brassicicola* and *Plectosphaerella cucumerina* (Ton & Mauch-Mani, 2004) on *A. thaliana*.

A number of studies have focused on the mechanisms underlying the resistance induced by BABA. Thus, BABA applied to tobacco at 10 mM led to the formation of reactive oxygen species, lipid peroxidation, induction of callose around lesions and an increase in the SA content of leaves (Siegrist *et al.*, 2000). Treatment with BABA has also been reported to lead to induction of PR proteins. Thus, BABA induced PR-1a, chitinase and glucanase in tobacco, tomato and pepper (Cohen *et al.*, 1999; Siegrist *et al.*, 2000), but not in Arabidopsis, cauliflower or tobacco (Cohen, 1994; Jakab *et al.*, 2001; Silué *et al.*, 2002). This suggests that induction of PR-proteins may not be the only mode of action of BABA which also leads to callose deposition, lignification and hypersensitivity in some plants (Cohen *et al.*, 1999; Siegrist *et al.*, 2000). Moreover, BABA is known to move systemically in tomato, tobacco and grapevines (Cohen *et al.*, 1999) and this may explain the systemic protection against pathogens observed in these and other plants (Cohen *et al.*, 1999; Siegrist *et al.*, 2000).

10.3.3.2 Syringolin

The syringolin peptide was originally isolated from the bacterium *P. syringae* pv. *syringae* (Wäspi *et al.*, 1998). It is one of the molecular determinants secreted by the bacterium and perceived by non-host plants like rice. Syringolin A was shown to induce resistance to *M. grisea* in rice when three-week-old plants were sprayed with a 40 mM solution 24 hours before inoculation (Wäspi *et al.*, 1998). In addition, syringolin A was shown to be perceived by wheat and to induce the accumulation of gene transcripts and resistance to powdery mildew infection if applied before inoculation (Wäspi *et al.*, 2001). Further, it eradicated powdery mildew on wheat if applied post-inoculation. This curative effect was accompanied by the induction of cell death and the re-activation of PR-related genes whose transcripts accumulated initially following powdery mildew inoculation but declined as infection progressed. The authors suggested that syringolin A counteracted the suppression of host defence reactions imposed by the pathogen on colonized cells (Wäspi *et al.*, 2001).

10.3.3.3 Harpin

Harpin is another example of an elicitor of bacterial origin. It was isolated from *E. amylovora*, the causal agent of apple and pear fire blight, and it induces HR and resistance in a variety of plants against a wide range of pathogens (Dong *et al.*, 1999). Although harpin is the active ingredient of the commercially available product Messenger®, which has been released in North America and Europe as an alternative to fungicides for various crop diseases, recent reports of the inducing activity of harpin have focused almost exclusively on the laboratory model *A. thaliana*. Growth of *P. syringae* pv. *tomato* and *P. parasitica* was inhibited in plants sprayed with harpin and inoculated five days thereafter (Dong *et al.*, 1999). A reduction in growth of *E. carotovora* subsp. *carotovora* was also observed in leaves infiltrated with harpin 24 hours before challenge inoculation (Kariola *et al.*, 2003). More recently, treatment of leaves of *A. thaliana* and tobacco with harpin at 15 $\mu g\ ml^{-1}$ led to HR-associated resistance to *P. parasitica* and TMV (Peng *et al.*, 2003).

10.3.4 Lipids

10.3.4.1 Oxylipins

Formation of oxygenated fatty acids, known collectively as oxylipins, is an early response of plant cells to both abiotic and biotic stress (Feussner & Wasternack, 2002). Indeed, oxylipins play diverse roles in plants as signal molecules for defence gene expression and as antimicrobial compounds (Blée, 2002; Farmer *et al.*, 2003). Many oxylipins are generated by the action of lipoxygenases (LOX), which in plants add molecular oxygen to pentadiene fatty acids like linoleic and linolenic acids. The products formed, fatty acid hydroperoxides, are subject to a diverse array of modifications leading to the generation of large numbers of other oxylipins. In plants, two carbon atoms in linoleic and linolenic acids are subjects for the action of LOX: C-13 and C-9 and the enzymes responsible are 13-LOXs and 9-LOXs, respectively (Feussner & Wasternack, 2002). Much is known about the products of 13-LOX action, since these include jasmonates, a family of potent biological regulators. Herbivore attack and the wounding that occurs as a result are thought to elicit the release of linolenic acid from membranes. This in turn is thought to lead to synthesis of jasmonic acid (JA), triggering the expression of defences against insect herbivores, like production of proteinase inhibitors (Glawe *et al.*, 2003). Jasmonates have been shown to mediate resistance responses to various pathogens (Norman-Setterblad *et al.*, 2000; Turner *et al.*, 2002) and the JA pathway has also been shown to be important for resistance in Arabidopsis to the biotrophic fungal pathogen *Erysiphe cichoracearum,* the bacterial pathogen *P. syringae* and the aphid *Myzus persicae* (Ellis *et al.*, 2002).

Exogenous application of jasmonates has been shown to induce resistance in a number of plants. So, JA and methyl jasmonate (MJ) were found to induce both local and systemic resistance against *Phytophthora infestans* in potato (Cohen *et al.*, 1993) and against *Pythium ultimum* in Norway spruce (Kozlowski *et al.*, 1994). In barley, Schweizer *et al.* (1993) demonstrated a protective effect of JA against powdery mildew, while MJ was shown to induce systemic resistance against powdery mildew in glasshouse studies and to provide control of mildew in the field (Mitchell & Walters, 1995). Subsequent studies on the barley–powdery mildew interaction demonstrated that the systemic resistance induced by MJ application was accompanied by increased activities of the defence-related enzymes PAL and peroxidase (Walters *et al.*, 2002). MJ has also been shown to induce resistance against soil-borne pathogens. For example, Buzi *et al.* (2004a) showed that soaking melon seeds in MJ provided complete protection against the gummy stem blight pathogen *Didymella bryoniae* and was associated with elevated activities of chitinase and peroxidase. Exposure of melon seeds to gaseous MJ provided less complete protection against *D. bryoniae, S. sclerotiorum* and *F. oxysporum* f. sp. *melonis,* and led to increased activity of chitinase but not peroxidase (Buzi *et al.*, 2004b). Both JA and MJ have been shown to reduce decay in grapefruit caused by the green mould *Penicillium digitatum,* and since neither compound exerted a direct antifungal effect on the fungus, the authors suggested that the jasmonates had enhanced the natural resistance of the fruit to the pathogen (Droby *et al.*, 1999).

Products of the 9-LOX pathway have only recently gained attention as potential defence compounds. For example, an anti-sense genomics approach has been used to demonstrate that 9-LOX is important for resistance in tobacco to *Phytophthora parasitica* (Rancé *et al.*, 1998), while more recently, Gobel *et al.* (2002) showed that a number of 9-LOX derived oxylipins accumulated in the interaction between *P. syringae* and the non-host potato. The accumulating oxylipins included two trihydroxy-oxylipins, 9,10,11-trihydroxyoctadecadienoic acid and 9,12,13-trihydroxyoctadecadienoic acid, as well as the divinyl ether compounds colneleic and colnelenic acids (Gobel *et al.*, 2002). These compounds started accumulating 12 hours after inoculation, and the authors suggested that they might be involved in the resistance response. This is not an unreasonable suggestion, since trihydroxy-oxylipins have been shown to possess antifungal activity (Masui *et al.*, 1989) and to induce local and systemic resistance to pathogen infection (Cowley & Walters, 2005). In the latter work, topical application of two trihydroxy-oxylipins at 30 μM induced both local and systemic resistance to powdery mildew infection in barley. Here, induction of both local and systemic resistance was associated with PAL activity, which increased significantly only following challenge inoculation of protected plants (Cowley & Walters, 2005). This suggests that treatment of barley with MJ primed plants to respond to attempted mildew infection.

10.3.5 Sugars

Cell-wall-derived polysaccharide fragments such as chitosan are not the only sugars to elicit induced resistance in plants. Trehalose is a non-reducing disaccharide commonly found in a wide variety of living organisms, including fungi, and is also associated with the protection of plants against different types of abiotic stresses (Drennan *et al.*, 1993). Trehalose has also been shown to induce resistance. Thus, a reduction in infection intensity of the powdery mildew fungus *Blumeria graminis* f. sp. *tritici* on wheat was observed after treatment with trehalose (Reignault *et al.*, 2001). Wheat plants grown under controlled conditions showed reductions in infection intensity of 50% and 95% respectively, following a single spray or three sprays of a trehalose solution (15 g l^{-1}), applied 48 hours prior to inoculation with powdery mildew (Reignault *et al.*, 2001).

10.3.6 Reactive oxygen species

Reactive oxygen species (ROS) are generated in plants during infection, and their role in host–pathogen interactions has been studied extensively. The importance of ROS in resistance to pathogens has led to increasing interest in products which can generate ROS, like Oxycom™ (see below). Generation of ROS may also be important in the effects of ozone on plant diseases. Ozone is known to influence disease development in plants, with reports of both increased and decreased pathogen infection (Fuhrer, 2003). Interestingly, however, in situations where exposure to ozone leads to reductions in disease development, Sandermann *et al.* (1998) has suggested that ozone-induced stress may lead to a burst of ROS, thereby activating SAR.

10.3.6.1 Oxycom™

Oxycom™ has been registered in North America for management of plant pathogens, especially those from the *Pythium* genus, downy mildews and powdery mildews. This commercial product contains ROS, as well as SA and other chemicals with fertilizer action. At the time of application, the two distinct components of the product have to be mixed together. Component A is a 5% v/v solution of peracetic acid containing 10–12% acetic acid and 20–22% hydrogen peroxide, while component B contains a mixture of plant nutrients, proprietary stabilizers and SA (Kim *et al.*, 2001). These authors reported that Oxycom™ can provide effective disease control on different crops, using an application rate of 1000–5000 ppm active ingredient and an application frequency ranging from five to 20 days, depending on the crop. For example, five applications of Oxycom™ at 14 day intervals provided significant control of *Bremia lactucae* on lettuce in the field. Oxycom™ has also been shown to induce resistance to *P. syringae* pv. *tabaci* in tobacco under controlled conditions (Yang *et al.*, 2002). In this study, plants were sprayed with a 5000 ppm solution of Oxycom™ three days before inoculation, leading to a 10-fold reduction in population growth of the bacterium. More recently, Blee *et al.* (2004) showed that treatment of tobacco with the mixture of ROS and SA in Oxycom™ provided greater protection against *P. syringae* pv. *tabaci* than either treatment alone.

10.3.6.2 Menadione sodium bisulphite (MSB)

A synthetic form of vitamin K_3, MSB is known to induce resistance in banana to the vascular wilt pathogen *F. oxysporum* f. sp. *cubense,* causal agent of Panama disease (Borges *et al.*, 2004). MSB has also been shown to enhance local and systemic resistance in oilseed rape to the stem canker pathogen *Leptushpaeria maculans* (Borges *et al.*, 2003). In this work, MSB induced the expression of ascorbate peroxidase, leading the authors to suggest that MSB augmented the production of ROS in treated plants.

10.3.7 Minerals and ions

10.3.7.1 Phosphates and phosphonates

Phosphate salts have been shown to induce systemic protection against anthracnose in cucumber caused by *Colletotrichum lagenarium* (Gottstein & Kuć, 1989), and later work demonstrated the broad spectrum of disease control achieved in cucumber using phosphates (Mucharromah & Kuć, 1991). In fact, phosphates have been shown to provide disease control on a range of hosts, including pepper, grapevines, rice and barley (Reuveni & Reuveni, 1995; Mandahar *et al.*, 1998; Reuveni *et al.*, 1998, 2000; Mitchell & Walters, 2004). It was suggested that basic phosphates applied to plants could sequester apoplastic calcium, altering membrane integrity and influencing the activity of apoplastic enzymes like polygalacturonases, thereby releasing elicitor-active oligogalacturonides from plant cell walls (Gottstein & Kuć, 1989; Walters & Murray, 1992). Indeed, subsequent work by Orober *et al.* (2002) showed that phosphate mediated resistance induction in cucumber was associated with localized cell death, preceded by a rapid generation of superoxide and hydrogen peroxide. These workers also detected local and systemic increases in levels of free and conjugated salicylic acid following phosphate application (Orober *et al.*, 2002). More recently, work on barley showed that application of phosphate to first leaves reduced powdery mildew infection by 89% in second leaves (Mitchell & Walters, 2004). Application of phosphate, as K_3PO_4, to first leaves led to significant increases in activities of phenylalanine ammonia-lyase (PAL), peroxidase and lipoxygenase in second leaves, and activities of these enzymes were increased further following pathogen challenge (Mitchell & Walters, 2004). Phosphates have also been shown to provide disease control under field conditions. Thus, K_2HPO_4 applied to rice as a 50 mM spray, reduced neck blast caused by the fungus *M. grisea* by 29–42%, with increases in grain yield of 12–32% (Mandahar *et al.*, 1998). Phosphate (K_3PO_4; 25 mM) applied to barley in a field trial reduced powdery mildew infection by up to 70% and gave an increase in grain yield of 12% compared to untreated controls (Mitchell & Walters, 2004). In cucumbers grown hydroponically, 20 ppm phosphate applied to the hydroponic solution reduced powdery mildew infection by 80–92%, with reductions of up to 91% in numbers of conidia produced on infected leaves (Reuveni *et al.*, 2000).

Phosphonate, the hydrolysis product of phosphoric acid (H_3PO_3), has also been shown to induce pathogen resistance in plants. Thus, Phytogard® (a formulation containing 58% potassium phosphonate, K_2HPO_3), was shown to provide protection against the downy mildew pathogen *P. parasitica* in cauliflower seedlings when used as a foliar spray or as a root treatment (Bécot *et al.*, 2000). In this work, foliar application of Phytogard® induced only local protection, while application to roots provided systemic protection, probably as a result of its translocation from root to shoot. Although Phytogard® reduced germination of *P. parasitica* spores, Bécot *et al.* (2000) argued that it induced resistance, since there was weak induction of β-1,3-glucanase and PR2 protein. Subsequent work showed that Phytogard® also induced resistance to *B. lactucae* in lettuce (Pajot *et al.*, 2001). However, while it completely inhibited spore germination, it had no effect on PR protein induction (Pajot *et al.*, 2001). Phosphonates are well known to possess powerful antifungal activity (Ouimette & Coffey, 1989), and interestingly, the fungicide Fosetyl-Al, marketed as Aliette® (active ingredient *O*-ethyl phosphonate), is known to exert both a direct effect on the pathogen and an indirect effect via stimulation of host defences (Nemestothy & Guest, 1990).

10.3.7.2 Silicon

The mineral studied most extensively as a resistance inducer is silicon. Although the mechanisms by which silicon reduces disease development are not yet fully understood, numerous studies have shown that the application of silicon-containing fertilizers can increase resistance in rice to several important pathogens, such as *M. grisea,* the brown spot pathogen, *Bipolaris oryzae*, and the causal agent of sheath blight, *R. solani* (Kim *et al.*, 2002). Seebold *et al.* (2001) conducted experiments under glasshouse conditions, where the soil was amended with calcium silicate at rates equivalent to 0, 2, 5 and 10 tonnes ha^{-1}. They found that incubation period, lesion size, rate of lesion extension, sporulation per lesion and diseased leaf area due to *M. grisea* were all changed by silicon treatment, resulting in reduced production of conidia and a slowing down of the rate of epidemic development. In field experiments, silicon was applied at 1000 kg ha^{-1}, either alone or combined with the tricyclazole fungicide edifenphos (Seebold *et al.*, 2004). This study showed clearly that the application of silicon allowed the use of reduced rates of fungicide to manage both leaf and neck blast. An active role for silicon in the resistance of wheat to powdery mildew has also been reported (Bélanger *et al.*, 2003). Here, under controlled conditions, the equivalent of approximately 3 tonnes ha^{-1} of silicon was added to the soil. Plants were protected against powdery mildew and exhibited cytological evidence of induced resistance. Thus, silicon-treated plants exhibited greater papilla formation and increased production of callose and phenolic material, leading the authors to suggest that silicon induced active localized cell defences against the powdery mildew fungus (Bélanger *et al.*, 2003). The effect of silicon was also investigated, either alone or in combination with a yeast biocontrol agent, *Cryptococcus laurentii*, on control of both blue mould (*Penicillium expansum*) and brown rot (*Monilia fructigena*) in sweet cherry (Qin & Tian, 2005). Dipping the fruits in a 1% silicon solution, in combination with the biocontrol agents at 1×10^7 cells ml^{-1}, provided significant control of both diseases. Moreover, silicon treatment was associated with increased activities of PAL, peroxidase and polyphenol oxidase in sweet cherry fruits (Qin & Tian, 2005).

10.3.8 Ultraviolet irradiation

Among the alternative agents for controlling post-harvest diseases, the use of ultraviolet-C light (UV-C) is often reported as a promising approach. Two mechanisms have been proposed to account for the reductions in storage rots by UV-C, a germicidal effect and induced host resistance (Stevens *et al.*, 1998). UV induces host resistance via hormesis, which is the elicitation of defence by a low dose of a radiation, like UV, X-rays and γ-rays (Stevens *et al.*, 1998). Resistance was induced against peach brown rot by reducing latent infection by *M. fructigena,* using a dose of 7.5 kJ m^{-2} of UV-C light (Stevens *et al.*, 1998). A number of subsequent studies has shown that low dose exposure to UV-C can control diseases of various crops, e.g. sweet potato root rot caused by *Fusarium solani* (Stevens *et al.*, 1999), the citrus green mould caused by *P. digitatum* (Arcas *et al.*, 2000), cabbage black rot caused by *X. campestris* pv. *campestris* (Brown *et al.*, 2001) and apple blue mould caused by *P. expansum* (de Capdeville *et al.*, 2002).

10.4 Conclusions

It is clear from the preceding sections that a bewildering array of biotic and abiotic agents is capable of inducing resistance to pathogen infection in plants. However, just a skim through this chapter should be enough to highlight the variation in levels of disease control achieved following topical application of inducing agents. Indeed, there are also many reports of induced resistance failing to provide disease control (Walters *et al.*, 2005). Understanding the sources of this variation in disease control and hopefully, as a consequence, being able to reduce the variation are crucial to the use of induced resistance in the field. Since induced resistance is a host response, its expression is likely to be influenced by, among other factors, the genotype and the environment. Indeed, there are reports that the magnitude of the induced resistance response is dependent on the plant cultivar used (see Walters *et al.*, 2005). But information in this area is scant. There needs to be greater emphasis on the means by which induced resistance can be incorporated into crop protection practice. Without such information, induced resistance is unlikely to enter mainstream crop protection.

10.5 Acknowledgements

The Scottish Agricultural College receives financial support from the Scottish Executive Environment and Rural Affairs Department.

10.6 References

Agostini JP, Bushong PM, Timmer LW, 2003. Greenhouse evaluation of products that induce host resistance for control of scab, melanose and Alternaria brown spot of citrus. *Plant Disease* **87,** 69–74.

Albersheim P, Valent BS, 1978. Plants when exposed to oligosaccharides of fungal origin defend themselves by accumulating antibiotics. *Journal of Cell Biology* **78,** 627–643.

Antoniw JF, White RF, 1980. The effects of aspirin and polyacrylic acid on soluble leaf proteins and resistance to virus infection in five cultivars of tobacco. *Phytopathologische Zeitschrift* **98,** 331–341.

Arcas MC, Botia JM, Ortuno AM, Rio JA, 2000. UV irradiation alters the levels of flavonoids involved in the defence mechanism of *Citrus aurantium* fruits against *Penicillium digitatum. European Journal of Plant Pathology* **106,** 617–622.

Aronson JM, 1981. Cell wall chemistry, ultrastructure and metabolism. In: Keen NT, Kosuge T, Walling LL, eds. *Physiology and Biochemistry of Plant–Microbial Interactions*. Rockville, MD: The American Society of Plant Biologists, 103–108.

Barber MS, Bertram RE, Ride JP, 1989. Chitin oligosaccharides elicit lignification in wounded leaves. *Physiological and Molecular Plant Pathology* **34,** 3–12.

Bécot S, Pajot E, Le Corre D, Monot C, Silué D, 2000. Phytogard (K_2HPO_3) induces localized resistance in cauliflower to downy mildew of crucifers. *Crop Protection* **19,** 417–425.

Bélanger RR, Benhamou N, Menzies JG, 2003. Cytological evidence of an active role of silicon in wheat resistance to powdery mildew (*Blumeria graminis* f. sp. *tritici*). *Phytopathology* **93,** 402–413.

Benhamou N, Bélanger RR, 1998. Benzothiadiazole-mediated induced resistance to *Fusarium oxysporum* f. sp. *radicis-lycopersici* in tomato. *Plant Physiology* **118,** 1203–1212.

Benhamou N, Theriault G, 1992. Treatment with chitosan enhances resistance of tomato plants to the crown and root rot pathogen *Fusarium oxysporum* f. sp. *radicis-lycopersici. Physiological and Molecular Plant Pathology* **41,** 33–52.

Ben-Shalom N, Ardi R, Pinto R, Aki C, Fallik E, 2003. Controlling gray mould caused by *Botrytis cinerea* in cucumber plants by means of chitosan. *Crop Protection* **22,** 285–290.

Blée E, 2002. Impact of phyto-oxylipins in plant defense. *Trends in Plant Science* **7,** 315–321.

Blee KA, Yang KY, Anderson AJ, 2004. Activation of defense pathways: synergism between reactive oxygen species and salicylic acid and consideration of field applicability. *European Journal of Plant Pathology* **110,** 203–212.

Borges AA, Borges-Perez A, Fernandez-Falcon M, 2004. Induced resistance to fusarial wilt of banana by menadione sodium bisulphite treatments. *Crop Protection* **23,** 1245–1247.

Borges AA, Cools HJ, Lucas JA, 2003. Menadione sodium bisulphite: a novel plant defence activator which enhances local and systemic resistance to infection by *Leptosphaeria maculans* in oilseed rape. *Plant Pathology* **52,** 429–436.

Bostock RM, 2005. Signal crosstalk and induced resistance: straddling the line between cost and benefit. *Annual Review of Phytopathology* **43,** 545–580.

Bostock RM, Sheaffer DA, Hammerschmidt R, 1986. Comparison of elicitor activities of arachidonic acid, fatty acids, and glucans from *Phytophthora infestans* in hypersensitivity expression in potato tuber. *Physiological and Molecular Plant Pathology* **29,** 349–360.

Brown JER, Lu TY, Stevens C, Khan VA, Lu JY, Wilson CL, Collins DJ, Wilson, MA, Igwegbe ECK, Chalutz E, Droby S, 2001. The effect of low dose ultraviolet light-C seed treatment on induced resistance in cabbage to black rot (*Xanthomonas campestris* pv. *campestris*). *Crop Protection* **20,** 873–883.

Buonaurio R, Scarponi L, Ferrara M, Sidoti P, Bertona A, 2002. Induction of systemic acquired resistance in pepper plants by acibenzolar-*S*-methyl against bacterial spot disease. *European Journal of Plant Pathology* **108,** 41–49.

Buzi A, Chilosi G, de Sillo D, Magro P, 2004a. Induction of resistance in melon to *Didymella bryoniae* and *Sclerotinia sclerotiorum* by seed treatments with acibenzolar-*S*-methyl and methyl jasmonate but not with salicylic acid. *Journal of Phytopathology* **152,** 34–42.

Buzi A, Chilosi G, Magro P, 2004b. Induction of resistance in melon seedlings against soil-borne fungal pathogens by gaseous treatments with methyl jasmonate and ethylene. *Journal of Phytopathology* **152,** 491–497.

Callow JA, 1984. Cellular and molecular recognition between plants and fungal pathogens. In: Linskens HF, Heslop-Harrison J, eds. *Encyclopedia of Plant Physiology, New Series, Vol. 17, Cellular Interactions*. Berlin: Springer, 212–237.

Cohen Y, 1994. 3-Aminobutyric acid induces systemic resistance against *Peronospora tabacina. Physiological and Molecular Plant Pathology* **44,** 273–288.

Cohen Y, Gisi U, Niderman T, 1993. Local and systemic protection against *Phytophthora infestans* induced in potato and tomato plants by jasmonic acid and jasmonic acid methyl ester. *Phytopathology* **83,** 1054–1062.

Cohen Y, Reuveni M, Baider A, 1999. Local and systemic activity of BABA (DL-3-aminobutyric acid) against *Plasmopara viticola* in grapevines. *European Journal of Plant Pathology* **105,** 351–361.

Cole D, 1999. The efficacy of acibenzolar-*S*-methyl, an inducer of systemic acquired resistance, against bacterial and fungal diseases of tobacco. *Crop Protection* **18,** 267–273.

Collmer A, Keen NT, 1988. The role of pectic enzymes in pathogenesis. *Annual Review of Phytopathology* **24,** 383–409.

Cowley T, Walters D, 2005. Local and systemic effects of oxylipins on powdery mildew infection in barley. *Pest Management Science* **61,** 572–576.

Dann E, Diers B, Byrum J, Hammerschmidt R, 1998. Effect of treating soybean with 2,6-dichloroisonicotinic acid (INA) and benzothiadiazole (BTH) on seed yields and the level of disease caused by *Sclerotinia sclerotiorum* in field and greenhouse studies. *European Journal of Plant Pathology* **104,** 271–278.

de Capdeville G, Wilson CL, Beer SV, Aist JR, 2002. Alternative disease control agents induce resistance to blue mold in harvested 'Red Delicious' apple fruit. *Phytopathology* **92,** 900–908.

Dempsey DA, Shah J, Klessig DF, 1999. Salicylic acid and disease resistance in plants. *Critical Reviews in Plant Sciences* **18,** 547–575.

Dong H, Delaney TP, Bauer DW, Beer SV, 1999. Harpin induces disease resistance in Arabidopsis through the systemic acquired resistance pathway mediated by salicylic acid and the *NIM1* gene. *The Plant Journal* **20,** 207–215.

Dong H, Li W, Zhang D, Tang W, 2003. Differential expression of induced resistance by an aqueous extract of killed *Penicillium chrysogenum* against *Verticillium* wilt of cotton. *Crop Protection* **22,** 129–134.

Drennan PM, Smith MT, Golsworthy D, van Staden J, 1993. The occurrence of trehalose in the leaves of the desication-tolerant angiosperm *Myrrothammus flabellifolius* Welw. *Journal of Plant Physiology* **142,** 493–496.

Droby S, Porat R, Cohen L, Weiss B, Shapiro B, Philosoph-Hadas S, Meir S, 1999. Suppressing green mold decay in grapefruit with postharvest jasmonate application. *Journal of the American Society for Horticultural Science* **124,** 184–188.

Dutton MV, Rigby NM, MacDougall AJ, 1997. Induced resistance to *Erwinia carotovora* ssp. *atroseptica* through the treatment of surface wounds of potato tubers with elicitors. *Journal of Phytopathology* **145,** 163–169.

El Ghaouth A, Arul J, Grenier J, Asselin A, 1992. Antifungal activity of chitosan on two postharvest pathogens of strawberry fruits. *Phytopathology* **82,** 398–402.

Elkemo H, Stensvand A, Tronsmo AM, 2003. Induced resistance as a possible means to control diseases of strawberry caused by *Phytophthora* spp. *Plant Disease* **87,** 345–350.

Ellis C, Karafyllidis I, Turner JG, 2002. Constitutive activation of jasmonate signalling in an *Arabidopsis* mutant correlates with enhanced resistance to *Erysiphe cichoracearum. Molecular Plant–Microbe Interactions* **15,** 1025–1030.

Farmer EE, Almeras E, Krishnamurthy V, 2003. Jasmonates and related oxylipins in plant responses to pathogenesis and herbivory. *Current Opinion in Plant Biology* **6,** 1–7.

Feussner I, Wasternack C, 2002. The lipoxygenase pathway. *Annual Review of Plant Physiology and Plant Molecular Biology* **53,** 275–297.

Fuhrer J, 2003. Agroecosystem responses to combination of elevated CO_2, ozone, and global climate change. *Agriculture, Ecosystems and Environment* **97,** 1–20.

Gális I, Smith JL, Jameson PE, 2004. Salicylic acid-, but not cytokinin-induced, resistance to WCIMV is associated with increased expression of SA-dependent resistance genes in *Phaseolus vulgaris. Journal of Plant Physiology* **161,** 459–466.

Glawe GA, Zavala JA, Kessler A, Van Dam NM, Baldwin IT, 2003. Ecological costs and benefits correlated with trypsin protease inhibitor production in *Nicotiana attenuata. Ecology* **84,** 79–90.

Gobel C, Feussner I, Hamberg M, Rosahl S, 2002. Oxylipin profiling in pathogen-infected potato leaves. *Biochimica et Biophysica Acta* **1548,** 55–64.

Godard J-F, Ziadi S, Monot C, Le Corre D, Silué D, 1999. Benzothiadiazole (BTH) induces resistance in cauliflower (*Brassica oleracea* var *botrytis*) to downy mildew of crucifers caused by *Peronospora parasitica. Crop Protection* **18,** 397–405.

Görlach J, Volrath S, Knauf-Beiter G, Hengy G, Beckhove U, Kogel KH, Oostendorp M, Staub T, Ward E, Kessmann H, Ryals J, 1996. Benzothiadiazole, a novel class of inducers of systemic acquired resistance, activates gene expression and disease resistance in wheat. *The Plant Cell* **8,** 629–643.

Gottstein HD, Kuć J, 1989. Induction of systemic resistance to anthracnose in cucumber by phosphates. *Phytopathology* **79,** 176–179.

Hammerschmidt R, Smith-Becker JA, 1999. The role of salicylic acid in disease resistance. In: Agrawal AA, Tuzun S, Bent E, eds. *Induced Plant Defenses Against Pathogens and Herbivores: Biochemistry, Ecology, and Agriculture*. St. Paul, MN: APS Press, 37–53.

Hofgaard IS, Ergon A, Wanner LA, Tronsmo AM, 2005. The effect of chitosan and Bion on resistance to pink snow mould in perennial ryegrass and winter wheat. *Journal of Phytopathology* **153,** 108–119.

Iriti M, Faoro F, 2003. Benzothiadiazole (BTH) induces cell-death independent resistance in *Phaseolus vulgaris* against *Uromyces appendiculatus. Journal of Phytopathology* **151,** 171–180.

Iriti M, Rossoni M, Borgo M, Faoro F, 2004. Benzothiadiazole enhances resveratrol and anthocyanin biosynthesis in grapevine, meanwhile improving resistance to *Botrytis cinerea. Journal of Agricultural ad Food Chemistry* **52,** 4406–4413.

Jackson AJ, Walters DR, Marshall G, 1994. Evaluation of *Penicillium chrysogenum* and its antifungal extracts as potential biological control agents against *Botrytis fabae* on faba beans. *Mycological Research* **98,** 1117–1126.

Jakab G, Cottier V, Toquin V, Rigoli G, Zimmerli L, Métraux JP, Mauch-Mani B, 2001. β-Aminobutyric acid-induced resistance in plants. *European Journal of Plant Pathology* **107,** 29–37.

Karban R, Kuć J, 1999. Induced resistance against pathogens and herbivores: an overview. In: Agrawal AA, Tuzun S, Bent E, eds. *Induced Plant Defenses Against Pathogens and Herbivores: Biochemistry, Ecology, and Agriculture*. St. Paul, MN: APS Press, 1–16.

Kariola T, Palomaki TA, Brader G, Palva ET, 2003. *Erwinia carotovora* subsp. *carotovora* and *Erwinia*-derived elicitors HrpN and PehA trigger distinct but interacting defense responses and cell death in Arabidopsis. *Molecular Plant–Microbe Interactions* **16,** 179–187.

Kendra FD, Hadwiger LA, 1984. Characterisation of the smallest chitosan oligomer that is maximally antifungal to *Fusarium solani* and elicits pisatin formation in *Pisum sativum*. *Experimental Mycology* **8,** 276–281.

Kim YC, Blee KA, JR, Anderson AJ, 2001. Oxycom under field and laboratory conditions increases resistance responses in plants. *European Journal of Plant Pathology* **107,** 129–136.

Kim SG, Kim KW, Park EW, Choi D, 2002. Silicon-induced cell wall fortification of rice leaves: a possible cellular mechanism of enhanced host resistance to blast. *Phytopathology* **92,** 1095–1103.

Kishore GK, Pande S, Podile AR, 2005. Chitin-supplemented foliar application of *Serratia marcescens* GPS5 improves control of late leaf spot disease of groundnut by activating defence-related enzymes. *Journal of Phytopathology* **153,** 169–173.

Kozlowski G, Buchala A, Métraux J-P, 1994. Methyl jasmonate protects Norway spruce [*Picea abies* (L.) Karst.] seedlings against *Pythium ultimum* Trow. *Physiological and Molecular Plant Pathology* **55,** 53–58.

Latunde-Dada AO, Lucas JA, 2001. The plant defence activator acibenzolar-*S*-methyl primes cowpea (*Vigna unguiculata* (L.) Walp.) seedlings for rapid induction of resistance. *Physiological and Molecular Plant Pathology* **58,** 199–208.

Lawton MA, Lamb CJ, 1987. Transcriptional activation of plant defence genes by fungal elicitor, wounding and infection. *Molecular Cell Biology* **7,** 335–341.

Linden JC, Stoner RJ, Knutson KW, Gardner-Hughes CA, 2000. Organic disease control elicitors. *Agro-Food Industry Hi-Tech* **September/October**, 32–34.

Liu HX, Wang BG, Luo YB, Jiang WB, 2005. Improving disease resistance in peach fruit during storage using benzo-(1,2,3)-thiadiazole-7-carbothionic acid *S*-methyl ester (BTH). *Journal of Horticultural Science and Biotechnology* **80,** 736–740.

Louws FJ, Wilson M, Campbell HL, Cuppels DA, Jones JB, Shoemaker PB, Sahin F, Miller SA, 2001. Field control of bacterial spot and bacterial speck of tomato using a plant activator. *Plant Disease* **85,** 481–488.

Lyon GD, Newton AC, 1999. Implementation of elicitor mediated induced resistance in agriculture. In: Agrawal AA, Tuzun S, Bent E, eds. *Induced Plant Defenses Against Pathogens and Herbivores: Biochemistry, Ecology, and Agriculture*. St. Paul, MN: APS Press, 299–318.

Malamy J, Carr JP, Klessig DF, Raskin I, 1990. Salicylic acid: a likely endogenous signal in the resistance response of tobacco to viral infection. *Science* **250,** 1002–1004.

Mandahar HK, Lyngs Jorgensen HJ, Mathur SB, Smedegaard-Petersen V, 1998. Resistance to rice blast induced by ferric chloride, di-potassium hydrogen phosphate and salicylic acid. *Crop Protection* **17,** 323–329.

Masui H, Kondo T, Kojima M, 1989. An antifungal compound 9,12,13-trihydroxy-*E*-10-octadecenoic acid from *Colocasia antiquorum* inoculated with *Ceratocystis fimbriata*. *Phytochemistry* **28,** 2613–2616.

Maxson-Stein K, He SY, Hammerschmidt R, Jones AL, 2002. Effect of treating apple trees with acibenzolar-*S*-methyl on fire blight and expression of pathogenesis-related protein genes. *Plant Disease* **86,** 785–790.

Métraux J-P, Signer H, Ryals J, Ward E, Wyss-Benz M, Gaudin J, Raschdorf K, Schmid E, Blum W, Inverardi B, 1990. Increase in salicylic acid at the onset of systemic acquired resistance in cucumber. *Science* **250,** 1004–1006.

Mitchell AF, Walters DR, 1995. Systemic protection in barley against powdery mildew infection using methyl jasmonate. *Aspects of Applied Biology* **42,** 323–326.

Mitchell AF, Walters DR, 2004. Potassium phosphate induces systemic protection in barley to powdery mildew infection. *Pest Management Science* **60,** 126–134.

Miyazawa J, Kawabata T, Ogasawara N, 1998. Induction of an acidic isozyme of peroxidase and acquired resistance to wilt disease in response to treatment of tomato roots with 2-furoic acid, 4-hydroxybenzoic hydrazide or salicylic hydrazide. *Physiological and Molecular Plant Pathology* **52,** 115–126.

Molloy C, Cheah L-H, Koolaard JP, 2004. Induced resistance against *Sclerotinia sclerotiorum* in carrots treated with enzymatically hydrolysed chitosan. *Postharvest Biology and Technology* **33,** 61–65.

Mucharromah E, Kuć J, 1991. Oxalate and phosphates induce systemic resistance against diseases caused by fungi, bacteria and viruses in cucumber. *Crop Protection* **10,** 265–270.

Nakashita H, Yasuda M, Nishioka M, Hasegawa S, Arai Y, Uramoto M, Yoshida S, Yamaguchi I, 2002a. Chloroisonicotinamide derivative induces a broad range of disease resistance in rice and tobacco. *Plant and Cell Physiology* **43,** 823–831.

Nakashita H, Yoshioka K, Yasuda M, Nitta T, Arai Y, Yoshido S, Yamaguchi I, 2002b. Probenazole induces systemic acquired resistance in tobacco through salicylic acid accumulation. *Physiological and Molecular Plant Pathology* **61,** 197–203.

Nemestothy GN, Guest DI, 1990. Phytoalexin accumulation, phenylalanine ammonia lyase activity and ethylene biosynthesis in Fosetyl-Al treated resistant and susceptible tobacco cultivars infected with *Phytophthora nicotianae* var. *nicotianae. Physiological and Molecular Plant Pathology* **37,** 207–219.

Newman MA, Roepenack-Lahaye EV, Parr A, Daniels MJ, Dow JM, 2002. Prior exposure to lipopolysaccharide potentiates expression of plant defenses in response to bacteria. *The Plant Journal* **29,** 487–495.

Norman-Setterblad C, Vidal S, Palva ET, 2000. Interacting signal pathways control defence gene expression in *Arabidopsis* in response to cell wall degrading enzymes from *Erwinia carotovora. Molecular Plant–Microbe Interactions* **4,** 43–438.

Oka Y, Cohen Y, 2001. Induced resistance to cyst and root-knot nematode in cereals by DL-β-amino-*n*-butyric acid. *European Journal of Plant Pathology* **107,** 219–227.

Orober M, Siegrist J, Buchenauer H, 2002. Mechanisms of phosphate-induced disease resistance in cucumber. *European Journal of Plant Pathology* **108,** 345–353.

Ouimette DG, Coffey MD, 1989. Comparative antifungal activity of four phosphonate compounds against isolates of nine *Phytophthora* species. *Phytopathology* **79,** 761–767.

Pajot E, Le Corre D, Silue D, 2001. Phytogard and DL-β-aminobutyric acid (BABA) induce resistance to downy mildew (*Bremia lactucae*) in lettuce (*Lactuca sativa* L.). *European Journal of Plant Pathology* **107,** 861–869.

Pearce RB, Ride JP, 1982. Chitin and related compounds as elicitors of the lignification response in wounded wheat leaves. *Physiological Plant Pathology* **20,** 119–123.

Peng J-L, Dong H-S, Dong HP, Delaney TP, Bonasera JM, Beer SV, 2003. Harpin-elicited hypersensitive cell death and pathogen resistance require the *NDR1* and *EDS1* genes. *Physiological and Molecular Plant Pathology* **62,** 317–326.

Perez L, Rodriguez ME, Rodriguez F, Roson C, 2003. Efficacy of acibenzolar-*S*-methyl, an inducer of systemic acquired resistance against tobacco blue mould caused by *Peronospora hyoscyami* f. sp. *tabacina. Crop Protection* **22,** 405–413.

Qin GZ, Tian SP, 2005. Enhancement of biocontrol activity of *Cryptococcus laurentii* by silicon and the possible mechanisms involved. *Phytopathology* **95,** 69–75.

Rancé I, Fournier J, Esquerré-Tugayé M-T, 1998. The incompatible interaction between *Phytophthora parasitica* var. *nicotianae* race 0 and tobacco is suppressed in transgenic plants expressing antisense lipoxygenase sequences. *Proceedings of the National Academy of Sciences USA* **95,** 6554–6559.

Reddy MVB, Arul J, Angers P, Couture L, 1999. Chitosan treatment of wheat seeds induces resistance to *Fusarium graminearum* and improves seed quality. *Journal of Agricultural and Food Chemistry* **47,** 1208–1216.

Reglinski T, Newton AC, Lyon GD, 1994a. Assessment of the ability of yeast-derived elicitors to control powdery mildew in the field. *Journal of Plant Diseases and Protection* **101,** 1–10.

Reglinski T, Newton AC, Lyon GD, 1994b. Induction of resistance mechanisms in barley by yeast derived elicitors. *Annals of Applied Biology* **124,** 509–517.

Reglinski T, Lyon GD, Newton AC, 1995. The control of *Botrytis cinerea* and *Rhizoctonia solani* on lettuce using elicitors extracted from yeast cell walls. *Journal of Plant Diseases and Protection* **102,** 257–266.

Reignault P, Cogan A, Muchembled J, Sahraoui ALH, Durand R, Sancholle M, 2001. Trehalose induces resistance to powdery mildew in wheat. *New Phytologist* **149,** 519–529.

Resende MLV, Nojosa GBA, Cavalcanti LS, Aguilar MAG, Silva LHCP, Perez JO, Andrade GCG, Carvalho GA, Castro RM, 2002. Induction of resistance in cocoa against *Crinipellis perniciosa* and *Verticillium dahliae* by acibenzolar-*S*-methyl (ASM). *Plant Pathology* **51,** 621–628.

Reuveni R, Dor G, Raviv M, Reuveni M, Tuzun S, 2000. Systemic resistance against *Sphaerotheca fuligenea* in cucumber plants exposed to phosphate in hydroponics system, and its control by foliar spray of mono-potassium phosphate. *Crop Protection* **19,** 355–361.

Reuveni R, Dor G, Reuveni M, 1998. Local and systemic control of powdery mildew (*Leveillula taurica*) on pepper plants by foliar spray of mono-potassium phosphate. *Crop Protection* **17,** 703–709.

Reuveni M, Reuveni R, 1995. Efficacy of foliar application of phosphates in controlling powdery mildew fungus on field grown winegrapes: effects on cluster yield and peroxidase activity in berries. *Journal of Phytopathology* **143,** 21–25.

Rohilla R, Singh US, Singh RL, 2001. Mode of action of acibenzolar-*S*-methyl against sheath blight of rice, caused by *Rhizoctonia solani* Kuhn. *Pest Management Science* **58,** 63–69.

Romero AM, Kousik CS, Ritchie DF, 2001. Resistance to bacterial spot in bell pepper induced by acibenzolar-*S*-methyl. *Plant Disease* **85,** 189–194.

Rose S, Parker M, Punja ZK, 2003. Efficacy of biological and chemical treatments for control of fusarium root and stem rot on greenhouse cucumber. *Plant Disease* **87,** 1462–1470.

Saikia R, Singh T, Kumar R, Srivastava J, Srivastava AK, Singh K, Arora DK, 2003. Role of salicylic acid in systemic resistance induced by *Pseudomonas fluorescens* against *Fusarium oxysporum* f. sp. *ciceri* in chickpea. *Microbiological Research* **158,** 203–213.

Sandermann H, Ernst D, Heller W, Langerbartels C, 1998. Ozone: an abiotic elicitor of plant defence reactions. *Trends in Plant Science* **3,** 47–50.

Sarathchandra RG, Niranjan Raj S, Shetty NP, Amruthesh KN, Shekar Shetty H, 2004. A chitosan formulation Elexa induces downy mildew disease resistance and growth promotion in pearl millet. *Crop Protection* **23,** 881–888.

Sathiyabama M, Balasubramanian R, 1998. Chitosan induces resistance components in *Arachis hypogaea* against leaf rust caused by *Puccinia arachidis* Speg. *Crop Protection* **17,** 307–313.

Schweizer P, Gees R, Mosinger E, 1993. Effect of jasmonic acid on the interaction of barley (*Hordeum vulgare* L.) with the powdery mildew *Erysiphe graminis* f. sp. *hordei. Plant Physiology* **102,** 503–511.

Seebold KW, Kucharek TA, Datnoff LE, Correa-Victoria FJ, Marchetti MA, 2001. The influence of silicon on components of resistance to blast in susceptible, partially resistant, and resistant cultivars of rice. *Phytopathology* **91,** 63–69.

Seebold KWJ, Datnoff LE, Correa-Victoria FJ, Kucharek TA, Snyder GH, 2004. Effect of silicon and fungicides on the control of leaf and neck blast in upland rice. *Plant Disease* **88,** 253–258.

Shailasree S, Sarosh BR, Vasanthi NS, Shetty HS, 2001. Seed treatment with β-aminobutyric acid protects *Pennisetum glaucum* systemically from *Sclerospora graminicola. Pest Management Science* **57,** 721–728.

Si-Ammour A, Mauch-Mani B, Mauch F, 2003. Quantification of induced resistance against Phytophthora species expressing GFP as a vital marker: β-aminobutyric acid but not BTH protects potato and Arabidopsis from infection. *Molecular Plant Pathology* **4,** 237–248.

Siegrist J, Orober M, Buchenauer H, 2000. β-aminobutyric acid-mediated enhancement of resistance in tobacco to tobacco mosaic virus depends on the accumulation of salicylic acid. *Physiological and Molecular Plant Pathology* **56,** 95–106.

Silué D, Pajot E, Cohen Y, 2002. Induction of resistance to downy mildew (*Peronospora parasitica*) in cauliflower by DL-beta-amino-n-butanoic acid (BABA). *Plant Pathology* **51,** 97–102.

Sparla F, Rotino L, Valgimigli MC, Pupillo P, Trost P, 2004. Systemic resistance induced by benzothiadiazole in pear inoculated with the agent of fire blight (*Erwinia amylovora*). *Scientia Horticulturae* **101,** 269–279.

Spletzer ME, Enyedi AJ, 1999. Salicylic acid induces resistance to *Alternaria solani* in hydroponically grown tomato. *Phytopathology* **89,** 722–727.

Stevens C, Khan VA, Lu JY, Wilson CL, Chalutz E, Droby S, Kabwe MK, Haung Z, Adeyeye O, Pusey LP, Tang AYA, 1999. Induced resistance of sweetpotato to Fusarium root rot by UV-C hormesis. *Crop Protection* **18,** 463–470.

Stevens C, Khan VA, Lu JY, Wilson CL, Pusey PL, Kabwe MK, Igwegbe ECK, Chalutz E, Droby S, 1998. The germicidal and hormetic effects of UV-C light on reducing brown rot disease and yeast microflora of peaches. *Crop Protection* **17,** 75–84.

Takenaka S, Nishio Z, Nakamura Y, 2003. Induction of defense reactions in sugar beet and wheat by treatment with cell wall protein fractions from the mycoparasite *Pythium oligandrum. Phytopathology* **93,** 1228–1232.

Terry LA, Joyce DC, 2004. Elicitors of induced disease resistance in postharvest horticultural crops: a brief review. *Postharvest Biology and Technology* **32,** 1–13.

Thuerig B, Felix G, Binder A, Boller T, Tamm L, 2006. An extract of *Penicillium chrysogenum* elicits early defense-related responses and induces resistance in *Arabidopsis thaliana* independently of known signalling pathways. *Physiological and Molecular Plant Pathology* **67,** 180–193.

Ton J, Mauch-Mani B, 2004. Beta-amino-butyric acid-induced resistance against necrotrophic pathogens is based on ABA-dependent priming for callose. *The Plant Journal* **38,** 119–130.

Turner JG, Ellis C, Devoto A, 2002. The jasmonate signal pathway. *The Plant Cell* S153–S164.
Tuzun S, Kloepper J, 1995. Practical application and implementation of induced resistance. In: Hammerschmidt R, Kuć J, eds. *Induced Resistance to Disease in Plants*. Dordrecht: Kluwer Academic, 152–168.
Umemura K, Tanimo S, Nagatsuka T, Koga J, Iwata M, Nagashima K, Amemiya Y, 2004. Cerebroside elicitor confers resistance to Fusarium disease in various plant species. *Phytopathology* **94,** 813–818.
Walters D, Cowley T, Mitchell A, 2002. Methyl jasmonate alters polyamine metabolism and induces systemic protection against powdery mildew infection in barley seedlings. *Journal of Experimental Botany* **53,** 747–756.
Walters D, Walsh D, Newton A, Lyon G, 2005. Induced resistance for plant disease control: maximising the efficacy of resistance elicitors. *Phytopathology* **95,** 1368–1373.
Walters DR, Murray DC, 1992. Induction of systemic resistance to rust in *Vicia faba* by phosphate and EDTA: effects of calcium. *Plant Pathology* **41,** 444–448.
Wäspi U, Blanc D, Winkler T, Ruedi P, Dudler R, 1998. Syringolin, a novel peptide elicitor from *Pseudomonas syringae* pv. *syringae* that induces resistance to *Pyricularia oryzae* in rice. *Molecular Plant–Microbe Interactions* **11,** 727–733.
Wäspi U, Schweizer P, Dudler R, 2001. Syringolin reprograms wheat to undergo hypersensitive cell death in a compatible interaction with powdery mildew. *The Plant Cell* **13,** 153–161.
White RF, 1979. Acetylsalicylic acid (aspirin) induces resistance to tobacco mosaic virus in tobacco. *Virology* **99,** 410–412.
Yang K-Y, Blee KA, Zhang S, Anderson AJ, 2002. Oxycom™ treatment suppresses *Pseudomonas syringae* infection and activates a mitogen-activated protein kinase pathway in tobacco. *Physiological and Molecular Plant Pathology* **61,** 249–256.
Yao H, Tian S, 2005. Effects of pre- and post-harvest application of salicylic acid or methyl jasmonate on inducing disease resistance of sweet cherry fruit in storage. *Postharvest Biology and Technology* **35,** 253–262.
Yoshioka K, Nakashita H, Klessig DF, Yamaguchi I, 2001. Probenazole induces systemic acquired resistance in Arabidopsis with a novel type of action. *The Plant Journal* **25,** 149–157.
Zhu YJ, Xiaohui Qiub X, Moore PH, Borth W, Hu J, Ferreira S, Albert HH, 2003. Systemic acquired resistance induced by BTH in papaya. *Physiological and Molecular Plant Pathology* **63,** 237–248.
Zimmerli L, Métraux J-P, Mauch-Mani B, 2001. β-Aminobutyric acid-induced protection of *Arabidopsis* against the necrotrophic pathogen *Botrytis cinerea. Plant Physiology* **126,** 517–523.

Chapter 11
Integration of induced resistance in crop production

Tony Reglinski[1], *Elizabeth Dann*[2] *and Brian Deverall*[3]
[1]HortResearch, Hamilton, New Zealand
[2]Department of Primary Industries and Fisheries, Queensland, Australia
[3]University of Sydney, Australia

11.1 Introduction

Ten years have passed since the first chemical resistance activator, 'Bion®', (acibenzolar-*S*-methyl, ASM), was registered as a plant 'tonic' and introduced to cereal cropping systems in Europe. Since then, many other chemical and microbial activators have been developed as a new generation of crop protectants has emerged to provide growers with additional options for disease management. The development of plant activators has been facilitated by co-ordinated research on a relatively small number of plant species that has unravelled some of the biochemical and molecular processes underlying the phenotypic expression of induced resistance. This has been accompanied by numerous field studies to demonstrate the practical application of induced resistance for disease control in crop production systems.

The use of the term 'activator' in the context of systemic acquired resistance (SAR) was put forward as follows by Kessmann *et al.* (1994) and used in the same way by Oostendorp *et al.* (2001). A chemical will be considered an 'activator' of the SAR response if the chemical induces resistance to the same spectrum of pathogens and induces expression of the same biochemical markers as in the biological model. Furthermore, the chemical should have no direct antimicrobial activity.

We have been much less strict in the use of 'activator' in this chapter because we wanted a broad and general term for the stimulating agents used in a wide variety of experiments and trials aimed at evaluating field performance of SAR, and have left considerations of underlying mechanisms for other chapters. We have thus used 'activator' in a broader sense than that intended by its proponents, covering also biological inducers of SAR and fungicides which may act directly on pathogens and also indirectly through SAR.

The efficacy of plant activators in field conditions has been variable compared to their performance in controlled glasshouse and laboratory conditions, and despite initial optimism and extensive research, the practical implementation of induced resistance in crop production systems has been slow. Variable performance of biologically based control strategies in the field is recognized to constitute a significant constraint for their practical implementation (Stewart, 2001; Shtienberg & Elad, 2002). The complexities of interactions between activators, plants and microbes in the field environment present an enormous challenge to achieving a consistent level of induced resistance. In this chapter, we consider the performance of plant activators in different crop production systems and discuss some of the factors that may affect efficacy. In addition, we discuss the potential to combine plant activators with fungicides, biocontrol agents, plant growth-promoting rhizobacteria and cultural control methods as practical means for reducing our dependence on traditional chemicals for plant disease control. This is very timely, given current global trends towards lower chemical inputs and greater interest in the adoption of more ecologically sensitive methods for crop production.

11.2 Induced resistance for disease control

Commercialization of activators, and the realization that induced resistance really may have a place in disease management, has resulted in increasing volumes of research and publications assessing efficacy of induced resistance under field or other commercial production conditions. Many of these studies arose from successful or promising glasshouse experimentation. In this section, we briefly outline examples where induced resistance has been particularly effective under field conditions, where efficacy in intractable host–pathogen systems under glass has warranted further field-testing, and where implementation of induced resistance has led to a demonstrated reduction in pesticide use. Induced resistance to post-harvest pathogens and in woody perennials such as fruit trees and conifers is less well known, and some examples will be presented here.

11.2.1 Successful implementation of induced resistance in field, glasshouse, forest and orchard production

The most intensively studied activators in the field have been ASM, 2,6-dichloro-isonicotinic acid (DCINA) and strains of plant growth-promoting rhizobacteria (PGPR). ASM shows particular promise for the control of foliar disease, although it can cause phytotoxicity if used repeatedly or at high concentrations. Efficacy can equal or better that of conventional fungicides, and opportunities exist for the use of ASM in integrated programmes and in situations where there is pressure to reduce fungicide inputs (discussed in detail in 11.4.1). Vallad & Goodman (2004) and da Rocha & Hammerschmidt (2005) have recently presented comprehensive reviews summarizing the field performance of ASM and DCINA across a broad range of crops, and discussion of these studies will not be repeated here. PGPR are applied as seed or root treatments and are more commonly associated with plant growth promotion and the suppression of soil-borne pathogens through direct antagonism (Kloepper, 1993). However, some PGPR are also capable of stimulating induced systemic resistance, making them a particularly attractive option in plant production (Van Loon *et al.*, 1998; Zehnder *et al.*, 2001; Kloepper *et al.*, 2004). PGPR-induced resistance is discussed in more detail in 11.4.3 and also in Chapter 8.

Other commercially developed plant activators that have demonstrated efficacy in the field include; Oxycom™ (Redox Chemicals Inc., Burley, ID), Milsana® (KHH BioScience Inc., Raleigh, NC), Elexa® (SafeScience, Boston, MA), and Messenger® (Eden Bioscience, Bothell, WA, USA). Field applications of Oxycom™ (one component being a 5% v/v stabilized solution of peracetic acid) effectively reduced foliar, berry and root diseases in lettuce, carrots and grapes (Kim *et al.*, 2001). Downy mildew of lettuce caused by *Bremia lactucae*, and diseases caused by *Pythium* sp. in lettuce and carrot, were significantly less severe after multiple Oxycom™ treatments compared with water controls and even the industry standard fungicide regime in some cases. Additionally, the numbers of forked carrot roots, an indicator of nematode damage, were lower in Oxycom™ treated plants. A combination of Oxycom™ and Microthiol® fungicide was more effective than Microthiol® alone in protecting grapes (berry clusters) from powdery mildew caused by *Uncinula necator*.

Milsana® Bioprotectant Concentrate is a registered plant activator for use on glasshouse grown ornamental (non-food) plants in the USA. Milsana® contains 5% of the ethanolic extract of giant knotweed (*Reynoutria sachalinensis*) and is active against fungal diseases, under glass and in field conditions, on a number of crops including cucumber (Daayf *et al.*, 1995; Fofana *et al.*, 2002), strawberry (Carlen *et al.*, 2004) and grapes (Schilder *et al.*, 2002; Schmitt *et al.*, 2002). In vineyard trials, application of Milsana® every 7–10 days reduced the incidence of powdery mildew and bunch rot (*Botrytis cinerea*) on grape berries to the same degree or better than sulfur and the copper containing agent FW 450 (Dow AgroSciences) (Schmitt *et al.*, 2002). A chitosan-based activator called Elexa® has also shown efficacy in grapes against downy mildew (*Plasmopara viticola*) and powdery mildew in field trials (Schilder *et al.*, 2002). A more recent formulation containing 4% chitosan called Elexa® 4 Plant Defense Booster is being marketed by Plant Defense Boosters Inc. (Syracuse, NY). Elexa® 4 PDB and Milsana® are both suitable for use in organic production systems.

Messenger® (a.i. harpin protein) has had mixed results as a crop protectant. It has good efficacy against blue mould in apples (de Capdeville *et al.*, 2003) but poor efficacy against target spot of tomato (Pernezny *et al.*, 2002) and grey mould in strawberry (Meszka & Bielenin, 2004). In recent studies on citrus, application of Messenger® was associated with retarded fruit maturation and delayed colour-break on a mid-season orange variety but not on a late-season Valencia orange (Graham & Leite, 2004). New formulations of harpin proteins including N-Hibit™, ProAct™ and MightyPlant™ have recently been registered for use across a broad range of crops and in home gardens. They are formulated as seeds or foliar treatments, and are proposed to improve crop growth, yield and quality (www.edenbio.com).

Induced resistance against pathogenic attack has not been widely studied in economically important orchard, forest, timber and landscape woody perennial species. Hubbes and co-workers at the University of Toronto, Canada, have carried out extensive investigations into the potential of induced resistance for the control of Dutch elm disease (DED) caused by *Ophiostoma novo-ulmi* (Hubbes, 2004). Hubbes & Jeng (1981) reported that four year old elm seedlings (*Ulmus americana*) acquired resistance to DED after pre-inoculation with the less aggressive *O. ulmi*. The identification of a proteinaceous elicitor from *O. ulmi* (Yang *et al.*, 1994) led to the development of a tree injection treatment for DED called ELMGuard (ArborSciences, Canada) (www.elmcare.com). In New Zealand, Chemcolour Industries have registered a product called TREET™ (a.i. 2-hydroxybenzoic acid) that is applied by stem injection and is proposed to enhance resistance against silverleaf (*Chondrostereum purpureum*) in pipfruit and stonefruit trees.

Application of chemical activators, by foliar spray or as a root drench, has resulted in elevation of disease resistance in some forestry and orchard crops. In *Pinus radiata* (Monterey Pine) seedlings, foliar application of salicylic acid (SA) or 5-chlorosalicylic acid (5CSA) induced resistance to inoculation with *Sphaeropsis sapinea* (Reglinski *et al.*, 1998) while treatment with chitosan induced resistance to *Fusarium circinatum*, the causal agent of pitch canker (Reglinski *et al.*, 2004). In the latter study, chitosan also induced systemic resistance to wound inoculation with *S. sapinea* in four year old *P. radiata*. ASM, applied as a foliar spray, reduced the incidence of root rot infection caused by *Phytophthora cinnamomi* in *P. radiata*, *Banksia integrifolia* and *Isopogon cuneatus* (Ali *et al.*, 2000). *Banksia attenuata* seedlings expressed elevated resistance to stem inoculation with *P. cinnamomi* one week after treatment with 0.5 mM benzoic acid, applied as either a soil drench or foliar spray (Williams *et al.*, 2003). A single soil drench application of ASM, DCINA or SA, 10 days prior to inoculation with *Colletotrichum gloeosporioides*, was sufficient to significantly reduce anthracnose disease in the foliage of four year old cashew (*Anacardium occidentale*) trees in a field experiment in Brazil (Lopez & Lucas, 2002). Four applications of ASM showed control efficacy against Japanese pear scab (*Venturia nashicola*) equal to that of the commercial fungicide in field trials (Ishii *et al.*, 1999).

11.2.2 Induced resistance for post-harvest disease control

Despite the fact that fruit maturation is accompanied by a decline in natural disease resistance (Prusky, 1996; Terry *et al.*, 2004), there is good evidence that inducible defences do have a role in post-harvest disease resistance. Terry & Joyce (2004) recently reviewed the use of various biotic and abiotic activators for post-harvest disease control in horticulture. Of particular interest is the potential for activators to suppress post-harvest disease, even when applied before flowering. There have been some remarkable examples of systemic protection against post-harvest diseases of fruit induced by foliar treatments with activators (ASM and harpin), and when activators were applied as a post-harvest spray or dip. Field trials on melons in Australia and China found effects of foliar ASM treatments on susceptibility of stored fruit to natural infections by *Alternaria* spp., *Fusarium* spp. and *Rhizopus* spp. but not *Tricothecium* spp. (Huang *et al.*, 2000). This result is particularly significant in showing that a single application of ASM to foliage before flowering partially protected melons harvested eight weeks later in the field and then held for a further 1–4 weeks in storage. In many experiments, foliar ASM applied pre-flowering followed by guazatine fungicide dipping of harvested melons gave greater protection than ASM or guazatine alone. The likelihood of ASM activating resistance in foliage and thus decreasing inoculum loads on developing fruit was discounted, as the foliage of untreated plants was apparently without any noticeable fungal infection. Additionally, Alternaria disease occurred initially where fruits were in contact with the soil, and thus did not arise from airborne inoculum. The authors suggested that either ASM or a secondary signal affects fruit-generating cells in the flower so that a long-lasting change in the fruit is initiated. It is also conceivable that signals for the stimulation of defences are constantly translocated from, and to, successive tissues, right through to fruit maturity.

These results have been confirmed and studies expanded in subsequent field and laboratory trials (R. McConchie, ACIAR website: http://www.aciar.gov.au). ReZist, silica, harpin (Messenger®) and ASM applied at various times prior to harvest all reduced post-harvest disease and enhanced melon quality. Chitinase and peroxidase activities were increased in leaves and fruit following these treatments. However, the beneficial effects were not observed in fruit that had received improper post-harvest care and handling, that is, incorrect temperatures and humidity which may have favoured the development of disease. Post-harvest dipping of melons in prochloraz, azoxystrobin and imazalil fungicides was also effective in reducing disease. It was suggested that recommendations for post-harvest disease control in melons would have to include adequate handling and management of conditions during transport and storage, in addition to pre- and post-harvest treatments with resistance activators and fungicides.

A recent study investigated whether post-harvest application of ASM to detached peach fruit (*Prunus persica* 'Jiubao') could affect severity of blue mould disease upon subsequent inoculation with *Penicillium expansum* (Liu *et al.*, 2005). Fruit were dipped for 5 minutes in solutions of ASM or sterile water, air-dried and maintained at constant temperature and relative humidity for 60 hours prior to inoculation with *P. expansum*. Disease incidence and severity were significantly reduced, and pathogen growth was delayed in ASM-treated fruit compared with water-treated control fruit. A similar study has been reported in three cultivars of apples treated with the harpin protein (de Capdeville *et al.*, 2003). Fruit were harvested and sprayed with harpin, or whole trees including fruit were sprayed with harpin either eight or four days before harvest. Fruit were wound inoculated with the blue mould pathogen, *P. expansum*, and stored in a commercial cold room at 0.5°C for 120 days. The incidence and severity of resulting disease were lower, and disease development was delayed in harpin-treated apples compared with fruit from formulation and water control treatments. The higher concentrations of harpin tested resulted in greater control, but there was no difference in control between the eight day and four day spray treatments.

Preliminary studies in mango (*Mangifera indica*) have shown that three soil drench applications to trees during the fruiting period or fruit immersion prior to harvest with ASM or soluble silicon could protect fruit from post-harvest anthracnose disease caused by *C. gloeosporioides* (Zainuri *et al.*, University of Queensland, Australia, unpublished results). Similarly, pre-harvest trunk injection of avocado with soluble silicon resulted in less severe anthracnose and increased shelf-life of fruit (J. Anderson *et al.*, Department of Primary Industries and Fisheries, Queensland, Australia, unpublished results). These studies are interesting because they provide evidence of protection when activators were applied after the initial infection with the pathogen had taken place. Anthracnose diseases in fruit generally arise from quiescent infections, i.e. *Colletotrichum* spp. infect early in the season and remain latent as appressoria (Muirhead & Deverall, 1981) or germinated appressoria (Coates *et al.*, 1993), until ripening occurs, when latency is broken and the fungus resumes growth, causing disease. Many important post-harvest diseases of fruits arise from latent field infections.

The species in these studies are known as 'climacteric' fruits, that is, they undergo a phase of increased respiratory activity during ripening, which commonly occurs after commercial harvest. Thus, it is feasible that the processes associated with active defence within harvested climacteric fruit may still be triggered by exposure to treatments shown to enhance resistance in whole plants. These encouraging reports of apparent resistance induced to quiescent pathogens and others of harvested fruits provide support for further rigorous testing under field and storage conditions with a view towards adoption of the technology in much broader applications than is currently the case. In some cases, there has been some doubt about whether post-harvest disease suppression was truly an induced resistance response or whether it was the result of the retardation of fruit ripening or direct anti-microbial effect of the activator on the target pathogen (Zainuri *et al.*, 2001). SA has been shown to retard the ripening of grapes (Kraeva *et al.*, 1998), and so the potential consequences of delayed harvest should be considered carefully before implementation of activators with hormonal activity.

11.3 Variable efficacy of induced resistance

The examples highlighted above provide evidence for successful field implementation of induced resistance into Integrated Crop Management (ICM) strategies. However, the inherent complexity of the field environment presents an enormous challenge to the use of plant activators, and efficacy can be highly variable. The variable or inconsistent efficacy of activators when used under field conditions is a major obstacle to the practical implementation of induced resistance for plant protection in agriculture. There is some evidence that induced resistance may be affected by light, temperature, soil type and soil nutrient level or availability, particularly nitrogen. Such parameters are known to affect the expression of constitutive defence, in the absence of resistance-inducing treatments. The examples discussed below illustrate the inconsistent or ineffective response of field-grown plants to resistance activators. Variability occurred among experiments and years or seasons of testing, cultivars tested, crop yields obtained and diseases to which control was sought. Additionally, some studies seeking to address how environmental or cultural factors may influence induced resistance will be presented.

11.3.1 Impact of cultivar and plant development on induced resistance

There is evidence of differential cultivar response to activator application. Barley exhibited a cultivar-specific response to treatment with *Bacillus subtilis* culture filtrates with the most marked effects on powdery mildew and crop yield observed in partially resistant cultivars (Steiner *et al.*, 1988). A similar cultivar-specific response was observed on barley after treatment with yeast extract (Reglinski *et al.*, 1994). In these studies, activator application affected disease/yield relationships suggesting that induced plants may exhibit both enhanced disease resistance and enhanced disease tolerance. Glasshouse and field experiments demonstrated that ASM treatments at least halved the incidence of bacterial wilt of tomato, caused by inoculations with *Ralstonia solanacearum*, in moderately resistant cultivars, but had little or no effect on incidence in susceptible cultivars (Pradhanang *et al.*, 2005). However, systemic colonization of stem tissue by *R. solanacearum* was lower after ASM treatment compared with untreated plants of both resistant and susceptible cultivars. This suggests that a defence response may still be activated in susceptible cultivars but was possibly overwhelmed by the high concentrations of bacteria used as inoculum in the study. Resistance to white mould in soybean, caused by *Sclerotinia sclerotiorum*, was consistently achieved by repeated applications of ASM or DCINA in field trials during 1993–1996 (Dann *et al.*, 1998). The greatest reduction in disease severity, up to 70%, and largest increases in seed yields, was demonstrated in Williams 82 and Elgin 87 cultivars, which are considered to be highly susceptible to the disease. The effects of activator treatments were not as large in the cultivars Corsoy 79 and NKS19-90, which are more resistant to white mould.

Field-grown potato 'Sebago' plants were treated with ASM as a foliar spray twice over one week at 60 days of crop growth (Bokshi *et al.*, 2003). There was no consistent reduction in dry rot in tubers from ASM-treated plants, when tubers were harvested four weeks after treatment, then wounded and inoculated with *Fusarium semitectum*. However, the severity of naturally infected leaf spotting diseases, mainly early blight caused by *Alternaria solani*, was reduced. In contrast to the field results, severity of dry rot in tubers of 'Coliban' collected from glasshouse-grown and ASM-treated plants was significantly reduced, as were early blight and powdery mildew foliar diseases caused by *A. solani* and *Erysiphe cichoracearum* respectively. In the glasshouse study, greater resistance to dry rot was observed when foliage was treated at 30 days of growth than at 60 days. β-1,3-glucanase activity was enhanced by ASM in leaves stems, tubers and stolons, but not in roots. It is possible that the application of ASM relatively late in plant growth in the field study, or the different cultivars used for field versus laboratory experiments, accounted for the lack of resistance to dry rot in field-raised tubers.

ASM was included in field trials with citrus crops evaluating efficacy of several fungicides in controlling two diseases. The incidence and severity of black spot on fruit caused by *Guignardia citricarpa* were reduced by up to 50% compared with untreated controls in mandarin 'Imperial' and orange 'Navel' when trees were treated three and four times, respectively, with ASM, commencing at petal fall (Miles *et al.*, 2004). In contrast, incidence and severity of brown spot on fruit caused by *Alternaria alternata* were not reduced in mandarin 'Murcott' when trees were sprayed 16 times with ASM commencing at flowering (Miles *et al.*, 2005). However, tank-mixing ASM with azoxystrobin provided significantly greater control of brown spot than azoxystrobin alone. The failure of ASM treatments alone to affect Alternaria brown spot disease in the field differs from a study in which lesion numbers on leaves of tangerine 'Dancy' were fewer in glasshouse-raised seedlings treated with ASM then inoculated with *A. alternata* (Agostini *et al.*, 2003). Similar levels of disease reduction were observed in leaves treated with other purported resistance activators, including Oxycom™ and Messenger®. The discrepancy could involve cultivar differences, environmental conditions and fruit versus leaf differences. The resistance response may have been saturated by the 16 applications of ASM in field-treated trees.

Field experiments on peanuts indicated that two out of 19 tested PGPR strains were effective at reducing late leaf-spot disease, caused by *Cercosporidium personatum*, but the effect did not persist through the season, and disease reduction was not as great or as long-lasting as that afforded by chlorothalonil fungicide treatment (Zhang *et al.*, 2001). There was no reduction in disease in plants previously treated with several resistance activators, including ASM, DCINA, SA, sodium salicylate and β-amino-butyric acid (BABA). However, one of the PGPR strains combined with BABA did offer significant protection compared with the untreated control but not for the complete season. This result is supported by glasshouse studies that showed that foliar sprays with BABA, but not the other chemical treatments at their tested concentrations, consistently and significantly reduced late leaf spot in two experiments. Soil drenches and foliar sprays, but not seed treatments, with seven of 19 tested PGPR strains were effective in only one of the glasshouse experiments. The authors concluded that resistance to disease in peanut might not be systemically inducible in the same way that it is induced by PGPRs, micro-organisms and chemical activators in other crops.

Activator efficacy may also be influenced by tissue age, since it is well documented that the disease resistance response, in some plant–pathogen interactions, is dependent upon developmental factors (Rupe & Gbur, 1995; Panter & Jones, 2002). In Arabidopsis, local responses to inoculation with *Pseudomonas syringae* pv. *maculicola* were more pronounced in young leaves than in older leaves (Zeier, 2005). Reactions typically associated with induced resistance including the oxidative burst, hypersensitive cell death, phenylalanine ammonia-lyase expression and SA accumulation were more intense and/or faster in the younger leaves. Similarly, systemic accumulation of SA and PR-1 and the expression of systemic acquired resistance were also greater in young leaves than old leaves. Interestingly, phenotypic expression of disease resistance was similar in the different aged leaves, suggesting that defence mechanisms other than those described above may contribute to disease resistance in older tissues. In studies on kiwifruit, topical application of SA elevated resistance to *S. sclerotiorum* in young leaves but not in mature fully expanded leaves (T. Reglinski *et al.*, unpublished results). These data illustrate the complexity of the defence network that contributes to plant disease resistance and present many interesting challenges for the practical implementation of activators in the field.

11.3.2 Impact of environmental and nutritional factors

Environmental parameters can have a significant impact on the expression of both constitutive and inducible resistance mechanisms (Falkhof *et al.*, 1988; Stout *et al.*, 1998; Wiese *et al.*, 2003; Dietrich *et al.*, 2004). One study investigated whether resistance to powdery mildew in wheat (*Blumeria graminis* f. sp. *tritici*) and barley (*B. graminis* f. sp. *hordei*), and rust in barley and green bean (*Puccinia hordei* and *Uromyces phaseoli*, respectively), could be induced by treating with a preparation from *Bacillus subtilis*, when plants were previously either maintained under constant temperature, light and humidity conditions in growth cabinets or subjected to altering conditions in the glasshouse or an outdoor nursery (Falkhof *et al.*, 1988). Resistance to the respective diseases could not be induced in any of the plant species maintained in growth cabinets, but infection densities were reduced by at least 40% in plants which were grown in the glasshouse or nursery prior to treatment with the activating preparation, that is, a decrease in efficacy of resistance under constant environmental conditions. Further experiments with barley demonstrated that temperature fluctuations and exposure to light were essential for an efficient response to the induction of powdery mildew resistance.

There are other indications of the importance of light in the expression of resistance and the occurrence of systemic resistance. For example, the development of hypersensitive cell death during an incompatible interaction and SA-induced PR-1 accumulation has been shown to be light- and phytochrome-dependent in Arabidopsis (Genoud *et al.*, 2002). Another study with Arabidopsis demonstrated that PAL induction, SA accumulation, PR1 expression and HR development were not induced by inoculation with avirulent *P. syringae* pv. *maculicola* (*Psm*), when plants were grown in the absence of light (Zeier *et al.*, 2004). Development of systemic resistance to virulent *Psm* normally occurs in response to inoculation with avirulent *Psm*. However, systemic resistance was not achieved when plants were maintained in the absence of light for the first (inducing) inoculation. Systemic resistance developed under medium and high (70 and 500 μmol photons m^{-2} s^{-1}, respectively) light conditions. Interestingly, resistance induced under high light was not accompanied by accumulation of SA or PR-1 in systemic tissue, demonstrating that systemic induced resistance can occur independently of these known markers, and that perhaps other as yet unidentified mechanisms are involved in the resistance to pathogens.

ASM-induced resistance to barley powdery mildew was differentially affected by soil types, levels of supplied N, P and K and organic amendments (Wiese *et al.*, 2003). Induced resistance was significantly reproducible in all four experiments when plants were grown in the ‘Kleinlinden’ soil, described as the ‘Bv’ horizon of a brown soil. The beneficial effect of ASM pre-treatment occurred inconsistently among experiments for the other three soil types in the study. Interestingly, the plants grown in 'Krofdorf' soil ('Ah' horizon of a brown soil), were much more resistant to disease than plants grown in the other soils, irrespective of whether they had been treated with ASM or not. The authors suggested that as the 'Krofdorf' soil was high in organic matter, it was likely to have a higher microbial activity than the 'Kleinlinden' soil, which contained almost no organic matter. This microbial activity may have enhanced disease resistance in barley, such that it was not further enhanced by ASM treatments. This demonstrates that basal levels of susceptibility or resistance can vary with something as simple as soil type. The two other soil types contained intermediate levels of organic matter, so the interaction with soil micro-organisms, and the 'inducibility' of resistance within plants, was influenced by other external factors, such as temperature, thus explaining the more variable results for plants growing in those soils. When the 'Kleinlinden' soil was amended with green manure or compost, there was inconsistent induction of resistance by ASM treatments. Application of varying rates of N, P or K as fertilizer did not consistently affect the induction of resistance by ASM in this study.

The effect of two levels of nitrogen fertilization (210 and 150 kg N ha^{-1}) on resistance to powdery mildew induced in barley by a *B. subtilis* culture preparation has been assessed under practical field conditions (Oerke *et al.*, 1989). Disease was more severe at the high level of applied N in plants not treated with *B. subtilis* preparation, and yields were unaffected or decreased. Induction of resistance by *B. subtilis* occurred at both fertilizer levels, but less effectively in plants at the high rate of N fertilizer, perhaps due to the greater underlying susceptibility at that rate. The greatest increases in yields occurred for 'Tapir' fertilized with the lower N rate and where powdery mildew was decreased by *B. subtilis*.

Dietrich *et al.* (2004) investigated the kinetics of chitinase, chitosanase and peroxidase activation by ASM in Arabidopsis grown with differing amounts of supplementary ammonium nitrate fertilizer. Constitutive activities of the enzymes increased almost linearly with increasing N rates. There was a significant interaction between levels of peroxidase and chitinase induction and N supply, i.e. they were induced more strongly at higher rates of N. A separate but related study reported increased resistance to *Psm* by ASM treatment (Dietrich *et al.*, 2005). It is likely that resistance would increase with increasing activities of the biochemical markers of resistance. For example, if plants are grown under limiting N conditions, then their defences may be compromised, and activators such as ASM may not be so effective. On the other hand, high levels of N fertilization have been shown to limit or restrict the induced resistance response in barley (Oerke *et al.*, 1989), so it seems a balanced N supply is critical and may be difficult to economically maintain under field conditions.

11.4 Compatibility of activators with other control methods

Biological limitations with induced resistance, such as lack of eradicant activity, have prompted the suggestion that plant activators should be considered as an additional option within the framework of an ICM strategy rather than as direct replacements for fungicides (Lyon & Newton, 1999). Indeed, the commercially available activators Elexa PDB® and Messenger® are recommended for use in rotation with conventional fungicides while the label for Bion®/Actigard® recommends that the product 'should be tank-mixed with other registered products with curative activity if disease is present at the time of application to ensure adequate disease control'. Initiatives such as EUREPGAP (European Retail Produce

Good Agricultural Practices) promote the use of ICM for the long-term improvement and sustainability of agricultural production. Adoption of these principles is mandatory in many crop production systems, for example, fruit and vegetable growers and packers shipping to participating stores in the UK are now required to provide EUREPGAP certification (see http://www.eurep.org/). In the following section, we discuss the compatibility of plant activators with other ICM options including fungicides, bactericides, biological control agents and plant growth-promoting rhizobacteria.

11.4.1 Fungicides

Fungicides remain vital for disease control in major crop production systems. Induced resistance is believed to contribute to the efficacy of several pesticides that were not specifically developed as activators, including probenazole (Iwata, 2001) and fosetyl-Al (Guest, 1984). Probenazole (3-allyloxy-1,2-benzisothiazole-1,1-dioxide, PBZ) has been used since 1975 to protect rice plants against the rice blast fungus (*Magnaporthe grisea*). However, there is strong evidence that probenazole operates as a plant activator since it has only weak antimicrobial activity against *M. grisea* (Watanabe, 1977) and has been shown to induce defence responses in rice (Midoh & Iwata, 1997) and Arabidopsis (Yoshioka *et al.*, 2001). PBZ is absorbed by the roots and then systemically translocated through the plant providing protection for 40–70 days (Iwata, 2001). Despite extensive use, there is no record of fungi developing resistance to this compound.

Concerns over residues in foods have led to restrictions on the use of certain chemicals and increased the need to find viable methods for reducing fungicide inputs. ASM has been evaluated in combination with fungicides for plant disease control. A tank mixture of ASM and azoxystrobin provided better control of powdery mildew (*B. graminis* f. sp. *tritici*) and leaf blotch (*Septoria tritici*) on wheat than that achieved by either component when applied separately (Stadnik & Buchenauer, 1999). Similarly, ASM improved efficacy of cypronidil against powdery mildew, but the addition of the plant activator offered no benefit in relation to grain yield compared with fungicide only. In Danish field trials, the use of ASM in an integrated control strategy for cereal diseases enabled a reduction in the number of fungicide applications in three of seven field trials (Jorgensen *et al.*, 1997).

Field management of scab (*Cladosporium oxysporum*) on passion fruit was significantly improved when ASM was tank mixed with azoxystrobin or when ASM was incorporated into an industry programme that included mancozeb, copper oxychloride and iprodione (Willingham *et al.*, 2002). The addition of ASM significantly reduced the incidence and severity of scab and also increased the percentage of marketable fruit compared with the industry standard programme. Furthermore, ASM + azoxystrobin significantly reduced the severity of alternata spot (*A. alternata*) on fruit compared with the industry schedule. However, ASM reduced the efficacy of the industry programme against alternata spot, and this indicates that integrated control will not necessarily be appropriate for all pathosystems. ASM alone reduced the severity of citrus black spot (*Guignardia citricarpa*) by *ca* 50%, compared with the untreated control, but did not enhance the efficacy of the industry standard programme when tank-mixed with fungicides (Miles *et al.*, 2004). However, arguably it is more important that plant activators facilitate a reduction in fungicide inputs without a loss in efficacy rather than enhance the efficacy of the management programme.

ASM enhanced the efficacy of trifloxistrobin against the white rust fungus (*Albugo occidentalis*) on spinach (Leskovar & Kolenda, 2002). Field studies demonstrated the benefits of ASM/fungicide combinations in two out of three seasons. Additionally, a combination of ASM with mefenoxam + copper hydroxide was more effective than mefenoxam + copper hydroxide alone on one occasion. However, in general ASM was more effective when combined with trifloxistrobin. The authors proposed that induced resistance was less effective as plants approached maturity because of an associated decline in internal signalling processes in senescing tissues. This observation is consistent with the notion that induced resistance requires an active plant response and highlights the importance of factoring application timing into the ICM framework.

The non-protein amino acid DL-3-amino-*n*-butanoic acid (β-aminobutyric acid, BABA), is a plant activator and has been shown to induce resistance against a broad range of fungal and bacterial plant pathogens (Jakab *et al.,* 2001; Cohen, 2002). Field experiments on Chardonnay and Cabernet Sauvignon grapevines demonstrated that BABA enhanced the activity of fosetyl aluminium (fosetyl-Al) and *N*-(trichloromethylthio)phthalamide (Folpet) for controlling downy mildew (*Plasmopara viticola*) (Reuveni *et al.*, 2001). Tank mixes containing BABA + fosetyl-Al and BABA + Folpet, at reduced rates, were as efficacious as metalaxyl-Cu (Ridomil®-Cu) or dimethomorph + mancozeb (Acrobat® Plus). A disease-management programme was proposed that integrated BABA with other fungicides in order to reduce intensive use of site-specific fungicides against *P. viticola*. This is of particular interest, since fungicide resistance is a concern for the control of downy mildew in vineyards (Leroux & Clerjeau, 1985). Spray combinations of BABA + mancozeb were also reported to exhibit synergistic activity against late blight (*Phytophthora infestans*) in potato and tomato and downy mildew (*Pseudoperonospora cubensis*) in cucumber (Baider & Cohen, 2003). Split-application experiments, where BABA was applied to the roots and mancozeb to the leaves, suppressed downy mildew sporulation on cucumber leaves in a synergistic manner. It was suggested that the combination of a plant activator and a fungicide could facilitate the use of lower dosages of fungicides on induced plants.

Besides consumer concerns over residues in foods, there is evidence that frequent fungicide application greatly increases the risk of the development of pathogen populations resistant to the active ingredients. The problem of resistance has increased since the advent of highly effective compounds with specific sites of action. The Fungicide Resistance Action Committee (FRAC) was formed over 20 years ago to provide fungicide resistance management guidelines to prolong the effectiveness of 'at risk' fungicides and to limit crop losses should resistance occur (www.frac.info). FRAC recommends against successive applications of the same fungicide but in favour of the use of mixtures or alternation of compounds with different modes of action. The integration of plant activators with fungicides has been suggested as a potential anti-resistance strategy (Ortega *et al.*, 1998; Gullino *et al.*, 2000). An interesting study by Ortega *et al.* (1998) evaluated the efficacy of plant activators on apple scab (*Venturia inaequalis*) in seedlings inoculated with strains of the pathogen that differed in their sensitivity to the triazole group of fungicides. The level of protection induced by either DCINA or 3,5-dichlorosalicylic acid against *V. inaequalis*, in the absence of fungicide treatment, varied by up to 50% depending upon the pathogen genotype. Furthermore, the efficacy of induced resistance was not correlated with strain sensitivity to triazoles. In combination experiments, flusilazole was more efficacious against *V. inaequalis* on DCINA-treated apple seedlings than untreated seedlings. The

EC_{50} value of flusilazole against a highly triazole-resistant strain shifted from 35.8 mg l^{-1}, on untreated seedlings, to 10.1 mg l^{-1} on DCINA-treated seedlings. The authors suggested that synergism between plant activators and flusilazole may enable a reduction in the number of fungicide applications, and possibly fungicide dose rate, in the field.

11.4.2 Bactericides

Several studies have investigated the use of plant activators against bacterial pathogens as replacements for or supplements to copper bactericides. ASM was evaluated over a four year period for the management of bacterial spot (*Xanthomonas axonopodis* pv. *vesicatoria*) and bacterial speck (*Pseudomonas syringae* pv. *tomato*) in glasshouse and field tomato production systems (Louws *et al.*, 2001). In three out of 10 experiments, copper hydroxide ($Cu(OH)_2$) plus ASM provided superior disease control to the standard bactericide programme. It was proposed that ASM might be particularly useful in fields where copper resistant pathogenic strains predominate. ASM protected young tomato plants against bacterial spot during transplant production under glass. However, the treated plants were visibly smaller than their untreated counterparts and showed a 50% reduction in average dry weight. The authors speculated that induction of resistance under crop stress conditions may occur at the expense of constitutive growth and recommended research to optimize the use of ASM during transplant production.

ASM and harpin were evaluated in combination with bacteriophages (Agriphage, OmniLytics, Inc., Salt Lake City, UT) to control bacterial spot in tomato transplants (Obradovic *et al.*, 2004). Both activator + bacteriophage combinations reduced bacterial spot more effectively than $Cu(OH)_2$ + mancozeb, with ASM + bacteriophage being the more effective. Curiously, the greater reduction in disease was not reflected by an increase in yield, and in some ASM-treated plots, yields appeared to be lower, though the difference was statistically insignificant. Conversely, harpin alone did not reduce bacterial spot severity but, when combined with bacteriophage, resulted in the highest fruit yields. In a followup study, necrotic spots typically observed on ASM-treated plants in response to bacterial challenge were not visible on plants treated with ASM in combination with bacteriophage (Obradovic *et al.*, 2005). This was attributed to the bacteriophage suppressing bacterial populations on the leaf surface to levels that would not induce a visible hypersensitive response.

Control of bacterial spot in pepper caused by *X. axonopodis* pv. *vesicatoria* (Romero *et al.*, 2001) and *X. campestris* pv. *vesicatoria* (Buonaurio *et al.*, 2002) was significantly improved when ASM was applied in combination with $Cu(OH)_2$. In field studies, ASM + $Cu(OH)_2$ had a significantly higher efficacy against bacterial spot than ASM or $Cu(OH)_2$ alone (Buonaurio *et al.*, 2002). Combinations of ASM applied with, or in rotation with, $Cu(OH)_2$ resulted in disease control equal to that obtained with weekly applications of $Cu(OH)_2$ + maneb (Romero *et al.*, 2001). However, there was an indication of a negative association between the intensity of ASM use and plant productivity (flower production, fruit set and maturity and crop yield) suggesting the need to consider potential trade-offs before frequent use of activators. Similarly, Gent & Schwartz (2005) reported that 10 weekly applications of ASM caused a reduction in onion bulb yield, but only in certain cultivars and in the absence of disease. This result may suggest a fitness cost to the plant by the ASM treatments, and this subject will be dealt with in more detail in Chapter 9. However, such intense use is not necessary, since four applications of ASM controlled Xanthomonas leaf blight (*X. axonopodis* pv. *alli*) at least as well as $Cu(OH)_2$ or $Cu(OH)_2$ + mancozeb, without affecting yield. It was proposed that the integration of ASM and bacterial antagonists with $Cu(OH)_2$ may eliminate the need for maneb and mancozeb in this pathosystem.

Copper-based bactericides are a standard control measure for citrus canker (*X. axonopodis* pv. *citri*), but there are concerns over the development of resistance in Xanthomonad populations (Rinaldi & Leite-Junior, 2000) and Cu accumulation in soils (Alva *et al.*, 1995). ASM and harpin have each been shown to demonstrate efficacy against fungal and bacterial diseases of citrus in glasshouse trials (Agostini *et al.*, 2003; Graham & Leite-Junior, 2004). The potential of using these compounds to augment the activity of copper oxychloride or copper hydroxide against citrus canker was evaluated in orchards trials. The copper-based formulations were highly effective when used alone, and neither ASM nor harpin enhanced their efficacy against citrus canker (Graham & Leite-Junior, 2004). However, there was an indication that the inclusion of plant activators in a spray programme may provide a means for reducing the number of applications of copper-based products without compromising disease control. This has potential environmental benefits with regard to reducing copper accumulation in soils.

ASM has been shown to induce resistance against fire blight (*Erwinia amylovora*) in apple (Brisset *et al.*, 2000; Maxson-Stein *et al.*, 2002) and has potential to augment the activity of bactericides currently used to control this disease. Weekly ASM application reduced fireblight incidence in a dose-dependent manner but was not as effective as streptomycin (Maxson-Stein *et al.*, 2002). However, the combined use of ASM and streptomycin was more effective than either treatment alone. An integrated approach was recommended where weekly ASM application was supplemented with streptomycin at critical times during bloom. It was proposed that the resulting reduction in streptomycin use, during the post-bloom period, would decrease the risk of the development of streptomycin-resistant strains of the pathogen.

11.4.3 Biological control agents

Many non-pathogenic saprophytes suppress the growth of plant pathogens through competition for nutrients, the production of inhibitory metabolites, and/or parasitism thereby naturally limiting the spread of plant disease in the environment (Elad, 2000; Hanson & Howell, 2004; Howell, 2003). While diverse microbes may contribute in this way to the biological control of plant pathogens, most research and development efforts have focused on isolates of three genera, Trichoderma, Bacillus and Pseudomonas. These biological control agents (BCAs) have been identified as another important component of ICM strategies, particularly for the control of soil-borne pathogens where conventional fungicides and plant activators have proven to be less effective. Some *Bacillus* spp. and *Pseudomonas* spp. are also referred to as plant growth-promoting rhizobacteria (PGPR) because of their intimate association with improved plant growth and health (Kloepper, 1993; Zehnder *et al.*, 2001; Chapter 8). Many BCAs and PGPR have also been reported to activate plant resistance against both soil-borne and airborne pathogens. Soluble chemicals produced by some PGPR, such as SA, as well as structural components of the micro-organism itself, such as membrane lipopolysaccharides, appear to play important roles in the induction of plant defences (Leeman *et al.*, 1995; De Meyer & Hofte, 1997; Bakker *et al.*, 2003).

Kloepper *et al.* (2004) recently suggested that the practical implementation of PGPR in agriculture and horticulture required adaptive research to identify ways of overcoming the innate variability of microbial-based disease management. It was suggested that more consistent and more effective disease control could be achieved by using microbial mixtures that include PGPR strains with different modes of action. Combinations of *Bacillus* spp. (Jetiyanon & Kloepper, 2002) and *Pseudomonas* spp. (De Boer *et al.*, 2003), containing antagonistic strains, and strains that activate plant resistance, have been shown to provide disease control superior to individual strains in a range of crops. Combined application of *Bacillus cereus* and chlorothalonil was effective in reducing disease severity in field tomatoes caused by *A. solani*, *P. infestans* and *Septoria lycopersici* (Silva *et al.*, 2004). Seed treatment with *Bacillus cereus* permitted a reduction in the frequency of fungicide applications, without any loss in efficacy, and afforded an increase in crop yield.

PGPR-induced resistance and chemically induced resistance, though phenotypically similar, have different regulatory pathways and therefore may be able to provide greater, more reliable and broader spectrum disease control when combined than when used individually (Van Wees *et al.*, 2000). A laboratory study by Chen *et al.* (1996) was one of the first attempts to evaluate interactions between chemically induced resistance and microbial biocontrol. Tobacco seedlings were treated with plant activators (DCINA or SA) and an antagonistic BCA (*B. cereus*), before inoculation with *Pythium torulosum*, *Pythium aphanidermatum* and *Phytophthora parasitica.* The combination treatments operated additively to provide greater suppression of damping off in tobacco seedlings than when either was used alone. The efficacy of antagonistic fluorescent *Pseudomonas* strains against damping off in cucumber was enhanced when the BCA was applied in combination with BABA as a soil drench (Vogt & Buchenauer, 1997). A slight additive effect was observed when using the BCA + BABA to control powdery mildew, but the control provided by the combination treatment was not significantly different from that achieved with BABA alone. The results obtained indicated that the combination treatment not only enhanced the level of disease control but also reduced the level of variability across the 10 independent experiments. Chitosan combines the ability to operate as a plant activator (Wilson *et al.*, 1994; El Ghaouth, 1997; Reglinski *et al.*, 2004) with direct antimicrobial activity (Ben Shalom *et al.*, 2003). The combined use of chitosan with an endophytic bacterium (*Bacillus pumilus*) has been shown to provide effective control of *Fusarium* wilt in tomato (Benhamou *et al.*, 1998).

In glasshouse studies, ASM and two PGPR-based products, BioYield™ flowable and Equity™ (Naturize Inc., Jacksonville, FL), were evaluated for their potential to control bacterial wilt (*Ralstonia solanacearum*) on tomatoes (Anith *et al.*, 2004). BioYield™ contains *Bacillus subtilis* Gb03 and *Bacillus amyloliquefaciens* IN937a, and Equity™ contains over 40 different microbial strains. ASM was initially applied as a foliar spray 14 days after germination. A second foliar application as well as a soil drench was applied five days before inoculation with *R. solanacearum*. PGPR treatments were applied twice, once as a seed treatment and then again as a soil drench seven days before inoculation. Combination treatments significantly reduced bacterial wilt incidence whereas, when used alone, the individual components were not effective. It was proposed that the suppression of pathogen inoculum by the PGPR reduced inoculum levels below a threshold where ASM could be effective.

Studies on grapes have demonstrated the benefits of combining compatible biocontrol treatments that occupy different environmental niches and/or have different modes of action. Plant activators Milsana® and Myco-Sin® (Schaette GmbH, Germany) and a bacterial antagonist (*Brevibacillus brevis*) were tested in combination for their ability to control powdery mildew, downy mildew and Botrytis in grapevines (Schmitt *et al.*, 2002). Application of the biocontrol treatments at 10-day intervals throughout the season reduced Botrytis incidence on grape berries to 29.8%, compared with 89.7% incidence on control plots. Disease control was comparable to that obtained using wettable sulfur and the copper-containing agent FW 450 (Dow AgroSciences, Indianapolis, IN). On Chardonnay grapevines, the activator 5-chlorosalicylic acid (5CSA) and the fungal antagonist *Ulocladium oudemansii* were used in combination to control Botrytis bunch rot (Reglinski *et al.*, 2005). Harvested bunches were incubated in high humidity chambers to encourage further Botrytis development. After 14 days incubation, Botrytis severity increased to 83% on untreated bunches compared with 37–41% on those treated with either 5CSA or *U. oudemansii*. Under these conditions, the combined use of 5CSA + *U. oudemansii* provided a significantly better control than each component alone and reduced bunch rot severity to less than 21%. More recent data indicate that the timing and targeting of each component are critical factors for optimizing efficacy (T. Reglinski & P. Elmer, unpublished results).

11.4.4 Cultural practice

Cultural practices (e.g. fertilization) that favour the development of strong plants will complement induced resistance, since general plant health will affect the ability of the plant to respond to activator treatments. This can range from accentuating the induced response to minimizing costs of induction, for example, the use of $Ca(NO_3)_2$ top dressing to overcome phytotoxicity in ASM-treated tomato (Cole *et al.*, 1999). As described earlier, nutrient balance in the soil and the use of particular fertilizers and organic amendments can have significant impacts on the induced resistance response (e.g. Stout *et al.*, 1998; Wiese *et al.*, 2003; Dietrich *et al.*, 2004). Moshe Reuveni & co-workers have published extensively on the induction of disease resistance by foliar application of phosphate fertilizers. Furthermore, it has been proposed that phosphate salts have the potential to contribute to an integrated disease-management programme, either in rotation with fungicides or in a tank mix with reduced rates of fungicide (Reuveni *et al.*, 1998a, b).

The benefits of using composts to maintain soil fertility and plant health have been known for centuries. Incorporation of various composted materials into soil and soil-less potting mixes has proven particularly effective for the suppression of soil-borne diseases (Hoitink *et al.*, 1997; Hoitink & Boehm, 1999). Furthermore, water-soluble compost extracts have shown potential to enhance plant health when applied as foliar sprays or drenches (Scheuerell & Mahaffee, 2002, 2004). Pythium and Phytophthora root rots are suppressed by the activity of a broad range of micro-organisms that naturally colonize composts (You & Sivasithamparam, 1995; Erhart *et al.*, 1999). Suppression of Rhizoctonia damping off is more variable and is dependent upon the presence of a narrower range of microbial antagonists (Scheuerell *et al.*, 2005). There are very few reports demonstrating the ability of composts to suppress foliar pathogens. Systemic disease resistance was induced in plants when they were grown in composts that were supplemented with specific rhizobacteria (Zhang *et al.*, 1998) and fungi (Pharand *et al.*, 2002; Krause *et al.*, 2003; Khan *et al.*, 2004). Disease suppression was associated with the enhancement of cytological (Pharand *et al.*,

2002) and biochemical (Zhang *et al.*, 1998) defences, but only following pathogen inoculation, suggesting that these composts potentiated the host resistance response. Direct elicitation of PR-gene expression by composts has been observed in leaves of Arabidopsis (Vallad *et al.*, 2003) and tomato (Kavroulakis *et al.*, 2005). In both cases, the plants also exhibited systemic resistance to foliar pathogens. The above examples demonstrate the potential for compost-amended media to activate systemic resistance in plants. However, the use of composts for this purpose is at a very early stage in development, and efficacy needs to be demonstrated across a range of crops before this could be considered as a viable option for disease control. The identification of specific biological, chemical and physical factors within composts that stimulate host resistance may ultimately lead to the development of tailor-made products for commercial crop production systems.

Other cultural practices and crop husbandry, such as pruning, are carried out to reduce the severity of fungal infection by removal of potential inoculum sources and by altering the canopy microclimate. However, leaf removal in grapevines dramatically reduced the formation of phenolic compounds in grape floral tissues (Keller *et al.*, 2000). This effect was possibly due to a reduction in the supply of photosynthate to the inflorescences. This is important with respect to Botrytis infection, since a low level of phenolic production during the critical bloom period may decrease overall resistance and increase primary infections of grapes. Preliminary studies in mango have shown that retention of sap by maintaining long (3 cm) peduncles at harvest can reduce the severity of post-harvest anthracnose disease (K. Hassan *et al.*, University of Queensland, Australia, unpublished results). It is likely that antifungal compounds present in sap, e.g. alkenylresorcinols, contribute to this observed decrease in disease. A greater understanding of the impact of such practices on activator-induced resistance is required.

11.5 Integration of plant activators in crop management

The vacuum created by the restricted use of traditional chemicals to satisfy regulatory requirements (e.g. EUREPGAP) and customer demands for more ecologically sensitive methods of food production will be the main drivers for the implementation of induced resistance in crop production. Plant activators have already shown potential to augment fungicide and bactericide activity, either as part of strategies against evolution of resistance to chemical control agents or simply as a means for reducing chemical inputs. However, given the fundamental differences in their modes of action, the implementation of activators as direct replacements for conventional chemicals will not always be appropriate, and factors affecting activator efficacy should be considered (Figure 11.1). Traditional chemicals generally directly target the pathogen and operate independently of the plant or the environment. Conversely, plant activators depend upon a rapid and intense plant response, and this in turn is affected by plant phenology and local environmental conditions. Therefore, the successful implementation of induced resistance in crop production demands a change in mindset that involves taking greater consideration of the influence of plant phenology and environmental conditions on activator efficacy. In addition, it is also important to consider the potential for additive and even synergistic activity between plant activators and other biologically based approaches. For example, the uses of microbial antagonists to suppress pathogen populations and activators to elevate resistance operate in different niches, but each contributes to reduced disease risk. These issues are addressed schematically in Figure 11.2.

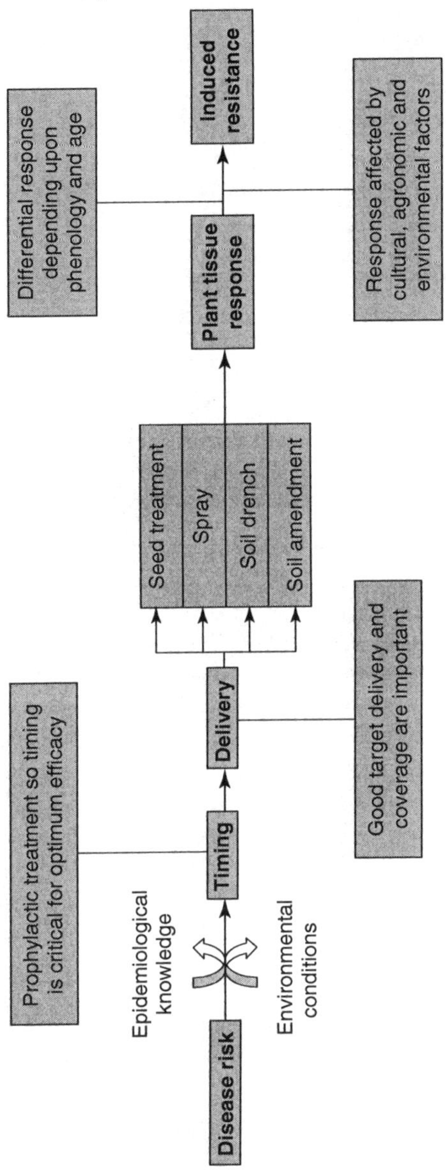

Figure 11.1 Factors influencing the efficacy of induced resistance.

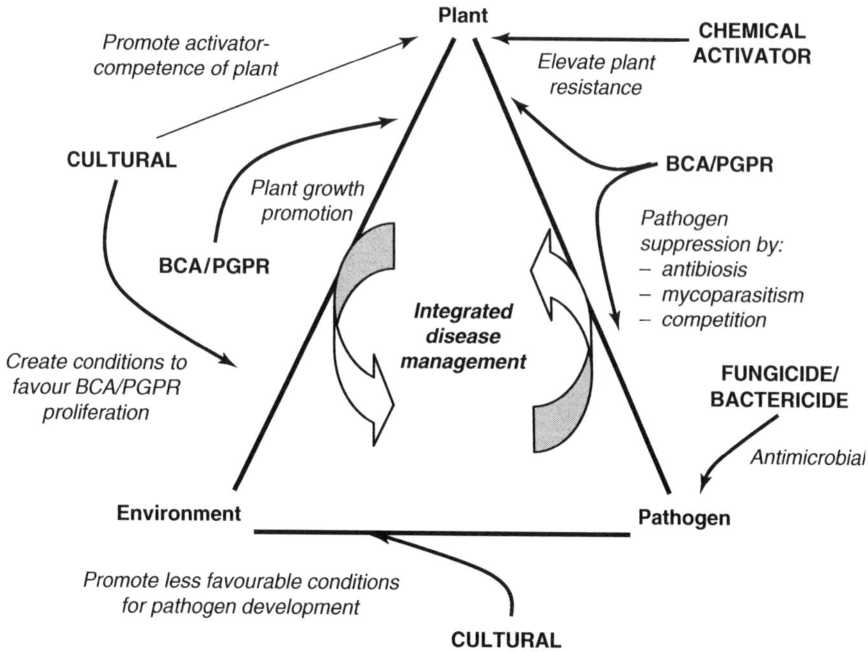

Figure 11.2 Interactions between chemical and microbial activators with other disease control measures in an integrated disease management system.

An altogether more holistic approach, akin to the philosophy underpinning organic production, is required in order to optimize induced resistance. In this section, we consider the opportunities for integrating induced resistance into crop production systems.

Induced resistance is likely to be most easily implemented in controlled production conditions, such as glasshouses, where environmental and cultural variables described earlier can be minimized. In nurseries and transplant production systems, there is opportunity to combine the use of microbial activators, as seed treatments or soil amendments, with aerial sprays of chemical activators to maximize the potential for induced resistance. These approaches, alongside the use of microbial antagonists and cultural control measures such as good hygiene, environmental manipulation to minimize leaf wetness and inclusion of composts, can provide an alternative to the frequent fungicide use that is common in nurseries. The use of microbial activators and BCAs has shown potential to be more effective than traditional chemicals against soil-borne disease. PGPR have been shown to enhance the rate and amount of seedling emergence and also to stimulate seedling growth in various crops (Chanway, 1997; Kloepper *et al.*, 2004; Chapter 8). This is of particular interest for nurseries where uneven emergence of seedlings necessitates over-sowing to meet production targets. Some PGPR and BCAs are marketed as growth promoters, even though they may well induce disease resistance or be antagonistic, simply because of the difficulty of demonstrating a causal relationship. The fortification of composts by the addition of specific PGPR and BCAs may enhance their disease-suppressive activity. These approaches are particularly applicable in nurseries because they are relatively simple to apply and should minimize the need for pesticide sprays.

Diseases can be a limiting factor to successful transplant production from the nursery; moreover, diseased transplants can serve as important sources of pathogen introduction into the field and community. Management of these diseases during transplant production commonly requires frequent chemical application, and the development of chemical-resistance within pathogen populations is a growing concern. The potential for activators applied in the nursery to protect transplants may be related to the longevity of the induced response. This may differ between monocotyledonous and dicotyledonous plants, since it has been reported that ASM-induced responses may be more persistent in monocots than dicots (Tally *et al.*, 1999). In operational terms, this may require activator application as seed treatments in nurseries followed by root dips before out-planting. However, microbial activators may be capable of more sustained activity than one-off chemical activator treatments and may not require repeat application. Diane Cuppels and co-workers (Canada) are evaluating the use of the biocontrol agents Mycostop®, Actinovate® and BlightBan®, in combination with chemical activators, to protect tomato and pepper seedlings and transplants against bacterial spot, anthracnose and early blight. PGPR inoculants have been used to enhance the health and survival of conifer seedlings in reforestation sites, and this approach could provide a cost-effective tool for enhancing plantation establishment (Reddy *et al.*, 1997; Chanway *et al.*, 2000). Growth promotion of spruce seedlings extended through the second growing season after out-planting (Chanway *et al.*, 2000). However, PGPR efficacy was site-specific, and there was an indication that the endophytic potential of the strains was an important contributory factor. The authors proposed that it might be necessary to match PGPR strain to out-planting site to optimize efficacy.

In perennial crops, orchards, vineyards and forestry plantations, induced resistance is more likely to be implemented by the application of chemical activators to foliage, flowers and fruit. An example of how induced resistance may be implemented in mango production is presented in Figure 11.3. For annual field crops, seed dressing with a chemical or biological activator may be a practical method of specifically targeting root-infecting fungal pathogens. A recent study in cotton has shown that ASM applied to seed prior to sowing in the field, or spraying into the furrow on top of the seeds at sowing, could reduce the severity of black root rot disease in seedlings raised in soils naturally infested with *Thielaviopsis basicola* (Mondal *et al.*, 2005). Induced resistance is achieved by temporal activation of host defences, so timing and delivery are critical to enable resistance elevation to coincide with periods of greatest disease risk during the growing season. Identification of potential risk periods will be assisted by more accurate disease forecasting used in conjunction with disease prediction models. Application timing should also consider plant physiological status since many defence mechanisms are developmentally regulated (Vavrina *et al.*, 2004; Whalen, 2005). The use of activators closer to harvest, or even as post-harvest treatments to climacteric fruit, may be particularly attractive in order to reduce fungicide and bactericide residues in fresh produce, assuming of course that the activators themselves leave no residues. However, ontogenetic constraints may limit the role of induced resistance in some crops as it has been suggested that inducible resistance will become less effective as tissues mature and approach senescence (Kessmann *et al.*, 1994). In grapes, the kinetics of berry ripening varied between berries on a cluster, between clusters on a vine and between different vines (Barnavon *et al.*, 2001). Thus, the ability of the grape berries to activate an effective induced response may show a similar variation.

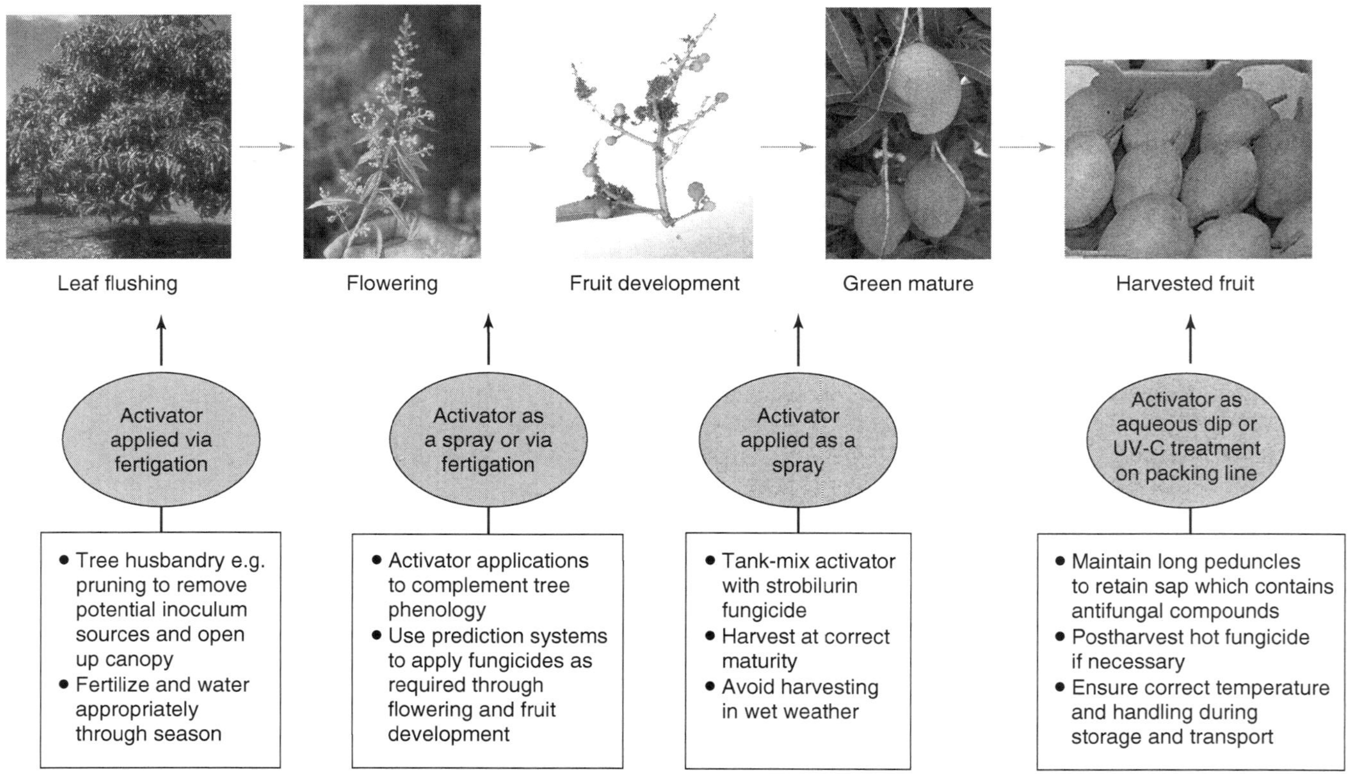

Figure 11.3 Theoretical example of how activators of induced resistance (text ovals) could be combined with existing mango production practices (text rectangles) through the season to provide effective protection against field and post-harvest diseases (Photos: T Cooke, I Bally & K Hassan).

11.6 Knowledge gaps

The previous sections of this chapter have established that induced resistance has a place in well designed and structured ICM programmes. Its potential for utilization in industries other than annual horticultural and broad acre cropping is expanding. Promising early trials demonstrate the applicability of induced resistance in the seedling, nursery and forestry industries and in post-harvest horticulture. New activators and novel methods for their application will continue to become available. Nevertheless, there are still many gaps in our knowledge which should be addressed by further directed research so that maximum benefit can be derived from the practical implementation of induced resistance into existing programmes.

The variable efficacy of activators is a major concern and is no doubt limiting the commercial adoption of induced resistance. The environmental and cultural factors affecting activator performance and plant resistance processes need to be determined. For example, how are plant defence mechanisms and induced resistance affected by normal seasonal climate changes and extremes of temperature, water, soil type including organic matter and nutrient/mineral contents, fertilization through the season, stages of plant development and plant ageing and cultivar selection, etc.? What is the best growth stage(s) for activator application? Are multiple applications necessary, and how are activators best applied? It is likely that optimal conditions and practices developed for one crop will not be transferable to another, so evaluation will need to be on a case-by-case basis. The choice of plant systems for experimentation and implementation will need to be guided by the importance of diseases, the availability and suitability of other control systems and economic assessments of inputs and derived benefits.

Additional knowledge of activators, particularly chemical activators such as ASM, would assist decisions on their practical use. For example, very little is known about the uptake and speed of movement of chemical activators, the sites to which they move and activate defences, and their breakdown and persistence. Similar sorts of issues need to be resolved for biological activators, such as PGPR. Of course, an activator must not be phytotoxic, or exert an overall 'fitness cost' to the extent where yields of commercial products are negatively impacted. There also need to be assurances that heightened defences, for example secondary metabolites where characterized, are not present in tissues at levels that may be harmful to the environment or to consumers.

The requirement for more detailed investigations into the underlying mechanisms associated with induced resistance is discussed in more detail elsewhere in this book. Advances in genomics and proteomics can also facilitate more efficient use of induced resistance in crop production. Such investigations may identify appropriate biochemical or molecular 'markers' of induced resistance or 'competency' to respond to activation, which may assist field implementation. For example, the development of biochemical or molecular techniques to measure 'plant resistance status' or competence for activation of resistance would provide a tool to allow more efficient use of activators with regards to timing and frequency of application. Evidence of activator–cultivar interactions (Reglinski *et al.*, 1994; Romero *et al.*, 2001) indicates that certain cultivars may be more competent than others for induced resistance. Activator competence could become an optional selection criterion in breeding programmes. Unfortunately, many modern high yielding crop cultivars appear to lack much of the natural resistance of old cultivars or related wild species. It is possible that breeding for high yield and other desirable traits has failed to retain genes that are essential for effective resistance. The technical feasibility of engineering broad spectrum and stable disease resistance is growing fast, and several transgenic plants exhibiting high levels of resistance to fungal and bacterial pathogens have been reported (Shah *et al.*, 1995; Vivier & Pretorious, 2002). However, constitutive expression of inducible resistance mechanisms has been accompanied by other undesirable phenotypic traits and fitness costs (Maleck *et al.*, 2002; Heidel *et al.*, 2004). The development of genetically modified (GM) crops with enhanced activator competence is perhaps a better option. However, public acceptance of GM is not widespread, and the introduction of engineered crops is more likely to be subject to political rather than scientific barriers.

There also needs to be more definitive studies conducted on the apparent durability of induced resistance, and on what selection effects there may be in populations of pathogens and saprophytic organisms in their natural environments. It has been suggested that plant activators would provide a durable method of disease control because they operate through the induction of multi-component plant defences and not directly against pathogens (e.g. Ruess *et al.*, 1996; Oostendorp *et al.*, 2001). However, Bousset & Pons-Kuhnemann (2003) reported that ASM exerted a selection pressure on the composition of barley mildew (*B. graminis* f. sp. *hordei*) populations when it was applied in combination with the fungicide ethirimol. This conclusion was based on the observation that pathogen population diversity was significantly lower in plots treated with ethirimol + ASM compared with that in plots treated with ethirimol alone. There was no difference in the rates of evolution of population diversity between ASM-treated and control plots, and this may suggest that, in the combination treatment, ASM induced an additional selection pressure among ethirimol-resistant isolates. These results were obtained in controlled laboratory conditions, and it was postulated that even greater variation in response to ASM may be expected among field populations where selection pressures are not constant. However, in contrast, it was recently reported that ASM-induced resistance to *X. axonopodis* pv. *vesicatoria* in Bell pepper was associated with a delay in the detection of race-change mutants and, where infection did occur, a reduction in disease severity (Romero & Ritchie, 2004). This suggests that plant activators may help to prolong the durability of major gene resistance, a very attractive proposition since genetic resistance is the preferred option for disease management. These results highlight the importance of understanding the interrelationships between plant activators, fungicides, varieties and pathogen populations in ICM systems.

11.7 Conclusions

Induced resistance is still a relatively new concept in disease management and, despite years of research, remains an under-utilized resource in disease management. It has been successfully adopted in many commercial production systems, and obviously has potential for integration into many more. It may be adopted most easily in situations where the variable efficacy discussed earlier in this chapter can be minimized, and where the use of activators can be integrated comfortably in ICM procedures rather than used as a stand-alone approach for disease control. Implementation of induced resistance will be advanced by co-ordination of considerably more fundamental and applied research.

The practical adoption of induced resistance in many countries will be largely driven by the withdrawal of traditional pesticides from sale and governed by product (activator) availability and choice, and also a demonstrated financial and/or environmental benefit.

The registration process and patenting in many countries are extensive and represent a significant economic challenge for the development of new activators. For this reason, it is likely that some potent plant activators, based on research trials, will never be exploited for practical application.

Good marketing and grower education are critical to the success of resistance activators. Growers must be able to make informed decisions about activator use. It is important that users understand the mode of action of activators, particularly if they are chemical formulations, that is, they target plant defences and have no direct effect on the pathogens. Users must also be aware of the limitations of induced resistance, for example the lack of eradicant activity, and that it needs time to work prior to pathogen infection taking place. Confidence in resistance activators will take a considerable time to develop, and may be easily destroyed by uninformed use. Therefore, within conventional crop production systems the transition from near total dependence on pesticides to ICM should be gradual. Timely transfer of research results to growers will allow more rapid implementation of practices that will enhance the control of important diseases, while improving the profitability of the industry.

11.8 References

Agostini JP, Bushong PM, Timmer LW, 2003. Greenhouse evaluation of products that induce host resistance for control of scab, melanose and Alternaria brown spot of citrus. *Plant Disease* **87,** 69–74.

Ali Z, Smith I, Guest DI, 2000. Combinations of potassium phosphonate and Bion (acibenzolar-*S*-methyl) reduce root infection and dieback of *Pinus radiata, Banksia intergrifolia* and *Isopogon cuneatus* caused by *Phytophthora cinnamomi. Australasian Plant Pathology* **29,** 59–63.

Alva AK, Graham JH, Anderson CA, 1995. Soil pH and copper effects on young 'Hamlin' orange trees. *Soil Science Society of America Journal* **59,** 481–487.

Anith KN, Momol MT, Kloepper JW, Marois JJ, Olson SM, Jones JB, 2004. Efficacy of plant growth-promoting rhizobacteria, acibenzolar-*S*-methyl, and soil amendment for integrated management of bacterial wilt on tomato. *Plant Disease* **88,** 669–673.

Baider A, Cohen Y, 2003. Synergistic interaction between BABA and mancozeb in controlling *Phytophthora infestans* in potato and tomato and *Pseudoperonospora cubensis* in cucumber. *Phytoparasitica* **31,** 399–409.

Bakker PAHM, Ran LX, Pieterse CMJ, van Loon LC, 2003. Understanding the involvement of rhizobacteria-mediated induction of systemic resistance in biocontrol of plant diseases. *Canadian Journal of Plant Pathology* **25,** 5–9.

Barnavon L, Doco T, Terrier N, Ageorges A, Romieu C, Pellerin P, 2001. Involvement of pectin methyl-esterase during the ripening of grape berries; partial cDNA isolation, transcript expression and changes in the degree of methyl-esterification of cell wall pectins. *Phytochemistry* **58,** 693–701.

Benhamou N, Kloepper JW, Tuzun S, 1998. Induction of resistance against Fusarium wilt of tomato by combination of chitosan with an endophytic bacterial strain – ultrastructure and cytochemistry of the host response. *Planta* **204,** 153–168.

Ben Shalom N, Ardi R, Pinto R, Aki C, Fallik E, 2003. Controlling gray mould caused by *Botrytis cinerea* in cucumber plants by means of chitosan. *Crop Protection* **22,** 285–290.

Bokshi AI, Morris SC, Deverall BJ, 2003. Effects of benzothiadiazole and acetylsalicylic acid on *β*-1,3-glucanase activity and disease resistance in potato. *Plant Pathology* **52,** 22–27.

Bousset L, Pons-Kuhnemann J, 2003. Effects of acibenzolar-S-methyl and ethirimol on the composition of a laboratory population of barley powdery mildew. *Phytopathology* **93,** 305–315.

Brisset M-N, Cesbron S, Thomson SV, Paulin J-P, 2000. Acibenzolar-*S*-methyl induces the accumulation of defense-related enzymes in apple and protects from fire blight. *European Journal of Plant Pathology* **106,** 529–536.

Buonaurio R, Scarponi L, Ferrara M, Sidoti P, Bertona A, 2002. Induction of systemic acquired resistance in pepper plants by acibenzolar-*S*-methyl against bacterial spot disease. *European Journal of Plant Pathology* **108,** 41–49.

Carlen C, Faby R, Karjalainen R, Pommier JJ, Steffek R, 2004. Control of air-borne diseases in strawberries with natural and synthetic elicitors. *Acta Horticulturae* **649,** 237–240.

Chanway CP, 1997. Inoculation of tree roots with plant growth promoting soil bacteria: An emerging technology for reforestation. *Forest Science* **43,** 99–112.

Chanway CP, Shishido M, Nairn J, Jungwirth S, Markham J, Xiao G, Holl FB, 2000. Endophytic colonization and field responses of hybrid spruce seedlings after inoculation with plant growth-promoting rhizobacteria. *Forest Ecology and Management* **133,** 81–88.

Chen J, Jacobson LM, Handelsman J, Goodman RM, 1996. Compatibility of systemic acquired resistance and microbial biocontrol for suppression of plant disease in a laboratory assay. *Molecular Ecology* **5,** 73–80.

Coates LM, Muirhead IF, Irwin JAG, Gowanlock DH, 1993. Initial infection processes by *Colletotrichum gloeosporioides* on avocado fruit. *Mycological Research* **97,** 1363–1370.

Cohen YR, 2002. ß-Aminobutyric acid-induced resistance against plant pathogens. *Plant Disease* **86,** 448–457.

Cole DL, Campbell PA, Latorre BA, 1999. The efficacy of acibenzolar-*S*-methyl, an inducer of systemic acquired resistance, against bacterial and fungal diseases of tobacco. *Crop Protection* **18,** 267–273.

da Rocha AB, Hammerschmidt R, 2005. History and perspectives on the use of disease resistance inducers in horticultural crops. *HortTechnology* **15,** 518–529.

Daayf F, Schmitt A, Belanger RR, 1995. The effects of plant extracts of *Reynoutria sachalinensis* on powdery mildew development and leaf physiology of long English cucumber. *Plant Disease* **79,** 577–580.

Dann EK, Diers B, Byrum J, Hammerschmidt R, 1998. Effect of treating soybean with 2,6-dichloroisonicotinic acid (INA) and benzothiadiazole (BTH) on seed yields and the level of disease caused by *Sclerotinia sclerotiorum* in field and greenhouse studies. *European Journal of Plant Pathology* **104,** 271–278.

De Boer M, Bom P, Kindt F, Keurentjes JJB, Van der Sluis I, Van Loon LC, Bakker PAHM, 2003. Control of Fusarium wilt of radish by combining *Pseudomonas putida* strains that have different disease-suppressive mechanisms. *Phytopathology* **93,** 626–632.

de Capdeville G, Beer SV, Watkins CB, Wilson CL, Tedeschi LO, Aist JR, 2003. Pre- and post-harvest harpin treatments of apples induce resistance to blue mold. *Plant Disease* **87,** 39–44.

De Meyer G, Hofte M, 1997. Salicylic acid produced by the rhizobacterium *Pseudomonas aeruginosa* 7NSK2 induces resistance to leaf infection by *Botrytis cinerea* on bean. *Phytopathology* **87,** 588–593.

Dietrich R, Ploss K, Heil M, 2004. Constitutive and induced resistance to pathogens in *Arabidopsis thaliana* depends on nitrogen supply. *Plant, Cell and Environment* **27,** 896–906.

Dietrich R, Ploss K, Heil M, 2005. Growth responses and fitness costs after induction of pathogen resistance depend on environmental conditions. *Plant, Cell and Environment* **28,** 211–222.

Elad Y, 2000. Biological control of foliar pathogens by means of *Trichoderma harzianum* and potential modes of action. *Crop Protection* **19,** 709–714.

El Ghaouth A, 1997. Biologically-based alternatives to synthetic fungicides for the control of postharvest diseases. *Journal of Industrial Microbiology & Biotechnology* **19,** 160–162.

Erhart E, Burian K, Hartl W, Stich K, 1999. Suppression of *Pythium ultimum* by biowaste composts in relation to compost microbial biomass, activity and content of phenolic compounds. *Journal of Phytopathology* **47**, 299–305.

Falkhof AG, Dehne HW, Schonbeck F, 1988. Dependence of the effectiveness of induced resistance on environmental conditions. *Journal of Phytopathology* **123,** 311–321.

Fofana B, McNally DJ, Labbe C, Boulanger R, Benhamou N, Seguin A, Belanger RR, 2002. Milsana-induced resistance in powdery mildew-infected cucumber plants correlates with the induction of chalcone synthase and chalcone isomerase. *Physiological & Molecular Plant Pathology* **61,** 121–132.

Genoud T, Buchala AJ, Chua NH, Métraux J-P, 2002. Phytochrome signalling modulates the SA-perceptive pathway in Arabidopsis. *The Plant Journal* **31,** 87–95.

Gent DH, Schwartz HF, 2005. Management of Xanthomonas leaf blight of onions with a plant activator, biological control agents, and copper bactericides. *Plant Disease* **89,** 631–639.

Graham JH, Leite-Junior RP, 2004. Lack of control of citrus canker by induced systemic resistance compounds. *Plant Disease* **88,** 745–750.

Guest DI, 1984. Modification of defence responses in tobacco and capsicum following treatment with Fosetyl-Al [Aluminium tris (*o*-ethyl phosphonate)]. *Physiological Plant Pathology* **25,** 125–134.

Gullino ML, Leroux P, Smith CM, 2000. Uses and challenges of novel compounds for plant disease control. *Crop Protection* **19,** 1–11.

Hanson LE, Howell CR, 2004. Elicitors of plant disease responses from biocontrol strains of *Trichoderma virens*. *Phytopathology* **94,** 171–176.

Heidel AJ, Clarke JD, Antonovics J, Dong X, 2004. Fitness costs of mutations affecting the systemic acquired resistance pathway in *Arabidopsis thaliana*. *Genetics* **168,** 2197–2206.

Hoitink HAJ, Stone AG, Han DY, 1997. Suppression of plant diseases by composts. *HortScience* **32,** 184–187.

Hoitink HAJ, Boehm MJ, 1999. Biocontrol within the context of soil microbial communities: A substrate-dependent phenomenon. *Annual Review of Phytopathology* **37**, 427–446.

Howell CR, 2003. Mechanisms employed by *Trichoderma* species in the biological control of plant diseases: The history and evolution of current concepts. *Plant Disease* **87,** 4–10.

Huang Y, Deverall BJ, Tang WH, Wang W, Wu FW, 2000. Foliar application of acibenzolar-*S*-methyl and protection of postharvest rock melons and Hami melons from disease. *European Journal of Plant Pathology* **106**, 651–656.

Hubbes M, 2004. Induced resistance for the control of Dutch elm disease. *Investigacion Agraria, Sistemas y Recursos Forestales* **13,** 185–196.

Hubbes M, Jeng RS, 1981. Aggressiveness of *Ceratocystis ulmi* strains and induction of resistance in *Ulmus americana*. *European Journal of Forest Pathology* **11**, 257–264.

Ishii H, Tomita Y, Horio T, Narusaka Y, Nakazawa Y, Nishimura K, Iwamoto S, 1999. Induced resistance of acibenzolar-*S*-methyl (CGA 245704) to cucumber and Japanese pear diseases. *European Journal of Plant Pathology* **105,** 77–85.

Iwata M, 2001. Probenazole – A plant defence activator. *Pesticide Outlook* **12,** 28–31.

Jakab G, Cottier V, Toquin V, Rigoli G, Zimmerli L, Metraux J-P, Mauch-Mani B, 2001. β-aminobutyric acid-induced resistance in plants. *European Journal of Plant Pathology* **107,** 29–37.

Jetiyanon K, Kloepper JW, 2002. Mixtures of plant growth-promoting rhizobacteria for induction of systemic resistance against multiple plant diseases. *Biological Control* **24,** 285–291.

Jorgensen LN, Jensen B, Smedegaard-Petersen V, 1997. Bion – A compound for disease control in cereal based on induced resistance. *Proceedings of 14th Danish Plant Protection Conference* **8,** 35–48.

Kavroulakis N, Ehaliotis C, Ntougias S, Zervakis GI, Papadopoulou KK, 2005. Local and systemic resistance against fungal pathogens of tomato plants elicited by a compost derived from agricultural residues. *Physiological & Molecular Plant Pathology* **66**, 163–174.

Keller M, Steel CC, Creasy GL, Bodson M, Verhoyen MNJ, 2000. Stilbene accumulation in grapevine tissues: Developmental and environmental effects. *Acta-Horticulturae* **514,** 275–286.

Kessmann H, Staub T, Hofmann C, Maetzke T, Herzog J, Ward E, Uknes S, Ryals J, 1994. Induction of systemic aquired disease resistance in plants by chemicals. *Annual Review of Phytopathology* **32,** 439–459.

Khan J, Ooka JJ, Miller SA, Madden LV, Hoitink HAJ, 2004. Systemic resistance induced by *Trichoderma hamatum* 382 in cucumber against Phytophthora crown rot and leaf blight. *Plant Disease* **88,** 280–286.

Kim YC, Blee KA, Robins J, Anderson AJ, 2001. OxycomTM under field and laboratory conditions increases resistance responses in plants. *European Journal of Plant Pathology* **107,** 129–136.

Kloepper JW, 1993. Plant growth-promoting rhizobacteria as biological control agents. In: Metting FB Jr, ed. *Soil Microbial Ecology – Applications in Agricultural and Environmental Management*. New York: Marcel Dekker, 255–274.

Kloepper JW, Ryu C, Zhang SA, 2004. Induced systemic resistance and promotion of plant growth by *Bacillus* spp. *Phytopathology* **94,** 1259–1266.

Kraeva E, Andary C, Carbonneau A, Deloire A, 1998. Salicylic acid treatment of grape berries retards ripening. *Vitis* **37**, 143–144.

Krause MS, De Ceuster TJJ, Tiquia SM, Michel FC, Jr., Madden LV, Hoitink HAJ, 2003. Isolation and characterization of rhizobacteria from composts that suppress the severity of bacterial leaf spot of radish. *Phytopathology* **93,** 1292–1300.

Leeman M, van Pelt JA, den Ouden FM, Heinsbroek M, Bakker P, Schippers B, 1995. Induction of systemic resistance against Fusarium wilt of radish by lipopolysaccharides of *Pseudomonas fluorescens*. *Phytopathology* **85,** 1021–1027.

Leroux P, Clerjeau M, 1985. Resistance of *Botrytis cinerea* Pers. and *Plasmopara viticola* (Berl. and de Toni) to fungicides in French vineyards. *Crop Protection* **4,** 137–160.

Leskovar DI, Kolenda K, 2002. Strobilurin+acibenzolar-*S*-methyl controls white rust without inducing leaf chlorosis in spinach. *Annals of Applied Biology* **140,** 171–175.

Liu H, Jiang W, Bi Y, Luo Y, 2005. Postharvest BTH treatment induces resistance of peach (*Prunus persica* L. cv. Jiubao) fruit to infection by *Penicillium expansum* and enhances activity of fruit defense mechanisms. *Postharvest Biology and Technology* **35,** 263–269.

Lopez AMQ, Lucas JA, 2002. Effects of plant defence activators on anthracnose disease of cashew. *European Journal of Plant Pathology* **108,** 409–420.

Louws FJ, Wilson M, Campbell HL, Cuppels DA, Jones JB, Shoemaker PB, Sahin F, Miller SA, 2001. Field control of bacterial spot and bacterial speck of tomato using a plant activator. *Plant Disease* **85,** 481–488.
Lyon GD, Newton AC, 1999. Implementation of elicitor-mediated induced resistance in agriculture. In: Agrawal AA, Tuzun S, Bent E, eds. *Induced Plant Defences Against Pathogens and Herbivores*, St. Paul, MN: APS Press, 299–318.
Maleck K, Neuenschwander U, Cade RM, Dietrich RA, Dangl JA, Ryals JA, 2002. Isolation and characterization of broad-spectrum disease resistant Arabidopsis mutants. *Genetics* **160,** 1661–1671.
Maxson-Stein K, He S-Y, Hammerschmidt R, Jones AL, 2002. Effect of treating apple trees with acibenzolar-*S*-methyl on fire blight and expression of pathogenesis-related protein genes. *Plant Disease* **86,** 785–790.
Meszka B, Bielenin A, 2004. Possibilities of integrated grey mould control on strawberry plantations in Poland. *Bulletin OILB/SROP* **27,** 41–45.
Midoh N, Iwata M, 1997. Expression of defense-related genes by probenazole or 1,2-benzisothiazole-3(2H)-one 1,1-dioxide. *Journal of Pesticide Science* **22,** 45–47.
Miles AK, Willingham SL, Cooke AW, 2004. Field evaluation of strobilurins and a plant activator for the control of citrus black spot. *Australasian Plant Pathology* **33,** 371–378.
Miles AK, Willingham SL, Cooke AW, 2005. Field evaluation of a plant activator, captan, chlorothalonil, copper hydroxide, iprodione, mancozeb and strobilurins for the control of citrus brown spot of mandarin. *Australasian Plant Pathology* **34,** 63–71.
Mondal AH, Nehl DB, Allen SJ, 2005. Acibenzolar-*S*-methyl induces systemic resistance in cotton against black root rot caused by *Thielaviopsis basicola. Australasian Plant Pathology* **34,** 499–507.
Muirhead IF, Deverall BJ, 1981. Role of appressoria in latent infection of banana fruits by *Colletotrichum musae*. *Physiological Plant Pathology* **19,** 77–84.
Obradovic A, Jones JB, Momol MT, Balogh B, Olson SM, 2004. Management of tomato bacterial spot in the field by foliar applications of bacteriophages and SAR inducers. *Plant Disease* **88,** 736–740.
Obradovic A, Jones JB, Momol MT, Olson SM, Jackson LE, Balogh B, Guven K, Iriarte FB, 2005. Integration of biological control agents and systemic acquired resistance inducers against bacterial spot on tomato. *Plant Disease* **89,** 712–716.
Oerke EC, Steiner U, Schönbeck F, 1989. The efficiency of induced resistance under practical culture conditions. V. Powdery mildew and grain yield of winter and spring barley in relation to nitrogen fertilization. *Journal of Plant Diseases and Protection* **96,** 140–153.
Oostendorp M, Kunz W, Dietrich B, Staub T, 2001. Induced disease resistance in plants by chemicals. *European Journal of Plant Pathology* **107,** 19–28.
Ortega F, Steiner U, Dehne HW, 1998. Induced resistance: A tool for fungicide resistance management. *Pesticide Science* **53,** 193–196.
Panter SN, Jones DA, 2002. Age-related resistance to plant pathogens. *Advances in Botanical Research* **38**, 251–280.
Pernezny K, Stoffella P, Collins J, Carroll A, Beaney A, 2002. Control of target spot of tomato with fungicides, systemic acquired resistance activators, and a biocontrol agent. *Plant Protection Science* **38,** 81–88.
Pharand B, Carisse O, Benhamou N, 2002. Cytological aspects of compost-mediated induced resistance against Fusarium crown and root rot in tomato. *Phytopathology* **92,** 424–438.
Pradhanang PM, Ji P, Momol MT, Olson SM, Mayfield JL, Jones JB, 2005. Application of acibenzolar-*S*-methyl enhances host resistance in tomato against *Ralstonia solanacearum*. *Plant Disease* **89,** 989–993.
Prusky D, 1996. Pathogen quiescence in postharvest diseases. *Annual Review of Phytopathology* **34,** 413–434.
Reddy MS, Funk LM, Covert DC, He DN, Pedersen EA, 1997. Microbial inoculants for sustainable forests. *Journal of Sustainable Forestry* **5,** 293–306.
Reglinski T, Elmer PAG, Taylor JT, Parry FJ, Marsden R, Wood PN, 2005. Suppression of Botrytis bunch rot in Chardonnay grapevines by induction of host resistance and fungal antagonism. *Australasian Plant Pathology* **34,** 481–488.
Reglinski T, Lyon, GD, Newton, AC, 1994. Assessment of the ability of yeast-derived resistance elicitors to control barley powdery mildew in the field. *Journal of Plant Disease and Protection* **101,** 1–10.
Reglinski T, Stavely FJL, Taylor JT, 1998. Induction of phenylalanine ammonia lyase activity and control of *Sphaeropsis sapinea* infection in *Pinus radiata* by 5-chlorosalicylic acid. *European Journal of Forest Pathology* **28,** 153–158.
Reglinski T, Taylor JT, Dick MA, 2004. Chitosan induces resistance to pitch canker in *Pinus radiata*. *New Zealand Journal of Forestry Science* **34,** 49–58.

Reuveni M, Harpaz M, Reuveni R, 1998a. Integrated control of powdery mildew on field-grown mango trees by mono-potassium phosphates, sterol inhibitor fungicides and the strobilurin Keroxim-methyl. *European Journal of Plant Pathology* **104,** 853–860.

Reuveni M, Oppenheim D, Reuveni R, 1998b. Integrated control of powdery mildew in apple trees by monopotassium phosphate fertilizer and sterol inhibitor fungicides. *Crop Protection* **17,** 563–568.

Reuveni M, Zahavi T, Cohen Y, 2001. Controlling downy mildew (*Plasmopara viticola*) in field-grown grapevine with beta-aminobutyric acid (BABA). *Phytoparasitica* **29,** 125–133.

Rinaldi DAMF, Leite-Junior RP, 2000. Adaptation of *Xanthomonas axonopodis* p.v. *citri* population to the presence of copper compounds in nature. *Proceedings of the International Society of Citriculture* **2,** 1064.

Romero AM, Kousik CS, Ritchie DF, 2001. Resistance to bacterial spot in bell pepper induced by acibenzolar-*S*-methyl. *Plant Disease* **85,** 188–194.

Romero AM, Ritchie DF, 2004. Systemic acquired resistance delays race shifts to major resistance genes in bell pepper. *Phytopathology* **94,** 1376–1382.

Ruess W, Mueller K, Knauf-Beiter G, Kunz W, Staub T, 1996. Plant activator CGA 245704: An innovative approach for disease control in cereals and tobacco. *Proceedings of the Brighton Crop Protection Conference 1996*, 53–60.

Rupe JC, Gbur JEE, 1995. Effects of plant age, maturity group, and environment on disease progress of sudden death syndrome of soybean. *Plant Disease* **79**, 139–143.

Scheuerell SJ, Mahaffee WF, 2002. Compost tea: Principles and prospects for plant disease control. *Compost Science & Utilisation* **10,** 313–338.

Scheuerell SJ, Mahaffee WF, 2004. Compost tea as a container medium drench for suppressing seedling damping-off caused by *Pythium ultimum. Phytopathology* **94,** 1156–1163.

Scheuerell SJ, Sullivan DM, Mahaffee WF, 2005. Suppression of seedling damping-off caused by *Pythium ultimum, P. irregulare*, and *Rhizoctonia solani* in container media amended with a diverse range of Pacific northwest compost sources. *Phytopathology* **95**, 306–315.

Schilder A, Gillet J, Sysak R, Wise J, 2002. Evaluation of environmentally friendly products for control of fungal diseases of grapes. In: *10th International Conference on Cultivation Technique and Phytopathological Problems in Organic Fruit-Growing and Viticulture*, Weinsberg, Germany, 163–167.

Schmitt A, Kunz S, Nandi S, Seddon B, Ernst A, 2002. Use of *Reynoutria sachalinensis* plant extracts, clay preparations and *Brevibacillus brevis* against fungal diseases of grape berries. In: *10th International Conference on Cultivation Technique and Phytopathological Problems in Organic Fruit-Growing and Viticulture*, Weinsberg, Germany, 146–151.

Shah DM, Rommens CMT, Beachy RN, 1995. Resistance to diseases and insects in transgenic plants: Progress and applications to agriculture. *Trends in Biotechnology* **13,** 362–368.

Shtienberg D, Elad Y, 2002. Is it possible to cope with the variability of biocontrol? *Bulletin IOBC/WPRS* **25,** 1–4.

Silva HSA, Romeiro RS, Carrer-Filho R, Pereira JLA, Mizubuti ESG, Mounteer A, 2004. Induction of systemic resistance by *Bacillus cereus* against tomato foliar diseases under field conditions. *Journal of Phytopathology* **152,** 371–375.

Stadnik MJ, Buchenauer H, 1999. Control of wheat diseases by a benzothiadiazole-derivative and modern fungicides. *Journal of Plant Disease and Protection* **106,** 466–475.

Steiner U, Oerke EC, Schonbeck F, 1988. The efficiency of induced resistance under practical conditions. IV. Powdery mildew and grain yield of winter barley cultivars with induced resistance and fungicide treatment. *Zeitschrift fur Pflanzenkrankheiten und Pflanzenschutz* **95,** 506–517.

Stewart A, 2001. Commercial biocontrol – reality or fantasy. *Australasian Plant Pathology* **30**, 127–131.

Stout MJ, Brovont RA, Duffey SS, 1998. Effect of nitrogen availability on expression of constitutive and inducible chemical defences in tomato, *Lycopersicon esculentum. Journal of Chemical Ecology* **24,** 945–963.

Tally A, Oostendorp M, Lawton K, Staub T, Bassi B, 1999. Commercial development of elicitors of induced resistance to pathogens. In: Agrawal AA, Tuzun S, Bent E, eds. *Induced Plant Defences Against Pathogens and Herbivores*, St. Paul, MN: APS Press, 357–369.

Terry LA, Joyce DC, 2004. Elicitors of induced disease resistance in postharvest horticultural crops: A brief review. *Postharvest Biology and Technology* **32,** 1–13.

Terry LA, Joyce DC, Adikaram NKB, Khambay BPS, 2004. Preformed antifungal compounds in strawberry fruit and flower tissues. *Postharvest Biology and Technology* **31,** 201–212.

Vallad GE, Cooperband L, Goodman RN, 2003. Plant foliar disease suppression mediated by composted forms of paper mill residuals exhibits molecular features of induced resistance. *Physiological & Molecular Plant Pathology* **63,** 65–77.

Vallad GE, Goodman RM, 2004. Systemic acquired resistance and induced systemic resistance in conventional agriculture. *Crop Science* **44,** 1920–1934.
Van Loon LC, Bakker PAHM, Pieterse CMJ, 1998. Systemic resistance induced by rhizosphere bacteria. *Annual Review of Phytopathology* **36,** 453–483.
Van Wees SCM, de Swart EAM, van Pelt JA, van Loon LC, Pieterse CMJ, 2000. Enhancement of induced disease resistance by simultaneous activation of salicylate- and jasmonate-dependent defense pathways in *Arabidopsis thaliana. Proceedings of the National Academy of Sciences USA* **97,** 8711–8716.
Vavrina CS, Roberts PD, Kokalis-Burelle N, 2004. Use of commercial systemic acquired resistance (SAR) inducers in the stand establishment of tomato; impact on plant growth, disease and nematode suppression. *Acta Horticulturae* **631,** 231–238.
Vivier MA, Pretorious IS, 2002. Genetically tailored grapevines for the wine industry. *Trends in Biotechnology* **20,** 472–478.
Vogt W, Buchenauer H, 1997. Enhancement of biological control by combination of antagonistic fluorescent *Pseudomonas* strains and resistance inducers against damping off and powdery mildew in cucumber. *Journal of Plant Disease and Protection* **104,** 272–280.
Watanabe T, 1977. Effects of probenazole (Oryzemate) on each stage of rice blast fungus (*Pyricularia oryzae Cavara*) in its life cycle. *Journal of Pesticide Science* **2,** 395–404.
Whalen MC, 2005. Host defence in a developmental context. *Molecular Plant Pathology* **6,** 347–360.
Wiese J, Bagy MMK, Schubert S, 2003. Soil properties, but not plant nutrients (N, P, K) interact with chemically induced resistance against powdery mildew in barley. *Journal of Plant Nutrition and Soil Science* **166,** 379–384.
Williams M, Senaratna T, Dixon K, Sivasithamparam K, 2003. Benzoic acid induces tolerence to biotic stress caused by *Phytophthora cinnamomi* in *Banksia attenuata. Plant Growth Regulation* **41,** 89–91.
Willingham SL, Pegg KG, Langdon PWB, Cooke AW, Peasley D, McLennan R, 2002. Combinations of strobilurin fungicides and acibenzolar (Bion) to reduce scab on passionfruit caused by *Cladosporium oxysporum. Australasian Plant Pathology* **31,** 333–336.
Wilson CL, El Ghaouth A, Chalutz E, Droby S, Stevens C, Lu JY, Khan V, Arul J, 1994. Potential of induced resistance to control postharvest diseases of fruits and vegetables. *Plant Disease* **78,** 837–844.
Yang D, Hubbes M, Jr., Jeng RS, Hubbes M, 1994. A glycoprotein isolated from culture filtrates of *Ophiostoma ulmi* as a mansonone-inducing elicitor on elm callus. *Mycological Research* **98,** 295–300.
Yoshioka K, Nakashita H, Klessig DF, Yamaguchi I, 2001. Probenazole induces systemic acquired resistance in Arabidopsis with a novel type of action. *The Plant Journal* **25,** 149–157.
You MP, Sivasithamparam K, 1995. Changes in microbial populations of an avocado plantation mulch suppressive to *Phytophthora cinnamomi. Applied Soil Ecology* **2,** 33–43.
Zainuri, Joyce D, Wearing AH, Coates L, Terry L, 2001. Effects of phosphonate and salicylic acid treatments on anthracnose disease development and ripening of 'Kensington Pride' mango fruit. *Australian Journal of Experimental Agriculture* **41,** 805–813.
Zehnder GW, Murphy JF, Sikora EJ, Kloepper JW, 2001. Application of rhizobacteria for induced resistance. *European Journal of Plant Pathology* **107,** 39–50.
Zeier J, 2005. Age-dependent variations of local and systemic defence responses in *Arabidopsis* leaves towards an avirulent strain of *Pseudomonas syringae. Physiological and Molecular Plant Pathology* **66,** 30–39.
Zeier J, Pink B, Mueller M, Berger S, 2004. Light conditions influence specific defence responses in incompatible plant–pathogen interactions: Uncoupling systemic resistance from salicylic acid and PR-1 accumulation. *Planta* **219,** 673–683.
Zhang S, Reddy MS, Kokalis-Burelle N, Wells LW, Nightengale SP, Kloepper JW, 2001. Lack of induced systemic resistance in peanut to late leaf spot disease by plant growth-promoting rhizobacteria and chemical elicitors. *Plant Disease* **85,** 879–884.
Zhang W, Han DY, Dick WA, Davis KR, Hoitink HAJ, 1998. Compost and compost water extract-induced systemic acquired resistance in cucumber and arabidopsis. *Phytopathology* **88,** 450–455.

Chapter 12
Exploitation of induced resistance: a commercial perspective

Andy Leadbeater and Theo Staub
Syngenta Crop Protection AG, Basel, Switzerland

12.1 Introduction

The crop protection industry is continually challenged to discover, develop and produce novel, more effective and safer disease control products. They help farmers to better prevent crop losses worldwide worth billions of dollars. Oercke *et al.* (1994) estimated losses due to plant diseases in the four major crops (rice, wheat, potato and maize) to be 10–16% of potential production, which translates into approximately US$64 billion in the years 1988–1990. The losses increased to US$84 billion if the top eight crops were included. This is not including the recent uprising of Asian rust of soybean, where losses are reported to be up to 80%, and the projected losses in the main growing regions of Brazil are 2.2 million tonnes ($487.3 million) (Yorinori *et al.*, 2005).

The common goal is the sustainable production of healthy crops and consistently good yields at high quality. The threats to food crops by plant pathogens are shifting constantly with the emergence of new diseases (like the recent upsurge of soybean rust in the Americas), in connection with new cropping systems or with the development of resistant pathogen strains that render present fungicides solutions ineffective. New disease control agents have to be effective and economical as well as safe for users, consumers and the environment. The Crop Protection Industry has a good record for introducing ever more effective products against fungal diseases with the discovery and introduction of such fungicide classes as the phenylamides, the DMIs and the strobilurins. It has been much less successful with the discovery of effective and safe products to control bacterial and viral pathogens.

It is widely known that the costs of discovering, developing and registering a new crop protection active ingredient are high and have increased in recent years. It is not sufficient simply to prove that an active substance is biologically effective and safe to crops: a large number of studies need to be carried out to ensure that products are safe to humans and the environment and do not cause any undesirable effects. The company concerned needs to carry out in-depth investigations into the science of the active substance (mode of action, uptake, translocation, metabolism, resistance risk, etc.) to ensure that the best product design and usage strategies can be implemented to maximize the agronomic benefits of the new compound. In addition, there are high costs related to formulation and production optimization.

A survey was carried out during 2002, on behalf of Crop Life America and the European Crop Protection Association (ECPA), of the leading global agrochemical companies to provide information on costs involved in the discovery, development and registration of a new conventional chemical crop protection active substance (McDougall, 2003). The results of this study showed that the overall costs for the discovery, development and registration of a new agrochemical product had risen during the period 1995–2000 from $152 million to $184 million. However, the lead time between the first synthesis of a new product and its commercialization increased from an average of 8.3 years to 9.1 years. Within the costs of product development in 2000, field trials were a very significant cost category, with a value of $25 million.

This survey confirms that the agrochemical industry invests a great deal of money each year in the field evaluation, development and management of new products to the market. Clearly, this level of investment can only be justified for markets large enough to ensure a return on investment to companies that is sufficient to justify the future investment into new compounds. To achieve a satisfactory long-term return on such an investment, sustainable use strategies to ensure their long-term effectiveness must be established at an early stage in the product life cycle.

If the key criteria of field performance, safety and economics are fulfilled, the new agent has a chance to be developed, and it is often of secondary importance how the candidate chemical works to achieve the protection. Therefore, in the past, the biochemical mode of action of new fungicide classes was often not known at the time the chemicals were selected for full development. In this chapter, we describe how industry dealt with the special case of systemically induced resistance where a biological concept was emerging, and products fitting that concept had to be found and developed. We include products that are introduced in practice and for which there is good evidence that they act through induced disease resistance. For earlier and more detailed reviews of the technical properties of chemical activators of plant defences against pathogens, see also Staub *et al.* (1992, 1997), Kessmann *et al.* (1994), Sticher *et al.* (1997), Tally *et al.* (1999) and Oostendorp *et al.* (2001).

12.2 Science and serendipitous discovery of resistance-inducing compounds

Traditionally, in the plant protection industry, new disease control agents have been discovered in mass screens designed to cover the spectrum of major diseases and including the temporal (protective, curative) and spatial (local vs. systemic) modes of action desired. It was usually only later that the biochemical mode of action of the most promising candidates was determined. For example, probenazole was introduced and used against the rice blast fungus (Watanabe, 1977) for some time before it was found to protect also against bacterial diseases of rice (Yamagouchi, 1998) and suspected of activating defence responses in the rice plant (Watanabe *et al.*, 1979). Only the rapid advance of the understanding of SAR and ISR mechanisms provided more detail on the possible mode of action of this chemical (Midoh & Iwata, 1996) that has been in use for 30 years. In this case, it was difficult to establish the activation of naturally occurring pathways, since they are not yet well understood in rice and monocots in general. Hence, some of the strongest evidence for the resistance inducing activity of probenazole comes from studies on model dicot plants such as parsley and Arabidopsis (Siegrist *et al.*, 1998). In addition, probenazole was shown to have some direct activity on the rice blast pathogen (Watanabe, 1977).

The screen leading to the discovery of Bion® (acibenzolar-*S*-methyl; ASM) was an extension of this biological screening approach (details see below). With the above criteria in mind, the concept of systemically induced resistance, as it was pioneered by Ross (1961) and expanded by Kuć (1982) with large research programmes on tobacco and cucurbits, seemed attractive as a model to discover novel disease control agents (den Hond, 1998). It showed a wide spectrum of protection which included fungi, bacteria and viruses as well as a long duration of protection which in several instances lasted the whole growing season. The protected, systemic parts of the plants showed no damage, and in some cases even stimulated growth. This is in contrast to locally induced resistance which was associated with tissue damage and localized production of phytoalexins, which was the emphasis of mainstream research in the early days of Kuć's work on systemically induced resistance.

The advance of biochemical and molecular knowledge of pathogens, plants and their interactions made alternative approaches in discovery of novel disease control agents possible. These include testing of chemicals on molecular targets *in vitro*, as is done in pharmaceuticals research or on transgenic plants and pathogens designed to signal activation or inhibition of certain pathways. Many of these approaches are biased in the sense that they are fixed on the intended targets or pathways while ignoring all other possible modes of action of a test chemical. Still another approach is the discovery of biochemical factors that are instrumental in stopping disease in biological models and their development or their use as lead structures for new products. A good example of this approach is harpin protein from bacteria that activates similar defence reactions in plants as the bacteria themselves (Dong *et al*., 1999).

12.3 Discovery of INAs and BTHs

The screening methods for resistance-inducing compounds were derived from the pioneering work of Kuć (1982) on systemically acquired resistance on cucumbers and tobacco. On both plants, he was able to show that infection of lower leaves with local lesion pathogens led to a systemic induction of resistance to a wide range of pathogen, including fungi, bacteria and viruses. For the initial mass screen, the cucumber model seemed more practical since it required much less space. The first isonicotinic acid derivatives (INAs) and benzothiadiazoles (BTHs) were identified in such screens using cucumber seedlings (den Hond, 1998). After localized application of the test chemicals, the whole plants were infected with the anthracnose fungus *Colletotrichum lagenarium*, and the protection pattern was compared with that of reference plants induced biologically by analogous localized infection with a local lesion pathogen. For the most promising candidates, their spectrum of protection, their direct activity *in vitro* and their physiological and biochemical effects on the plants were investigated on both cucumber and tobacco (Métraux *et al*., 1991). Several chemical groups showed activity patterns that were comparable to the biological model. Of these, the INAs offered most promise overall and were selected for field tests and possible development (Métraux *et al*., 1991).

12.4 Identification of BION® and other SAR activators

Using similar methods, many BTHs were identified, which showed a pattern of protection on cucumber and tobacco that matched that of the biological models and of the INAs

(Kessmann *et al.*, 1996; Kunz *et al.*, 1997). They showed the same spectrum of protection on cucumber and tobacco, and induced the same pattern of PR proteins. In addition, it was determined that neither the parent compounds nor their metabolites formed in the plants showed any *in vitro* activity against the pathogens. Through extensive field testing of many candidates, CGA 245704 was identified as a candidate for further development (Ruess *et al.*, 1996). During this testing process, it soon became clear that the transfer of laboratory and greenhouse results to the field was not as straightforward for these compounds as for fungicides. The activated protection and crop tolerance depended on many factors such as growth stage and growth conditions, more so than had been the experience with the more classical fungicide candidates with direct action against the pathogens. Moreover, the spectra of protection were found to be quite different from one plant species to another. However, benzothiadiazoles induced disease resistance in many crop plants, with the protection spectra often including fungal, bacterial and viral pathogens. The discovery of this large crop spectrum on which the benzothiadiazoles are active suggested that the resistance mechanisms induced by these chemicals are conserved widely throughout the plant kingdom. On the basis of its wide spectrum of plant pathogen control on many crops and its novel mode of action, CGA 245704 (ASM) was selected for commercial development in the early 1990s (Ruess *et al.*, 1996). It has been successfully introduced in Europe and other countries under the trade name Bion®, and in the USA as Actigard®.

More recently, tiadinil was introduced as an activator of SAR for the control of rice blast. It did not show direct activity against the rice blast pathogen *Magnaporthe grisea*, and in model studies on tobacco it showed a biological and biochemical induction pattern that was consistent with SAR induction (Yasuda *et al.*, 2004). As with probenazole, the elucidation of the resistance inducing activity of tiadinil on rice is difficult due to the lack of a good monocot model for SAR and other induced resistance pathways (see Figure 12.1 for the chemical structures of a variety of SAR inducers).

12.5 The role of basic studies in the discovery of BION® and other SAR/ISR products

In parallel with the discovery and development of INAs and BTHs, molecular biology made rapid progress in elucidation of the signal pathways involved in systemically acquired resistance induced by both biological and chemical agents (Ryals *et al.*, 1996). It was possible to profile more rapidly and more precisely the mode of action of the different chemical candidates identified in the biological screens. The description of this progress of the underlying mechanisms occupies large parts of the rest of this book. In the selection and development process, use could be made of these novel techniques to monitor efficiently the physiological and molecular state of the plants during laboratory or field tests. For Bion®, molecular studies on tobacco and Arabidopsis also showed that it activates the SAR pathway by mimicking the activity of salicylic acid, which is a key signal molecule for biological SAR activation on these model plants (Gaffney *et al.*, 1993; Friedrich *et al.*, 1996; Lawton *et al.*, 1996). Although evidence is good that Bion® also induces disease resistance in monocots (Görlach *et al.*, 1996), the analogy to the biological system is lacking because no such system is available comparable to Arabidopsis or tobacco in the dicots (Kessmann *et al.*, 1996).

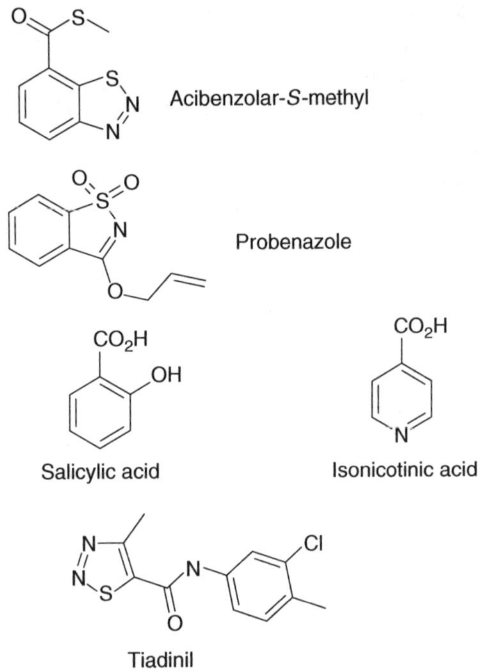

Figure 12.1 Active ingredients shown to be inducers of systemic acquired resistance.

Such molecular profiling of gene activation which is now possible with many hundreds of genes on microchips allows rapid determination of whether a new candidate with suspected activity for resistance induction is in fact activating resistance mechanisms and whether the mechanisms are of a known or novel nature. Also, for the early discovery and selection process, methods could be developed that offered faster and more detailed information on the activation profiles at the RNA, protein synthesis or enzyme activity level. For example, with reporter genes, it is possible to measure very efficiently the activation of selected proven resistance induction pathways without having to challenge the plant with pathogens.

12.6 Identification of harpin

Harpin is a naturally occurring bacterial protein, 403 amino acids in size, present in a number of species of plant pathogenic bacteria. It is necessary for pathogenicity on host plants and for the HR response on non-host plants, The first harpin protein was isolated at Cornell University from the bacterium *Erwinia amylovora* (Wei *et al.*, 1992). Early in the characterization of harpin from *E. amylovora*, it was discovered that foliar applications of harpin could elicit disease resistance in plants without causing visible HR reactions and, surprisingly, increase plant growth. Based on these findings, harpin was developed into the commercial product Messenger®. In *E. amylovora*, the causative agent of fire blight disease in pear and apple, the *hrp* gene cluster was dissected, and a single protein was identified that elicited hypersensitive responses in non-host plants and was necessary for pathogenesis. This protein was given the name harpin and the corresponding gene designated *hrpN*. This was the first example of such a protein and gene identified from any bacterial species. For commercial production of the protein, the DNA sequence coding for Harpin aβ was put into a weakened strain of *Escherichia coli* (*E. coli* K-12) commonly used experimentally and commercially. This genetically modified *E. coli* K-12 produces large amounts of Harpin aβ, which is then isolated and purified from the bacterial growth medium. Commercially produced harpin protein is identical to the protein that occurs in nature. *E. coli* K-12 is considered to be a non-pathogenic, nutritionally deficient bacterium which is unable to grow in the environment. Harpin is concentrated from the growth medium of the genetically modified *E. coli*, and the bacterial cells are killed and removed from the marketed product. Messenger™ was registered by the EPA in April 2000 (Jones, 2001). Since then, a second product containing Harpin aβ protein (ProAct™) has been launched by the Eden Bioscience Corporation. Both products are marketed as 'Plant Health Regulators' that enhance crop growth, quality and yield as well as possessing features such as improved plant stand and vigour, reduced transplant shock and suppression of nematode egg production.

On Arabidopsis, harpin activates the synthesis of marker proteins for both the salicylic acid dependent SAR pathway via the stimulation of salicylic acid synthesis and the ethylene/jasmonic acid dependent pathway (Peng *et al.*, 2003). A protein from Arabidopsis (HrBP1) was identified in a yeast two hybrid system that binds harpin and that might be the target site for the harpin activity. HrBP1 like proteins were found in several crops, which might explain the wide spectrum of plants on which harpin is active. Harpin seems to mimic local lesion pathogens in the biological SAR systems and stimulate the accumulation of salicylic acid, while Bion® acts downstream in the SAR signal pathway and mimics salicylic acid itself.

12.7 The commercial development of an induced resistance product

As has been described earlier, there are several potential commercial benefits to developing and marketing products which protect plants from pathogens using induced resistance. Because such compounds rely on the activation of the plant's natural broad spectrum defence mechanisms, they may offer a solution to many problems in plant protection which have not yet been solved satisfactorily such as bacteria and viruses. Even if the levels of protection achieved with resistance activators against some of these difficult to control problems are not better than current conventional solutions, the product safety profile or other considerations can be significantly better (e.g. compared with anti-bacterials which are important to preserve in human medicine). Some of them are active at very low use rates, making them economically attractive to growers, industry and environmental agencies alike. Because there is no direct effect on the target pathogen, and because the plant defence mechanism induced is likely to be complex in nature, the risk of resistance arising to the inducer can be considered rather low, leading to a sustainable use of such a product. In the case of probenazole, no resistant pathogenic strains have arisen, in spite of three decades of intensive commercial use. In addition, combinations of resistance activators with conventional products can produce interesting synergistic effects (Staub *et al.*, 1997) and can reduce the risk of resistance against the latter (Romero *et al.*, 1998).

Because they act on physiological processes in the plant, there can be positive growth effects in addition to protection against harmful pathogens. In the development of ASM (Bion®), it was found that its use in potatoes can be associated with an increased proportion of desirably sized seed potatoes. It also improved stress resistance in cabbage, resulting in fewer 'split heads' – both mean a considerable increase in the marketable yield. Work with harpin has shown pronounced growth promotion effects in a range of crops in addition to disease control benefits. Induced resistance products are also able to have a surprising spectrum of activity in a crop. A series of trials with seed potatoes showed that Bion® is a very effective SAR inducer against viral diseases but also provides protection against aphids and in other cases against nematodes (Tally *et al.*, 1999).

For resistance activators without direct activity on pathogens, the challenges in such areas as dose or rate definition are more difficult than with conventional plant protection products. Many trials were carried out on rate titration, and it became difficult to identify the lowest effective rate – it appears that once the plant is 'activated', higher rates of applications do not improve disease protection or residual activity. There appears to be some variability in the efficacy of SAR due to variability in the physiological state of plants (Guedes *et al.*, 1980).

There are also several issues that have to be considered in the development of a product acting only via induced resistance. Because it is not the product but the plant species itself which determines the SAR spectrum to be activated, the results from one plant species cannot be transferred automatically to others. The product is not like a fungicide which can always combat powdery mildew regardless of the crop; for example, a very efficient SAR reaction can be induced in cereals against powdery mildew, but not equally efficiently in cucurbits against powdery mildew. With ASM, the lack of a good biological model experiment for monocots for use in greenhouse trials meant that less basic research had been carried out in these crops, and the activity in monocots had to be better understood. Activated plants need a certain amount of time before their defence mechanisms are fully operational (between four and seven days following use of ASM). This must be taken into account when determining timing of the first application and means that a truly protective application must be made to the crop. The period of protection given by the product, and therefore the interval after which repeat application needs to be made, varies from crop species to species – it must therefore be evaluated on a crop by crop basis (SAR generally lasts significantly longer in monocots than in dicots). Product formulation or adjuvant utilization can be very important in this context (Reglinski *et al.*, 1994). SAR can be variable and sometimes inconsistent, especially in comparison with a good conventional plant protection product. This may need to be solved through the use of programmes including conventional products, or mixtures with conventional chemistry. Because the compounds act on plant systems, the questions of crop tolerance and of metabolic costs of the activation process to the crop need very careful evaluation. During the development of ASM, it became clear that visual phytotoxicity could be caused in some situations. However, it was only in a very few species that the phytotoxicity was so high that development of the product was avoided (ornamental species such as *Pelargonium* and *Begonia*). In most other situations, through adjustment of the product rate and timing, phytotoxicity could be reduced to acceptable levels or avoided so that overall the use of the product was beneficial rather than detrimental. It was found that phytotoxicity effects were most pronounced when crops were in any case under conditions of environmental stress, and use rates were rather high. (Interestingly, at low use rates, beneficial effects of applying the product to crops under stress conditions, for example drought, could often be seen). As with resistance activation, unacceptable crop effects had to be determined on a crop by crop basis and the use defined accordingly (Tally *et al.*, 1999). Although this is not always the case (for example with probenazole), many target markets for such a product are rather 'niche' in nature (small), and this leads to problems in justifying the development and regulatory costs in many potential opportunity markets.

12.8 Innovation in registration?

Because products acting only via induced resistance in plants have no direct effect on the target pathogen (i.e. the active substance in the product is not a fungicide, insecticide, bactericide, etc.), this results in some potential beneficial positions in the regulatory requirements for commercialization. In some countries, such products do not fall under the direct legislation for the registration of crop protection products, for example in Germany where the category 'Pflanzenstärkungsmittel' (plant strengthener) exists. This means that lower data requirements are needed to bring a product which falls into this category to the market, resulting in faster times to market. An example of an SAR product currently authorized in Germany under this legislation is Messenger® (harpin). Bion® was first authorized in Germany under this legislation, in June 1996.

At the same time, an issue that can arise in the registration of such products is that the efficacy, while useful, may not be as reliable or at the same level as a conventional product. In countries where a minimum level of protection (control) is required for registration of a fungicide, for example, this may become an issue if the induced resistance product does not reach the threshold and therefore cannot be registered (as a conventional fungicide type product). To try to avoid this issue, an innovative approach was taken in the registration of ASM with the proposal to create a new crop protection product category – Plant Activators. The ISO commission accepted the use of the term 'Plant Activator' as a new product category for products working solely via SAR, with ASM accepted as the first representative of this category. The definition of the category is 'A substance that protects plants by activating their defence mechanisms against pests or diseases'.

The Swiss regulatory authority introduced a new product class 'stimulator of natural defence' and entered it as a separate entry in the category 'plant protection substances'. The European Commission recognized that the regulatory procedures based upon synthetic chemical active substances have been regarded as a barrier to the commercialization of alternative products. They defined a 'plant strengthener' as a substance or micro-organism shown to protect plants against harmful organisms by activating the defences of the plant through: stimulating resistance/defence mechanisms in the plant, or the competition of the plant strengthener with harmful organisms for space and food substances in the phyllosphere or rhizosphere. Although they state that 'plant strengtheners' are plant protection products, and therefore covered by articles 2(1.1) and 2(1.2) of Directive 91/414/EEC concerning the placing of plant protection products on the market, up to 2001 few applications for approval of such products had been submitted. Several reasons were given for non-submission: uncertainty whether the Directive included 'plant strengtheners', the high cost of an application for a 'plant strengthener' in relation to limited use/markets, the nature of the products and expected low risk profile of most of the plant strengtheners. The Commission and Member States, however, stated it to be a matter of importance that plant strengtheners are authorized within the EU as plant protection products. To address this situation, a specific guidance document (Sanco/1003/2000) was composed as a helpful instrument in the process of authorization of 'plant strengtheners' and lays down a tiered approach for plant strengtheners with a low risk profile. If strengtheners have a high risk profile or are based on microbials, the dossier requirements of Annex II/III are fully applicable.

In the majority of countries, therefore, a product such as ASM, although it acts solely via SAR, falls under the conventional crop protection product legislation. Bion®/Actigard® has been registered in many countries within this legislation and achieved Annex 1 inclusion in the European Union registration process in October 2001, having achieved all the stringent requirements relating to human and environmental safety, as well as efficacy. Acibenzolar-*S*-methyl, as Actigard®, achieved full registration in the USA in August 2000. In the USA, ASM is classified as a plant activator, with label recommendations for use on brassica crops, tomatoes, spinach and tobacco, for protection against fungal, bacterial and viral diseases depending on the crop.

In the registration process of several countries, most notably in the European Union, there is a requirement to evaluate the resistance risk of active ingredients and products. This is achieved in conventional plant protection products by a series of studies investigating the biochemical mode of action against the target pest, the natural inherent variation in sensitivity of pest populations to the active ingredient, the impact of chemical- and radiation-induced mutations on pest sensitivity, etc., and would normally include a baseline evaluation of pest populations in the most important markets. With SAR inducing compounds, all the above are clearly very difficult to achieve (because the compound under test has no or very low inherent direct activity against the target pathogen), and so this regulatory hurdle has to be addressed largely by other observations. Several mechanisms appear to be activated simultaneously against the pathogen attack (Görlach *et al.*, 1996) – this feature is believed to limit the risk of development of pathotypes insensitive to the activated defence response of SAR based products. ASM, probenazole and tiadinil are classified by the Fungicide Resistance Action Committee as unique sub-groups within group P (host plant defence induction) 'resistance not known'.

12.9 Commercial experiences with induced resistance products

There are today a number of commercial products available whose effect is solely by induced resistance (Table 12.1). These include Oryzemate® (probenazole), Bion®/Actigard® (acibenzolar-*S*-methyl) and Messenger® (harpin). Oryzemate® has been used successfully in Asian rice production for a number of years, mainly against the blast pathogen (*M. grisea*). It remains today one of the most important products for the protection of rice against blast disease in Japan. It is noteworthy that no resistance has developed against this product in over 20 years of intensive use and is therefore a valuable component of disease management programmes in rice in Japan. ASM, as Bion® and Actigard®, is successfully registered and sold worldwide in a wide range of crops including tomatoes, tobacco, pears, bananas, lettuce and other leafy vegetables, nuts and cucurbits.

Induced resistance products have been received quite successfully into markets, depending on the crops and the expectations of the users. In some cases, where the protection of

Table 12.1 Commercial products with good evidence for inducing disease resistance in plants.

Trade names	Chemical name	Mode of action	Key biological properties	Reference
BION, ACTIGARD	Acibenzolar-*S*-methyl	Mimics SA in natural SAR	Broad spectrum including fungi, bacteria and viruses on many crops	Kessmann *et al.* (1996), Ruess *et al.* (1996)
MESSENGER, ProAct	Harpin protein	Mimics local lesion in natural SAR (depends on SA production)	Enhanced crop growth, quality and yield, suppression of nematode egg production	Jones (2001)
V-GET	Tiadinil	Mimics SA	Controls rice blast	Yasuda *et al.* (2004)
ORYZEMATE	Probenazole	Induces various PR proteins and lipids on rice (may depend on SA)	Fungal and bacterial protection of rice and some vegetables	Watanabe *et al.* (1979), Yoshioka *et al.* (2001)
OXYCOM A & B	Reactive oxygen and plant stimulant (acetic acid, hydrogen peroxide, plant nutrients, proprietary stabilizers and salicylic acid)	Stimulation of various defence genes involving a MAPK pathway	Increased plant cell wall strength and improved root health – reduction of nematodes, bacteria in a range of crops	Yang *et al.* (2002)
ELEXA 4	Chitosan	Unknown, may require SA	Fungal protection of fruit, vegetables, ornamentals, cereals, turf and rice	Manufacturer literature (www.plantdefenseboosters.com/elexa.html)
IODUS 40	β-1,3-Glucan	Stimulation of induced resistance (SA and JA pathways)	Fungal diseases of various crops	Manufacturer literature (www.goemar.com)

crops is at a similar level to that given by conventional products (e.g. tobacco blue mold with ASM), the products have been well accepted and used. In other situations where there are few if any alternatives (e.g. bacterial control in vegetables such as tomatoes and fireblight control in pome fruit), the acceptance and success of the product have also been good. In intensive production of arable crops, however, where there is a wide range of reliable and effective conventional products (e.g. European cereals) the success of these products has been limited. Used alone, the 'efficacy' was found to be insufficient or variable, and at the time of introduction of some products, the market was competing with new, highly potent chemicals such as the strobilurins for disease control. However, as discussed earlier in this chapter, one of the features of induced resistance products appears to be the limited risk of target pathogen resistance occurring, so this lower 'efficacy' relative to conventional highly potent fungicides or bactericides may be considered sufficient and valuable in the long-term to ensure sustainable crop protection. Future strategies for use of resistance inducers in such crops may be to 'support' conventional chemical products in mixtures to bring another mode of action and exploit fully the frequent synergisms observed in such combination. In addition, such supportive spray schedules (mixtures or alternations) with conventional chemistry can achieve reliable protection of crops and yield benefits not least as an additional tool in crop management and in the management of resistance of pathogens to conventional fungicide partners.

An example of the practical benefit of such a mixture product as described above is the combination of ASM with mancozeb, which is sold for use on various vegetable crops in Asia (Bion M®). This is a combination of 1 g of ASM plus 48 g of mancozeb, which is usually applied to deliver in the region of 1–2.5 g of ASM + 48–120 g of a.i. mancozeb per 100 l of spray solution. The activity of this mixture has been found to be at least equivalent to the standard rate of mancozeb, which is usually two to three times higher than the rate used in the mixture. The addition of a very small quantity of plant activator to a conventional fungicide in this case allows a drastic reduction in the rate of the fungicide partner to be achieved, with the benefit of reduced environmental load of the conventional fungicide among others.

ASM is also sold for use on bananas under the tradename BOOST® to support the disease management programme against black Sigatoka disease (*Mycosphaerella fijiensis*). The defence mechanism in bananas is activated when BOOST® is applied, and the spray interval of conventional fungicides can be extended. This not only provides excellent disease control but leads to significant reductions in inputs such as fungicides and oil.

For some fungicides, with primary direct activity on the pathogens secondary effects via induction of inducing activity have been claimed. For carpropamide ([1*RS*, 3*SR*]-2,2-dichloro-*N*-1-(4-chlorophenyl)ethyl-1-ethyl-3-methyl-cyclopropane-carboxamide), the main protective effect against *Maynaporthe grisea* on rice is based on the inhibition of fungal melanin biosynthesis, while it has been proposed that the long lasting activity after single treatments originates from the cyclopropane part of the molecule which may act as a plant activator (Thieron *et al.*, 1998). An analogue of this cyclopropane part, 2,2-dichloro-3,3-dimethylcyclopropane carboxylic acid (WL 28325), had been known for more than 20 years as a specific and systemic research compound against rice leaf blast. It showed low direct fungitoxicity against the blast pathogen, and treated plants respond more quickly and in a resistant manner to infection (Langcake *et al.*, 1983). When melanin precursors were fed to carpropamid treated plants, fungal melanin production in *M. grisea* was restored, but this did not result in full loss of protection against rice blast. By contrast, the related fungicide tricyclazole, which lacks the cyclopropane moiety and which interferes at a different step in melanin biosynthesis, lost its protective activity upon restoration of melanin biosynthesis by feeding of the appropriate melanin precursor (Thieron *et al.*, 1998).

There are also reports of some strobilurin fungicides inducing resistance in tobacco against tobacco mosaic virus and the wildfire pathogen *Pseudomonas syringae* pv. *tabaci* (Herms *et al.*, 2002). Here, and in many other cases of suspected induced resistance, the evidence is lack of direct action *in vitro* and a faster response of the plant to infection. However, wherever pre-challenge markers of resistance are lacking, it is difficult to separate the effects of a treatment on the plant from those on the pathogen. There are many reports of fungistatic fungicides producing phenocopies of resistance reactions on plants that are indistinguishable from the reactions produced by genetic resistance (Ward 1984) or by plant activators. For fosetyl-Al, activation of plant resistance had been proposed, based on its weak *in vitro* activity against the target pathogens before it was shown that resistant mutants of *Phytophthora capsici* selected *in vitro* were also insensitive on treated plants (Fenn & Coffey, 1985). Recently, claims have been made by the manufacturer that the powdery mildew fungicide proquinazid, as well as having direct fungicidal effects, also acts via induced resistance to control the disease in cereals, for example. With all the molecular tools available today, it is much easier to demonstrate key pre-challenge effects that indicate activation or priming of resistance pathways in plants by chemicals.

There are many 'niche' markets where a very high biological potential of resistance inducers has been identified and proved, but where the economics of product development and marketing are questionable. It is in these opportunity markets where maximum use of reduced risk programmes, reduced data requirements and mutual recognition of registrations need to exist and be utilized to bring benefits to growers.

12.10 Conclusions

The phenomenon of induced resistance in plants offers new possibilities for practical exploitation in crop protection against diseases: use of key genes in breeding programmes, and application of biological, biochemical or synthetic agents that activate one or more of the biochemical pathways leading to systemically activated plant defence.

The large protection spectra that are characteristic of most induced resistance agents are very crop specific but often include pathogens such as bacteria and viruses, against which no (or rather ineffective) products were available so far. These characteristics create potential market opportunities for such products above and beyond what can be achieved by more conventional products, for example fungicides, and are also attractive propositions for the market place because of the generally very low rates of activator required, the low risk of target pathogen resistance, and the benefits possible in reducing environmental costs and loads of conventional fungicides and bactericides.

Disease control with plant activators seems more crop-specific as well as more dependent on growth and environmental conditions than that with traditional fungicides. This creates additional challenges to the development and eventual marketing of such products; however, due to the nature of induced resistance products, regulatory hurdles can be reduced and greater use made of grower/industry groups to bring solutions for more niche markets.

Most promise comes from combining resistance inducers with traditional disease control measures in an intelligent way to exploit their strengths and the synergism that is often observed with such combinations. Where fungicides are prone to a buildup of resistant pathogen populations, combinations with activators of plant resistance can extend the useful life and the reliability of such fungicides. Evidence also exists that use of plant activators can slow down the ability of pathogens to mutate to overcome host plant resistance.

The rapid advance of our knowledge of the molecular mechanisms underlying the biological and chemical resistance activation in plants should enable a more rapid and reliable development of resistance activators in the future. In addition, this knowledge has opened new opportunities for the use of key genes from the resistance induction signal pathways in crop breeding programmes.

12.11 References

den Hond F, 1998. Systemic acquired resistance: a case of innovation in crop protection. *Pesticide Outlook* **9,** 18–23.

Dong H, Delaney TP, Bauer DW, Beer SV, 1999. Harpin induces disease resistance in Arabidopsis through the systemic acquired resistance pathway mediated by salicylic acid and the NIM1 gene. *The Plant Journal* **20,** 207–215.

Fenn ME, Coffey MD, 1985. Further evidence for the direct mode of action of fosetyl-Al and phosphorous acid. *Phytopathology* **75,** 1064–1068.

Friedrich L, Lawton K, Ruess W, Masner P, Specker N, Gut Rella M, Meier B, Dincher S, Staub T, Uknes S, Métraux J-P, Kessmann H, Ryals J, 1996. A benzothiadiazole derivative induces systemic acquired resistance in tobacco. *The Plant Journal* **10,** 61–70.

Gaffney T, Friedrich L, Vernooij B, Negretto D, Nye G, Uknes S, Ward E, Kessmann H, Ryals J, 1993. Requirement of salicylic acid for the induction of systemic acquired resistance. *Science* **261,** 754–756.

Görlach J, Volrath S, Knauf-Beiter G, Hengy G, Beckhove U, Kogel KH, Oostendorp M, Staub T, Ward E, Kessmann H, Ryals J, 1996. Benzothiadiazole, a novel class of inducers of systemic acquired resistance, activates gene expression and disease resistance in wheat. *The Plant Cell* **8,** 629–643.

Guedes MEM, Richmond S, Kuć J, 1980. Induced systemic resistance to anthracnose in cucumber as influenced by the location of the inducer inoculation with *Colletotrichum lagenarium* and the onset of flowering and fruiting. *Physiological and Molecular Plant Pathology* **17,** 229–233.

Herms S, Seehaus K, Koehle H, Conrath U, 2002. A strobilurin fungicide enhances the resistance of tobacco against tobacco mosaic virus and *Pseudomonas syringae* pv. *tabaci. Plant Physiology* **130,** 120–127.

Jones J, 2001. Harpin. *Pesticide Outlook* **12,** 134–135.

Kessmann H, Oostendorp M, Staub T, Goerlach J, Friedrich L, Lawton K, Ryals J, 1996. CGA 245704, Mode of action of a new plant activator. *Brighton Crop Protection Conference – Pests and Diseases*, 961–966.

Kessmann H, Staub T, Hofmann C, Maetzke T, Herzog J, Ward E, Uknes S, Ryals J, 1994. Induction of systemic acquired resistance in plants by chemicals. *Annual Review of Phytopathology* **32,** 439–459.

Kuć J, 1982. Induced immunity to plant diseases. *BioScience* **32,** 854–860.

Kunz W, Schurter R, Maetzke T, 1997. The chemistry of benzothiadiazole plant activators. *Pesticide Science* **50,** 275–282.

Langcake P, Dartwright D, Ride JP, 1983. The dichloropropanes and other fungicides with indirect mode of action. In: Lyr H, Polter C, eds. *Systemische Verbindungen und antifungale Verbindungen*. Berlin: Akademie-Verlag, 199–210.

Lawton KA, Friedrich L. Hunt M, Weymann K, Delaney TP, Kessmann H, Staub T, Ryals J, 1996. Benzothiadiazole induces disease resistance in *Arabidopsis* by activation of the systemic acquired resistance signal transduction pathway. *The Plant Journal* **10,** 71–82.

McDougall P, 2003. *The cost of new agrochemical product discovery, development and registration in 1995 and 2000*. Report D/03/EJ/12344. Brussels: European Crop Protection Association and CropLife America.

Métraux J-P, Ahl-Goy P, Staub T, Speich J, Steinemann A, Ryals J, Ward E, 1991. Induced systemic resistance in cucumber in response to 2,6-dichloro-isonicotinic acid and pathogens. In: Hennecke H & Verma DPS, eds. *Advances in Molecular Genetics of Plant–Microbe Interactions* **1**. Dordrecht: Kluwer Academic, 432–439.

Midoh N, Iwata M. 1996. Cloning and characterization of a probenazole-inducible gene for an intercellular pathogenesis-related protein in rice. *Plant Cell Physiology* **37,** 9–18.

Oercke E-C, Dehne H-W, Schoenbeck F, Weber A, 1994. *Crop Production and Crop Protection – Estimated Losses in Major Food and Cash Crops.* Amsterdam: Elsevier.

Oostendorp M, Kunz W, Dietrich B, Staub T, 2001. Induced resistance in plants by chemicals. *European Journal of Plant Pathology* **107,** 19–28.

Peng J-L, Dong H-S, Dong H-P, Delaney TP, Bonasera JM, Beer SV, 2003. Harpin-elicited hypersensitive cell death and pathogen resistance require NDR1 and EDS1 genes. *Physiological and Molecular Plant Pathology* **62,** 317–326.

Reglinski T, Newton AC, Lyon GD, 1994. Assessment of the ability of yeast-derived elicitors to control barley powdery mildew in the field. *Journal of Plant Disease and Protection* **101,** 1–10.

Romero AM, Kousik CS, Ritchie DF, 1998. Systemic acquired resistance delays race shifts to major resistance genes in pepper. *Proceedings of the Annual Meeting of the American Phytopathological Society, Las Vegas* (Abstract).

Ross FA, 1961. Systemic acquired resistance induced by localized virus infection in plants. *Virology* **14,** 340–358.

Ruess W, Mueller K, Knauf-Beiter G, Staub T, 1996. Plant activator CGA-245704: An innovative approach for disease control in cereals and tobacco. *Brighton Crop Protection Conference – Pests and Diseases,* 53–60.

Ryals JA, Neuenschwander UH, Willits MG, Molina A, Steiner H, Hunt MD, 1996. Systemic acquired resistance. *The Plant Cell* **8,** 1809–1819.

Siegrist J, Muhlenbeck S, Buchenauer H, 1998. Cultured parsley cells, a model system for the rapid testing of abiotic and natural substances as inducers of systemic acquired resistance. *Physiological and Molecular Plant Pathology* **53,** 223–238.

Staub T, Ahl-Goy P, Kessmann H, 1992. Chemically induced disease resistance in plants. In: Lyr H and Polter C, eds. *Proceedings of the 10th International Symposium on Systemic Fungicides and Antifungal Compounds.* Stuttgart: Ulmer-Verlag, 239–249.

Staub T, Ruess W, Neuenschwander U, Oostendorp M, 1997. Les stimulateurs des defenses naturelles des plantes: perspectives d'utilisation pour la protection contre les maladies. *Proceedings of the 5th International Conference on Plant Diseases.* Paris: ANPP, 77–86.

Sticher L, Mauch-Mani B, Métraux J-P, 1997. Systemic acquired resistance. *Annual Review of Phytopathology* **35,** 235–270.

Tally A, Oostendorp M, Lawton K, Staub T, Bassi B, 1999. Commercial development of elicitors of induced resistance to pathogens. In: Agrawal AA, Tuzun S, Bent E, eds. *Induced Plant Defenses Against Pathogens and Herbivores.* St. Paul, MN: APS Press, 357–369.

Thieron M, Pontzen R, Kuahashi Y, 1998. Carpropamid: a rice fungicide with two modes of action. *Pflanzenschutz-Nachrichten Bayser* **51,** 257–278.

Ward EWB, 1984. Suppression of metalaxyl activity by glyphosate: Evidence that host defenses contribute to metalaxyl inhibition of *Phytophthora megasperma* f. sp. *glycine*. *Physiological Plant Pathology* **25,** 381–386.

Watanabe T, 1977 Effects of probenazole (Oryzemate®) on each stage of rice blast fungus (*Pyricularia oryzae* Cavara) in its life cycle. *Journal of Pesticide Science* **2,** 395–404.

Watanabe T, Sekizawa Y, Shimura M, Suzuki Y, Matsumoto K, Iwata M, Mase S, 1979. Effects of probenazole (Oryzemate®) on rice plants with reference to controlling rice blast. *Journal of Pesticide Science* **4,** 53–59.

Wei Z-M, Laby RJ, Zumoff CH, Bauer DW, He SY, Collmer A, Beer SV, 1992. Harpin elicitor of the hypersensitive response produced by the plant pathogen *Erwinia amylovora*. *Science* **257,** 85–88.

Yamagouchi I, 1998. Activators of systemic acquired resistance. In: Hudson D & Miyamoto J, eds. *Fungicidal Activity.* London: Wiley, 193–219.

Yang K-Y, Blee KA, Zhang S, Anderson AJ, 2002. Oxycom™ treatment suppresses *Pseudomonas syringae* infection and activates a mitogen-activated protein kinase pathway. *Physiological and Molecular Plant Pathology* **61,** 249–256.

Yasuda M, Nakashita H, Yoshida S, 2004. Tiadinil, a novel class of activators of systemic acquired resistance, induces defense gene expression and disease resistance in tobacco. *Journal of Pesticide Science* **29,** 46–49.

Yorinori JT, Frederick RD, Costamilan LM, Bertagnolli PF, Hartman GE, Godoy CV, Nunes J, 2005. Epidemics of soybean rust (*Phakopsora pachyrhizi*) in Brazil and Paraguay from 2001 to 2003. *Plant Disease* **89,** 675–677.

Yoshioka K, Nakashita H, Klessing DF, Yamagouchi I, 2001. Probenazole induces systemic acquired resistance in Arabidopsis with a novel type of action. *The Plant Journal* **25,** 149–157.

Chapter 13
Induced resistance in crop protection: the future, drivers and barriers

Gary Lyon[1], *Adrian Newton*[1] *and Dale Walters*[2]
[1]Scottish Crop Research Institute, Dundee, UK
[2]Scottish Agricultural College, Edinburgh, UK

13.1 Introduction

Much has been published on induced resistance, e.g. Lyon *et al.* (1995), Oostendorp *et al.* (2001), Heil & Bostock (2002), Durrant & Dong (2004), Fobert & Després (2005) and many others, though these reviews focus largely on molecular aspects of this phenomenon through studying the regulation of genes or the biochemical responses associated with it.

It is clear that there are a number of issues to be resolved if induced resistance is going to become a widely accepted solution within integrated disease control strategies. Concerns may be based on the perceived or real problems, that yield will be compromised (i.e. lower), that disease control will be less effective than with conventional pesticides, that they are unreliable, possibly due to unspecified environmental interactions, and that multiple applications of elicitors will be necessary. This book has attempted to discuss some of these issues and, where possible, indicate the future direction of research in this area.

Though initial research on elicitors was based on their ability to induce resistance to disease, there is increasing evidence that they may have wider applications and be useful as inducers of resistance to abiotic stresses such as drought. Furthermore, compounds such as *cis*-jasmone, reported to induce resistance to pests such as aphids (Bruce *et al.*, 2003), may also stimulate the plant to attract natural aphid enemies such as parasitoids (Birkett *et al.*, 2000). Thus, elicitors may become an increasingly desirable asset in future, performing several different roles which may make the difference in determining whether they become a normal component of crop management.

13.2 Strategies to increase efficacy and durability in the field

There is strong evidence that some induced resistance can be activated in field grown plants without the presence of pathogens. For example, genes such as SAR8.2, regarded as a marker for systemic acquired resistance, are up-regulated not only by BTH, BABA and salicylic acid, but also by abiotic stresses such as drought, sodium chloride and low temperature (Lee & Hwang, 2003). Simulated acid rain has been shown to induce genes associated with the salicylic acid signalling pathway (Lee *et al.*, 2006), again providing evidence of potential interactions of induced resistance and environmental factors.

Knowledge that induced resistance is already occurring in the field is important in trying to predict what remaining potential there is to induce higher levels of resistance. That some commercial elicitors such as Bion® are able to make a useful contribution to disease control in selected situations suggests that this potential exists, but we need to better understand why it is so quantitatively variable and frequently sub-optimal in the field.

The deployment of elicitors in an integrated crop management system has previously been discussed (Lyon & Newton, 1999), though it has not yet been widely incorporated into practice. There are a relatively small number of fungicides that have been developed and marketed as such, but which in addition to direct anti-microbial activity also appear able to induce some pathogen resistance-related responses. Such discoveries have been fortuitous rather than specifically screened for, but open up the possibility of dual action crop protectants which may be more robust and durable than single mode-of-action pesticides.

Formulation of agrochemicals is an important consideration in the development of successful products. Larger agrochemical companies have a wealth of expertise in this area while smaller ventures may struggle here, although the formulation requirements for elicitors may differ considerably from those for fungicides which target pathogens directly.

13.3 What research is required to make induced resistance work in practice?

Some workers have assumed that elicitors can be used as a direct replacement for existing pesticides without the need to modify any other component of the agricultural system. Evidence to date suggests that this is unlikely to be a very effective strategy and that induced resistance may need to be considered as part of a new approach including using cultivars specifically chosen for appropriate elicitor response or disease tolerance, these being possibly genetically modified (GM) plants, perhaps used in spray combinations with existing pesticides, changes to farm environment scheme payments to encourage less total pesticide use through better targeting, and even an acceptance that in some cases more disease may be the necessary price to pay for a reduction in pesticide usage. Plant pathologists have focused largely on understanding the nature of induced resistance, and little extensive work has yet been undertaken to incorporate this knowledge into practical, integrated packages that can be applied to agricultural or horticultural production situations. Plant breeders have not carried out selection on the basis of response to elicitors (see Section 13.4), nor are they likely too unless it can be proven to be correlated with other beneficial traits such as increased tolerance to disease or specific pathogen resistance expression per se.

A greater understanding of biotic and abiotic stresses that can affect induction or inhibition of induced resistance-related genes is required before the problem of their unreliability is solved. Stress factors include environmental conditions, such as humidity, temperature, varying regimes of temperature and light versus constant conditions, effect of saprophytic microbes, effect of other chemicals, wind, drought, cultivars, crop management regimes, nutrients, etc. These responses need to be quantified, which may be difficult to do by directly measuring their effect on resistance, but it may be possible by quantifying the expression of SAR-related genes (e.g. using quantitative RT-PCR). The use of reporter genes linked to promoters putatively associated with induced resistance such as for PR1 genes has been tried with limited success in the past but does suffer from the disadvantage that there may be different molecular patterns associated with different types of induced resistance and that it would be beneficial not to preclude some molecular responses which may be unique to certain triggers or environments. Many believe that PR1 is a good indicator of induced resistance, particularly in Arabidopsis and tobacco, though Molina *et al.* (1999) suggested that PR gene expression is not a reliable indicator of induced resistance in cereals and showed that wheat PR1 genes that were pathogen inducible did not respond to salicylic acid, BTH or isonicotinic acid. Are there going to be better genes to use as 'indicators', and can these be used in a quantitative manner to screen for more effective inducers/primers?

Elicitor activity is clearly distributed among many different types of substance, and there is always scope for screening more. Recent work on compounds such as the cyclodextrins (Bru *et al.*, 2006) offers the possibility that there could be breakthroughs on new classes of compounds which might dramatically change our concepts. Compounds not regarded as inducers of resistance, such as the polycyclic aromatic compound phenanthrene, have been shown to induce PR1 and to inhibit the expansin gene EXP8 (Alkio *et al.*, 2005). These compounds are regarded as potentially phytotoxic causing oxidative damage to plants. They are therefore distinctly different from the compounds described in Chapter 2 that are known to enhance resistance, but they provide evidence that a wider range of compounds may interact with resistance responses, thus leaving little opportunity for 'resistance inducers' to further enhance such pathways, or indeed they may be blocking such responses. However, finding out why our existing compounds do not achieve their apparent potential in the field could be even more productive in the longer-term. Perhaps we need also to consider different types of screens to detect compounds able to prime plants to respond more quickly to subsequent infection but which, themselves, do not trigger a high and resource costly expression of resistance.

Kogel & Langen (2005) stated that some resistance inducing chemicals, including BTH for instance, may also induce plant genes which are not directly related to plant defence. The full extent of this is not yet known but does raise the possibility of explaining, in part, some of the more complex responses to elicitors which have been noted. However, it does emphasize the need for integrated approaches, not only to crop protection in practice, but also at fundamental discovery and development research levels. For example, crop protectants are frequently developed in specialist facilities where the primary focus is on a range of target organisms. From a molecular perspective, this may not be an appropriate strategy, as there are clearly many molecular mechanisms common to both biotic and abiotic stresses. Understanding these mechanisms may also provide explanations of some of the observations made in the field. For example, Petersen *et al.* (2000) reported that AtMPK4 activity is required to repress systemic acquired resistance. Interestingly, AtMPK4 is up-regulated by cold and salt stress, thereby providing a molecular explanation for one aspect of the interaction between SAR and the environment. A similar interaction between pathogen and abiotic stress is apparent in tobacco, as SIPK is activated by both salicylic acid and osmotic stress (Hoyos & Zhang 2000; Mikołajczyk *et al.*, 2000). Better characterized signal transduction pathways will highlight the extent to which crosstalk can occur and whether it can be manipulated successfully or whether it will hinder the successful application of elicitors.

The information derived from understanding which genes are up- or down-regulated in response to elicitors now needs to be followed up by looking at the amounts and activity of proteins. Some initial work on a proteomics approach to studying plant response to elicitors has recently been published by Chivasa *et al.* (2006). We need to distinguish more clearly those plant responses that are specific to induced resistance from those that are activated by a large number of different 'treatments' (stresses and compounds). For example, in Arabidopsis some of the glutathione-*S*-transferases such as At4g02520 and At1g02920 seem to be up-regulated by a very large number of treatments (see the DRASTIC gene expression database at www.drastic.org.uk).

Salicylic acid appears to be involved in signalling not only in SAR but also in response to various abiotic stresses such as cold (Janda *et al.*, 1999). Janda *et al.* (1999) showed that treatment of hydroponic maize (*Zea mays*) with salicylic acid increased the tolerance to low temperature stress. Thus, while we are frustrated by the lack of consistency in disease control when elicitors are used in the field, their future may be more to do with inducing resistance to other sorts of abiotic stress.

As discussed in Chapter 9, there is the issue that induced resistance may be associated with allocation costs and trade-offs. Research in this area, at least in terms of induced resistance to pathogens, is in its infancy. However, concerns over allocation costs and trade-offs is important from an agricultural perspective, since farmers and growers are unlikely to be keen on induced resistance if its use leads to a penalty in terms of grain yield and quality (Walters & Boyle, 2005). The recent work of van Hulten *et al.* (2006) showed that induced resistance by priming offers an efficient approach to disease control in Arabidopsis in the presence of pathogen pressure, since under these conditions, the benefits of priming-mediated resistance outweighed the costs. Whether the same is true for different crop species and different inducing agents requires further experimentation.

13.4 Can we breed plants with enhanced responsiveness to inducers?

An important question is: how could plant breeders select for elicitor-responsive breeding lines, and would they get different responses using different elicitors? Would marker-assisted selection be a viable option? Undoubtedly, there is variation between cultivars in the extent to which resistance can be induced, but how much variability is available in breeding material? Do we have appropriate parental lines, and is there more or less variation in distant relatives? Do wild species have more inducible resistance? Is there more variability in some crops, species, genera or families than others? Will there be more variability in response to different inducers in different species? Will enhanced response to one inducer be correlated with response to others? Would selection be for increased response of the whole response pathway or just a faster (or stronger) response of part of the pathway? Will these inducer-responsive plants have changed drought and cold tolerance as a side effect, enabling breeders to use particular elicitors to select for more tolerant genotypes? Thus, there are many more questions that can be considered for which evidence is still sparse. Answering these questions will be important in our efforts to bring induced resistance into the mainstream of crop protection.

13.5 The potential for GM plants containing SAR-related genes

There are many publications, particularly on Arabidopsis, listing genes induced by a variety of elicitors. These lists include many potential signalling genes which are candidates for GM transformations. Of course, there is potential for using different types of genes. One could use 'late stage' response genes which themselves may have direct antimicrobial activity. These genes could be either constitutively induced to provide continuous protection or regulated by a pathogen-responsive promoter to ensure more targeted and therefore resource-efficient up- or down-regulation only in response to infection. One could also use elicitor-responsive promoters to regulate genes so that resistance could be more effectively activated in response to spray application of chemical elicitors. Transformation with transcription factors could well be the most effective way forward ensuring an effective response to infection. Of general concern is that there may be a cost involved in constitutive resistance, but van Hulten *et al.* (2006) showed that priming plants is likely to be less costly to the plant than direct triggering of resistance, and this concept may be as applicable whether we are considering application of chemical elicitors or genetic modification. It will be vital to field test any GM plants under a wide variety of environmental conditions to look for pleiotropic effects such as improved abiotic stress tolerance.

13.6 Political, economic and legislation issues

There is an increasingly strong view, at least in Europe, that disease protection strategies should be as environmentally benign as possible, with little or no negative impact on non-target organisms and no impact on human health. This is leading to pressure on growers, often via supermarkets, to reduce chemical inputs. Such pressures are helping to encourage the growth of organic food production which may attract a premium price ensuring adequate financial returns for growers. It is clear, therefore, that if there was political pressure, particularly via financial incentives, to move away from traditional types of pesticides to those that acted, at least in part, through activation of induced resistance, then there would be greater encouragement to solve existing problems with induced resistance. Currently, public opinion in Europe seems to be against the use of GM plants, even when they can be demonstrated to have the potential to reduce pesticide inputs. This contrasts markedly with the acceptance of GM technology per se in many other parts of the world, and thus one can expect GM plants using genes associated with induced resistance to be used outwith Europe first.

In addition, farmers and growers may want to avoid emergence of resistance to existing plant protection products. It is important that disease control is reliable, and so we need to think of using elicitors, together with fungicides, in disease control programmes. This should allow farmers to use fewer fungicide sprays and have more options for managing fungicide usage to improve product longevity through avoiding resistance selection. However, we need to know when to apply elicitors – early in crop growth to 'prime' the plant, with fungicide applied later or on appearance of any disease? Clearly, further research is needed on these more 'practical', but nonetheless important, issues.

13.7 Conclusion

The title of this chapter highlights the future, drivers and barriers related to induced resistance and its implementation in crop protection practice. The future of crop protection certainly has a place for induced resistance as it is a mechanism already exploited in breeding. As a criterion for the development of crop protectants, it may increase in importance in multi-function products. However, the future for crop protectants based only on resistance induction depends on understanding and overcoming the efficacy problems attributable to environmental interactions. The main drivers remain durability of disease control while minimizing environmental impact. Indeed, induced resistance offers great potential and advantages over conventional crop protectants and the deployment of simple pathogen recognition genes. The main barrier remains our lack of knowledge of the mechanisms underlying induced resistance and how they interact with environmental variables, and subsequently how we might overcome these problems. While much excellent research has been carried out thus far, more is needed to answer key questions.

13.8 Acknowledgements

We are grateful to The Scottish Executive Environment and Rural Affairs Department (SEERAD) for continued support and funding.

13.9 References

Alkio M, Tabuchi TM, Wang X, Colón-Carmona A, 2005. Stress responses to polycyclic aromatic hydrocarbons in Arabidopsis include growth inhibition and hypersensitive response-like symptoms. *Journal of Experimental Botany* **56**, 2983–2994.

Birkett MA, Campbell CAM, Chamberlain K, Guerrieri E, Hick AJ, Martin JL, Matthes M, Napier JA, Pettersson J, Pickett JA, Poppy GM, Pow EM, Pye BJ, Smart LE, Wadhams GH, Wadhams LJ, Woodcock CM, 2000. New roles for *cis*-jasmone as an insect semiochemical and in plant defense. *Proceedings of the National Academy of Sciences USA* **97**, 9329–9334.

Bru R, Sellés S, Casado-Vela J, Belchí-Navarro S, Pedreño MA, 2006. Modified cyclodextrins are chemically defined glucan inducers of defense responses in grapevine cell cultures. *Journal of Agricultural and Food Chemistry* **54**, 65–71.

Bruce TJA, Martin JL, Pickett JA, Pye BJ, Smart LE, Wadhams LJ, 2003. *cis*-Jasmone treatment induces resistance in wheat leaves against the grain aphid, *Sitobion avenae* (Fabricius) (Homoptera: Aphididae). *Pest Management Science* **59**, 1031–1036.

Chivasa S, Hamilton JM, Pringle RS, Ndimba BK, Simon WJ, Lindsey K, Slabas AR, 2006. Proteomic analysis of differentially expressed proteins in fungal elicitor-treated *Arabidopsis* cell cultures. *Journal of Experimental Botany* **57**, 1553–1562.

Durrant WE, Dong X, 2004. Systemic acquired resistance. *Annual Review of Phytopathology* **42**, 185–209.

Fobert PR, Després C, 2005. Redox control of systemic acquired resistance. *Current Opinion in Plant Biology* **8**, 378–382.

Heil M, Bostock RM, 2002. Induced systemic resistance (ISR) against pathogens in the context of induced plant defences. *Annals of Botany* **89**, 503–512.

Hoyos ME, Zhang S, 2000. Calcium-independent activation of salicylic acid-induced protein kinase and a 40-kilodalton protein kinase by hyperosmotic stress. *Plant Physiology* **122**, 1355–1364.

Janda T, Szalai G, Tari I, Páldi E, 1999. Hydroponic treatment with salicylic acid decreases the effect of chilling injury in maize (*Zea mays* L.) plants. *Planta* **208**, 175–180.

Kogel K-H, Langen G, 2005. Induced disease resistance and gene expression in cereals. *Cellular Microbiology* **7**, 1555–1564.

Lee SC, Hwang BK, 2003. Identification of the pepper SAR8.2 gene as a molecular marker for pathogen infection, abiotic elicitors and environmental stresses in *Capsicum annuum*. *Planta* **216**, 387–396.

Lee Y, Park J, Im K, Kim K, Lee J, Lee K, Park J-A, Lee T-K, Park D-S, Yang J-S, Kim D, Lee S, 2006. Arabidopsis leaf necrosis caused by simulated acid rain is related to the salicylic acid signaling pathway. *Plant Physiology and Biochemistry* **44**, 38–42.

Lyon GD, Newton AC, 1999. Implementation of elicitor mediated induced resistance in agriculture. In: *Induced plant defenses against pathogens and herbivores*. Agrawal AA, Tuzun S, Bent E, eds. St. Paul, MN: APS Press, 299–318.

Lyon GD, Reglinski T, Newton AC, 1995. Novel disease control compounds: the potential to immunize plants against infection. *Plant Pathology* **44**, 407–427.

Mikońłajczyk M, Awotunde OS, Muszyńłska G, Klessig DF, Dobrowolska G, 2000. Osmotic stress induces rapid activation of a salicylic acid-induced protein kinase and a homolog of protein kinase ASK1 in tobacco cells. *The Plant Cell* **12**, 165–178.

Molina A, Görlach J, Volrath S, Ryals J, 1999. Wheat genes encoding two types of PR-1 proteins are pathogen inducible, but do not respond to activators of systemic acquired resistance. *Molecular Plant–Microbe Interactions* **12**, 53–58.

Oostendorp M, Kunz W, Dietrich B, Staub T, 2001. Induced disease resistance in plants by chemicals. *European Journal of Plant Pathology* **107**, 19–28.

Petersen M, Brodersen P, Naested H, Andreasson E, Lindhart U, Johansen B, Nielsen HB, Lacy M, Austin MJ, Parker JE, Sharma SB, Klessig DF, Martienssen R, Mattsson O, Jensen AB, Mundy J, 2000. Arabidopsis MAP kinase 4 negatively regulates systemic acquired resistance. *Cell* **103**, 1111–1120.

van Hulten M, Pelser M, van Loon LC, Pieterse CMJ, Ton J, 2006. Costs and benefits of priming for defense in Arabidopsis. *Proceedings of the National Academy of Sciences USA* **103**, 5602–5607.

Walters DR, Boyle C, 2005. Induced resistance and allocation costs: what is the impact of pathogen challenge? *Physiological and Molecular Plant Pathology* **66**, 40–44.

Index

Printed and bound in the UK by
CPI Antony Rowe, Eastbourne